A World from Dust

A World from Dust

How the Periodic Table Shaped Life

Ben McFarland

Illustrations by Gala Bent

and

Mary Anderson

OXFORD

UNIVERSITY PRESS

OXFORD
UNIVERSITY PRESS

Oxford University Press is a department of the University of Oxford. It furthers
the University's objective of excellence in research, scholarship, and education
by publishing worldwide. Oxford is a registered trade mark of Oxford University
Press in the UK and certain other countries.

Published in the United States of America by Oxford University Press
198 Madison Avenue, New York, NY 10016, United States of America.

© Oxford University Press 2016

First Edition published in 2016

Cataloging-in-Publication data is on file at the Library of Congress
ISBN 978–0–19–027501–3

1 3 5 7 9 8 6 4 2
Printed by Sheridan, USA

To those who always welcome me home:

Laurie (Chapters 5 and 8),
Sam, Aidan, Brendan, and Benjamin (Chapters 3 and 12),

and Bethany (Chapter 12)

CONTENTS

ACKNOWLEDGMENTS

The most valuable and scarce resource needed for this book was time. I was generously given time through two sources: the BioLogos Foundation's ECF grant and conferences, organized by Kathryn Applegate and Deborah Haarsma, and the award of sabbatical time from Seattle Pacific University, organized by Dean Bruce Congdon.

All my time would have amounted to much less without the time my colleagues, students, and friends devoted to reading and remarking on manuscript drafts. Many thanks to Eric Hanson, Jenny Tenlen, One Pagan, Larry Funck, Ben Wilcox, Kevin Hilman, Thane Erickson, Jeff Brown, and Mike Korpi. Also, thanks to the students of CHM 3410 Survey of Physical Chemistry and BIO/CHM 4363 Biochemistry Survey who read and commented on early (and far too lengthy) chapters. A special thanks to Alex Garcia, who was in *both* courses.

Thanks to Jeremy Lewis and Anna Langley for help in the editing process.

Thanks to Chip MacGregor and Richard Dahlstrom for generous and valuable advice.

Finally, thank you to my family for listening to me think aloud on these topics at random moments, and for all your constant manifestations of love and support.

AUTHOR'S NOTE

This book is about the nature of history and the history of nature. Because I understand the world in terms of chemistry, it is also a book about elements and molecules and the animals, vegetables, and minerals built from them. It is natural history told by a chemist.

The first three chapters introduce several chemical concepts that will be more or less familiar, depending on your own experience with chemistry. If you need extra reinforcement for a word or concept, try a quick Internet search. I drop a few details intended to help with such searches from time to time. Also, the periodic table (Figure 0.1) is your road map throughout the book. On this table, similar elements are stacked on top of each other, and elements get bigger as you scan downward, almost as if gravity has pulled the heavier ones down.

Because the nature of history is a philosophical topic, how you look at it is shaped by prior convictions. Stephen Jay Gould's book *Wonderful Life* is also about the nature of history, and in my first and last chapters I discuss how Gould's interpretations were shaped by his prior convictions. However, I don't intend to throw stones, because my own prior convictions undoubtedly shaped this book.

Here are five of mine: I studied protein chemistry in grad school for five years; my favorite color is green; I teach undergraduates at a liberal arts institution, so bridging the gap between the arts and sciences is crucial to me (at the very least to keep my friends in other departments); I presented science projects in the science fair every year from 7th through 12th grades; and my family and I attend Bethany Community Church in Seattle, where I learned to teach a diverse group of people who were not undergraduates and could leave at any moment—and sometimes did.

Of those categories, the last one is theological, which is an unusual category to find at the beginning of a popular science book. I am interested in bridging the perceived gap between science and faith, or, in more neutral categories, between fate and free will or order and disorder.

Some readers may attribute my conclusions about the nature of history to this last aspect of my own personal history. Actually, my faith is ambiguous on the question of the nature of history. In this book, I emphasize the chemical order that shapes the chaos of mutation and flow, restricting the paths of history—but others of the same faith emphasize the contingency and freedom of creation, and the welter of possible paths that history could have taken, and the untold possible choices a person faces each day. Faith alone does not dictate my conclusion about the nature of history, but rather, it supports my ultimate conclusion that the conversation between these voices should continue in a balanced dialogue. More on that in Chapter 12.

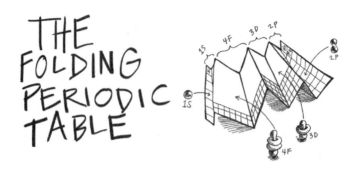

THE FOLDING PERIODIC TABLE

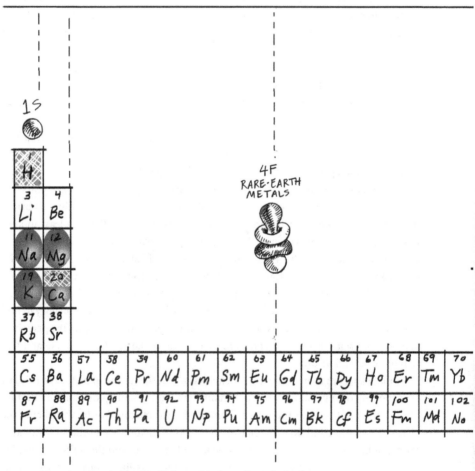

FIG. 0.1 The origami periodic table. Refer to this throughout the book as your road map. The "origami" is explained in Chapter 3.

= ELEMENTS FOR BIOCHEMICAL BALANCE.

 = ELEMENTS FOR BIOCHEMICAL BUILDING.

 = ELEMENTS FOR BIOCHEMICAL CATALYSIS.

 = ELEMENTS WITH CAPACITY FOR
BOTH BUILDING + BALANCE

2P

3D
TRANSITIONAL
METALS

																2 He
										5 B	6 C	7 N	8 O	9 F	10 Ne	
										13 Al	14 Si	15 P	16 S	17 Cl	18 Ar	
21 Sc	22 Ti	23 V	24 Cr	25 Mn	26 Fe	27 Co	28 Ni	29 Cu	30 Zn	31 Ga	32 Ge	33 As	34 Se	35 Br	36 Kr	
39 Y	40 Zr	41 Nb	42 Mo	43 Tc	44 Ru	45 Rh	46 Pd	47 Ag	48 Cd	49 In	50 Sn	51 Sb	52 Te	53 I	54 Xe	
71 Lu	72 Hf	73 Ta	74 W	75 Re	76 Os	77 Ir	78 Pt	79 Au	80 Hg	81 Tl	82 Pb	83 Bi	84 Po	85 At	86 Rn	
103 Lr	104 Rf	105 Db	106 Sg	107 Bh	108 Hs	109 Mt	110 Ds	111 Rg	112 Cn	113 Uut	114 Uuq	115 Uup	116 Uuh	117 Uus	118 Uuo	

The body of this book is shaped so that anyone can read it, meaning there are many details and debates to be found beneath the surface. References are cited at the end that allow readers to dive into the details and limitations of the individual studies. References are cited by quotations from the text in the end matter (rather than by footnotes) to avoid interrupting the flow of the narrative. Because this book is written for non-scientists, I leave it to the reader to investigate the methods or limitations of individual studies depending on individual expertise and interest. Study guides associated with this book, for use in general chemistry, biochemistry, or physical chemistry courses, will lead students through the details of selected studies.

As Darwin said of natural selection, "I believe in the truth of the theory, because it collects under one point of view, and gives a rational explanation of, many apparently independent classes of facts" (p. 16). What natural selection did for Darwin, chemistry does for me. In my view, a chemical sequence and chemical order shape the chaos of biology and history in surprising, yet rational ways, explaining many facts. It is a long story, but it coheres with chemical logic, and it shows that the nature of history is ordered by chemistry. That has reshaped the way I look at every blade of grass and every rock on the beach. I hope you enjoy thinking about this old subject in this new way.

1

ARSENIC LIFE?

a puzzling biochemical press conference // the natural laboratory at Mono Lake // a global scientific test // phosphorus builds, arsenic kills // how to mop up arsenic // magnesium's fitting relationship // a less wonderful life is more predictable

GEOLOGY SHAPES CHEMISTRY AT MONO LAKE

In December 2, 2010, at 11:16 a.m., I received the first of three emails from students in my biochemistry class, all asking if I had heard the news. A press conference at 11 a.m. had announced that scientists had discovered a bacterium that uses arsenic instead of phosphorus in its DNA. Soon there was a hashtag for this: #arseniclife. We were excited and a little puzzled. I had just lectured about how phosphorus was uniquely useful to DNA. I shrugged and mumbled something about how textbooks can be rewritten.

Today, the dust has settled—and the textbook reads the same as ever. DNA is made of phosphorus, never arsenic. That December press conference was followed by two full years of multiple experiments in labs around the world. It confirmed what the textbook said all along, yet the story was well worth it. The "arsenic life" story was never just about microbiology. It's about science itself, how we know things, and the nature of natural history.

Everyone should know this story. It will temper expectations when the next press-conference-induced hashtag makes its way halfway around the world while science is still lacing up its boots. More than that, it shows something deep about what kind of world we live in, something underreported because it is so intricate and comes from so many different places. There is a hidden order that makes some sense of biology and even sociology, and that hidden order is chemistry.

All life, from a lakewater bacterium to the neurons firing in your brain as you read this, is hemmed in. It is free to randomly adapt to its surroundings with nearly infinite creativity, but its overall path is as constrained as if it were walking on the deck of a ship crossing the ocean. The ultimate movement, on the scale of billions of years, is shaped by chemical rules.

One of these rules is that phosphorus makes good DNA, while arsenic does not. To reach this conclusion, we have to start where the arsenic life story started. The home of the purported arsenic-using bacteria that caused all this trouble is in a remote spot of California called Mono Lake.

Mono Lake sits at the bottom of a basin just east of Yosemite, at the foot of Mount Conness. It is one of the oldest lakes in North America, peaceful and uniquely beautiful, with towers of rugged tufa that look like they belong on an alien moon. It is also poisonous. Its arsenic-laced waters put the surrounding ecosystem under constant chemical threat.

Mono Lake is an unusual place, hosting unusual life, which even uses unusual atoms. Its unusual chemistry comes from its unusual geology. From above, Mono Lake is an asymmetric shape, its west shore squarish and angular, its east a round arc. This sharp western edge is cracked by a geological fault, traced by Highway 395. West of this fault, the Sierra Nevada mountains are pushed up by deep geological forces, while to the east, the lake is pulled down. The mountains to the west catch the moisture so that Mono Lake itself is arid and rain-shadowed. This is where the Nevada desert begins.

Highway 395's fault still bubbles with potential volcanoes. The pressure from underneath raises the entire Sierra Nevada range by about a millimeter a year, tilting it up and away from the Mono Lake basin. Sometimes the pressure is released explosively. A volcano 250 years ago suddenly formed a new island in the middle of lake.

The water in Mono Lake is a concentrated, liquid form of the Sierra Nevadas. Like the Dead Sea, Mono Lake sits at the bottom of a bowl of rock, with many streams running in but no streams running out. Once an atom arrives at Mono Lake, it cannot leave unless it is light enough to evaporate up, liquid enough to seep down, or lucky enough to be eaten by an animal that can walk or fly away. Water carries heavy rock atoms down from the surrounding hills and they are trapped.

At Mono Lake, the water is as much mountain as it is lake. The dissolved rock, especially the calcium, makes Mono Lake's water very "hard." It is so hard that when it evaporates, towers of tufa rock are left behind (Figure 1.1). The classic album cover to Pink Floyd's *Wish You Were Here* was taken at Mono Lake. It shows a diver's legs projecting out of blue water with wrinkled towers of sand-colored rock all around, as if his splash was turned to stone. Sometimes the water in lakes like this can get so "hard" that a bird sitting on the water too long will calcify. The rock creeps up and coats the bird's feathers, eventually overtaking and ossifying the entire bird, like something from the works of Edgar Allan Poe.

Tufa towers are built from chemistry, and the work is done by time. You can make your own by mixing baking soda, table salt, Epsom salts, and Borax in a gallon of water, then adding calcium chloride. Your bucket will include six elements important to rocks (and to life): sodium, chlorine, magnesium, sulfur, calcium, and carbon. Only the boron (added as Borax) does not play a major role in this book. After mixing, all you need is the patience to wait for the water to evaporate. Over months, inexorably, the calcium will link together with carbonate from the baking soda to make limestone tufa.

As the calcium turns into tufa, the other chemicals stay dissolved in the lake water. The first two, sodium and chlorine, come from table salt and dissolve well in water. Too heavy to evaporate, sodium and chlorine remain trapped in Mono Lake and make it twice as salty as the ocean. Most of the remaining rocky atoms are

FIG. 1.1 At Mono Lake, did the unusual chemistry that formed the Tufa towers also reshape the elements in microbial DNA? Also, note how magnesium interacts with DNA's phosphates.

surrounded by shells of oxygen atoms in water—carbon as carbonate (that was the baking soda), sulfur as sulfate (Epsom salts), boron as borate (Borax), even arsenic as arsenate (not included here because I presume you don't want poison in your bucket).

Much of this book is spent thinking about what these chemicals are doing at the atomic level, and there are two major differences when you shrink to this scale. The first is that *everything is constantly moving*. A river on a still day may look serene, but at the biological level it is filled with animals moving, plants waving, and currents flowing. This motion is magnified at the tiniest levels, where molecules wiggle in place or zip about like bees in a bottle.

Second, when molecules fit together, *they only care about two things: shape and charge*. Shape is familiar—all atoms are spheres that can stack together like a supermarket display of oranges—but charge is unusual. Unless you work with wires or rub your feet across shag carpeting, you don't normally sense charge imbalances at our macro level. But at the nanometer level, charge moves things around. Each atom is made of heavy protons with a positive charge and light electrons with negative charge. When these charges are symmetric and balanced, the overall charge is neutral, but when they fall askew, a chain of domino effects can start, and chemistry can happen.

Here it's truer than ever that opposites attract, because positive and negative charges pull together. If the pulling forces put two negative electrons between two positive protons, the negatives tie the positives together with a chemical bond. A group of likewise bonded atoms is a molecule. If opposite charges are on two different molecules, the two will pull together. The chemistry of Mono Lake—all chemistry, in fact—is formed from patterns of positive–negative interactions.

Each "-ate" molecule noted earlier is a different element coated with three or four negative oxygen atoms. These are strong negative charges and can pull on neighboring molecules. Sometimes a water molecule comes too close, so that the "-ate" molecule's negative oxygen attracts the water's positive hydrogen, and yanks it right off. Take away an H^+ from H_2O water, and OH^- hydroxide molecule is left behind.

So when rock dissolves in water, the rocky "-ate" molecules make enough OH^- hydroxide to change the water chemistry. This change is measured by the pH scale. If almost all of the water is intact, the pH is 7 and the solution is balanced and neutral. If water is pulled apart to make OH^- hydroxide, the pH is above 7 and the solution is basic. If instead it's pulled to make H^+ hydrogens, the pH is below 7 and the solution is acidic.

One rule of thumb is that dark granite is acidic, while light limestone is basic. The light-colored spires in Mono Lake indicate basic conditions, and the lake's pH is around 10. It has as much hydroxide "base" as milk of magnesia and is only a pH step or two away from ammonia. These basic solutions feel slightly slippery, almost thick, as if you could dive in and leave behind a splash of stone.

So Mono Lake is no garden spot, but neither is it dead. Local geology made the lakewater hard and basic, which shapes the local biology. Fish cannot survive in Mono Lake's high-pH water, but smaller, more nimble organisms can adapt and get everything they need.

In spring, the lake turns green with life as lake algae gorge on the swollen streams' runoff. The rivulets that feed the lake bring rock molecules such as phosphate for food. Sunlight gives algae power to grow, so they will piece small molecules together into larger molecules that life can use: sugars, fats, and proteins. Brine shrimp eat the algae, and birds eat the brine shrimp—not to mention the black flies that swarm, breed, and thrive on the lakeshore, which from a human perspective may be a little too productive.

But not all rocks are good for life. The high levels of dissolved phosphate in the spring runoff are accompanied by high levels of dissolved arsenic in the oxygenated form of arsenate, which is also concentrated in the lake. Life has learned to live with the arsenic. It was this extreme ecosystem that attracted scientists curious about the biochemistry formed from this poisoned geology.

BACTERIA THAT DEFY ARSENIC

This brings us to the December 2010 press conference about arsenic life. Ever since the 1980s, the era of the cold fusion debacle, scientists have learned to be suspicious of science by press conference. Still, this had something that cold fusion didn't: peer review and publication in a prominent journal. A group of scientists, including members of the NASA Astrobiology Institute, published evidence in the journal *Science* that bacteria from Mono Lake could use arsenic instead of phosphorus.

The question asked in the 2010 *Science* paper was provocative: Did the extreme environment of Mono Lake force life to build with poisonous elements? In particular, could a bacterium in Mono Lake have performed the biochemical alchemy of living on arsenic, turning a poisonous sword into a productive plowshare?

Felisa Wolfe-Simon is listed in the first, best spot on the 2010 *Science* paper that claimed that bacteria can pull this off. She grew bacteria in the lab from Mono Lake mud. Day after day, she would reduce the phosphate levels while keeping the arsenic around, in order to prove that the bacteria were not only *tolerating* the arsenic but actually *using* it. Day after day, she came into the lab and the bacteria were still defiantly growing.

Wolfe-Simon removed the phosphate because of a pattern in the periodic table. There, the symbol for arsenic (As) is directly beneath the symbol for phosphorus (P). Each column in the periodic table contains a family of elements that have a similar electron arrangement and therefore often do the same chemistry. Phosphorus is used by all life to build DNA and cell membranes. Once the phosphate was removed, the bacteria would be forced to borrow the next-best thing from the environment instead: arsenate.

The fact that elements in a column of the periodic table are chemically similar is incredibly useful in the lab. If you want to tweak a molecule, try switching out some of its elements for others in the same column. They often will bond the same but will have different shapes that change the chemistry just a little. In 2006, chemists built a new superconductor from lanthanum, oxygen, iron, and phosphorus. If phosphorus worked, the researchers reasoned, why not try its chemical cousin one box down, arsenic? They did, and it worked, if anything, even better.

In superconductors the two can substitute, so why not in DNA? Before finding the Mono Lake bacteria, Wolfe-Simon asked the same question in a 2009 paper titled "Did nature also choose arsenic?" She speculated that a "shadow biosphere" of arsenate-using organisms may have played a part in life's origins—in other words, an alien biochemistry.

The problem is that for all known organisms, phosphorus is essential for life, while arsenic is only essential for death. In this case, their chemical family resemblance explains why arsenic is such a dangerous poison (and useful operatic plot device). Today both molecules are found surrounded with oxygens, as the "-ate" molecules phosphate and arsenate. As the periodic table would allow us to predict, their shapes are the same, and their sizes are also practically the same: if phosphate is a handball, then arsenate is a racquetball. If you can mistake one for the other when rummaging around inside a gym bag, then the cell likewise can grab an arsenate when it intends to use a phosphate.

Phosphate and arsenate may look the same, but they do not act the same. If you ingest too much arsenic, you will experience headache, confusion, and sleepiness. Your body tries to expel it out of, ahem, one end or the other. Many different parts of your body break down at once. Your stomach and muscles ache and convulse, and your kidneys malfunction. Everything that grows will misfire, whether it's your hair falling out in clumps or your fingernails turning white. Enough arsenic causes coma and death.

The difference between life-giving phosphate and life-stealing arsenate is all in the timing. Phosphate latches onto other molecules and stays bound for days, but arsenate binds and then drops off in a matter of seconds. Phosphate is the studious engineer that builds bonds to last, while arsenate is much more easily distracted. Arsenate's bigger size makes for longer bonds allowing just enough room for water. Water, pressing in from all directions, squeezes into arsenate and breaks its bonds. In a test tube, arsenate chemicals break down in seconds. Arsenic simply does not cohere with the body's proteins and metabolites—it sticks but does not stick around.

Arsenic is toxic when it is used by the body in the place of phosphate. It is like an old post-it that falls off too fast. Phosphate sends essential messages for muscle contraction and growth throughout the body, which arsenate disconnects. When the phosphate messages die, so does the cell. Rapidly living cells (hair and kidneys) are the first to feel the pinch, but all cells use phosphorus, so all cells are in danger. Heart cells without energy shut down and die.

Arsenate also carries a second kind of danger. Because it is bigger than phosphorus, arsenic has more room to carry extra electrons. In a cell, arsenate absorbs electrons like an electronic sponge, and this sabotages the cell's electron balance. Arsenate sheds its electrons randomly, and these extra electrons react powerfully with molecules, especially oxygen. Random reactions shred the cell like interior shrapnel. In an experiment in which yeast cells were fed arsenic, when the electronic balance was disrupted, the DNA fragmented like a dropped egg. Arsenic also pushes proteins out of shape by interfering with their sulfur atoms.

From this perspective, it seems like folly to grow cells in arsenic-rich, phosphate-poor broth, but the proof is in the experiment. Wolfe-Simon's logic was also based on the periodic table, and in the fact that microbes thrive in Mono Lake, tolerating chemical conditions that destroy plants and animals in hours. A single-celled organism without complex systems may be able to heave a microbial sigh and turn in the absence of phosphate to the next-best thing, even if the next-best thing is normally poisonous. Could a microbe turn an element of death into one of life?

Wolfe-Simon's experiments showed that the microbes could live in high levels of arsenic and very low levels of phosphate. This was an exciting result, but still, just the beginning. Survival in arsenate does not mean the cell is made of arsenate. Atoms are too small to see with a microscope, so Wolfe-Simon and company looked at the cells themselves, and observed bacteria that looked inflated and shot through with huge holes filled with arsenate. Bacteria usually do this when faced with a toxin—they shove the toxin inside a bubble like they're cleaning house by shoving all the extra stuff in a closet. Just don't open the closet.

Wolfe-Simon and colleagues still had to show that the arsenate *replaced* phosphate in essential molecules. Inside the cell, phosphate sticks to proteins, coats cell membranes, and floats around as small energy-bearing molecules like adenosine triphosphate ("ATP," in which "TP" means *tri*-phosphate). Perhaps most important, phosphate forms the backbone of DNA, the molecule that tells the cell how to make all the other molecules. If Wolfe-Simon could find arsenate not just lightly stuck to, but permanently fixed in the backbone of DNA, that would close the case.

But the smoking gun wasn't found. The rest of Wolfe-Simon's paper contains several surprisingly weak experiments. Isolated DNA showed a faint shimmer of arsenate in a large DNA molecule, but the shimmer was so faint it could have been background noise. X-rays shot through the cells produced a complex shape of data. This did not look like plain old arsenate, so it might have been bound to *something*, but then again, it didn't clearly look like anything else either. The data were faint and the error bars shown bracketing the data were too large for many scientists' comfort, including my own. Instead of a firm conclusion, we had a mystery.

THE 2011–2012 MONO LAKE BACTERIA TOUR

Six months later, *Science* magazine took a step I had never seen before. It published eight separate critiques, each short and argumentative, all written by scientists questioning Wolfe-Simon's paper. Chemists cited arsenate's fragile bonding chemistry; biologists questioned the techniques; everyone asked why more detailed experiments weren't done. (Now, this is the kind of comment that can always be leveled at a paper—it is much easier to propose an experiment than to carry one out.)

The real hope of changing minds was not in print but in the lab. Wolfe-Simon and others sent bacteria around the world, and other scientists set to work. One researcher in Canada even blogged her experiments daily. In summer 2012, a year and a half after *Science* first published Wolfe-Simon's results, two papers appeared in *Science*, followed by two more in other journals, constituting a parade of evidence:

1. A lab from Switzerland found the Mono Lake bacteria grew at "very low" concentrations of phosphate but *not* at "very, very low" concentrations. Some trace phosphorus could have contaminated Wolfe-Simon's original media. This Swiss lab also used a technique called mass spectrometry that could essentially weigh individual molecules to look for heavy arsenic atoms. Arsenic levels in DNA were too low to be measured, meaning that the DNA was at least 99.99% phosphate.

2. Rosie Redfield and her lab were the live-blogging Canadian scientists. They also found that the Mono Lake bacteria would not grow in "very, very low" concentrations of phosphate, and their mass spectrometry results also came up empty. The bacteria acted the same in Vancouver as they did in Switzerland. Finally, Redfield stored the DNA in water for two months, and, anticlimactically, nothing happened. Every arsenate-linked molecule known to chemists reacts with water, but this DNA was as sturdy as normal phosphate-linked DNA, mostly likely because it *was* phosphate-linked DNA.

This is enough evidence to provide the verdict that the Mono Lake bacteria are not arsenate utilizing but are merely arsenate resistant. This showed that they are *not* arsenic eaters, but it didn't show what they *are*. How are they so good at discriminating phosphate handballs from arsenate racquetballs? Two more papers answered these questions:

3. Researchers from Israel used a complex and slow technique with X-rays that can see individual atoms. They examined phosphate-binding proteins outside the cell that find phosphate and let it into the cell. In an arsenate-rich environment, these phosphate-binding proteins would be challenged with thousands of arsenate imposters for every phosphate. This lab looked closely at the Mono Lake bacteria's phosphate-binding protein, to find out why it was especially good at accepting phosphate and binding arsenic.

Their pictures showed that the Mono Lake protein has a phosphate-binding hole. This hole is like a lock, and phosphate fits inside like a key. Phosphate's four oxygens are spread out from each other and stick out evenly from the central phosphorus. This tetrahedral structure is like a camera tripod with the top camera-holding arms extended fully up: the phosphorus is at the middle where the four arms come together and at the tip of each arm is an oxygen atom. The phosphate-binding site on the protein is a "lock" that mirrors phosphate's four negative oxygens with a slightly larger tetrahedral hole of four positively charged hydrogen atoms (Figure 1.2). The unbalanced charges in both the protein and the phosphate become balanced as opposite charges attract and tiny magnets snap the phosphate into place.

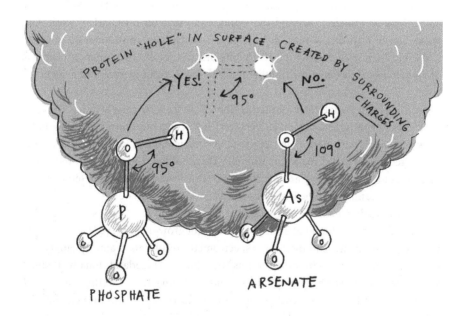

FIG. 1.2 Phosphate and arsenate molecules both adopt similar geometries, but a protein can distinguish between them by the O-H angle shown at the top of each molecule.

The Mono Lake protein has one twist, but it's a good one. On each tetrahedron, one of the oxygens has pulled an extra hydrogen from water. The angle for this hydrogen is 95 degrees for phosphate and it is 109 degrees for arsenate. This structural difference can discriminate between toxin and nutrient. The Mono Lake protein puts a negative oxygen atom that fits with a 95-degree angle for hydrogen but not a 109-degree angle. This one feature holds phosphate in place while arsenate slips out.

4. The last piece of evidence was making the numbers add up. Wolfe-Simon's very low phosphate level was not enough for a cell to live on, so how did they make up the difference? A lab from Miami found out a somewhat grisly secret: these bacteria scraped up phosphate by cannibalizing the ribosome, the most important molecule in the cell.

Ribosomes are protein-making molecules made from RNA (a molecule that looks just like DNA except for one oxygen-hydrogen pair). Every protein in a cell is assembled by a ribosome. A full quarter of the dry weight of a bacterium is in the ribosomes alone. About one-quarter of the mass of RNA is phosphate, making it a tempting target for a phosphate-starved bacterium. The Miami lab saw the Mono Lake bacterium breaking down its ribosomes for their phosphate backbones. Needless to say, this is not a sustainable strategy in the long term. The cell is chopping up its furniture to feed its furnace.

All four studies converged on a single conclusion: the Mono Lake bacterium *resists* using arsenate as much as possible. Life as we know it will accept no substitutes, and does not incorporate arsenate but finds new ways to reject it. This was not what Wolfe-Simon was looking for, which makes it even more convincing. The biology changes constantly, but it is molded around a solid, incontrovertible fact of chemistry: arsenate is not suitable for building DNA. Even in Mono Lake, arsenate DNA would be a house of cards thousands high, falling apart within seconds.

PHOSPHORUS: THE LAST ELEMENT STANDING

All this re-emphasizes something biologists have known for a long time, that phosphorus has an unavoidable association with life. In Mono Lake it makes algae bloom in the spring. This extends back in time. Old rocks, after glaciers pass through, have suspiciously high amounts of phosphate. It looks like advancing glaciers scraped phosphate out from rocks and fed it to the sea, seeding a spring-like burst of plant and microbe activity. More phosphorus made more life, as geology led to biology through chemistry. This has implications for all of biochemistry, because other elements that go well with phosphorus also go well with life.

In terms of the periodic table, phosphorus has biological advantages over arsenate below it, over nitrate above it, over sulfate to its right, and over silicate to its left. In fact, it has advantages over every other element on the periodic table in forming a medium-strength bond to negative charge. This makes it the best choice on the table for energy and information chemistry.

To understand the chemical advantage of phosphorus, let's shrink to the scale where we can watch atoms move, bond, and fly apart. Instead of an atmosphere, a microbe is surrounded by a blizzard of flowing water molecules. Everything skitters around like a butterfly unless it is tied down. Movement is mediated by the constant random jostling and flow of molecules. It's like trying to move through the crowd at a big outdoor festival. In this turmoil, the best shapes will fit together like three-dimensional puzzle pieces, propelled by random motion and guided by electric fields of charge. Stable arrangements of atoms that fit together with shape and charge will hold together better (like phosphate in the Mono Lake phosphate-binding protein).

In this context, a molecule needs three characteristics to be useful for building life's structures:

1. It must be *available* in water.
2. It must have a fitting *shape and charge* for its purpose.
3. Its *bonds should last* as long as they need to (and no longer).

Both arsenate and phosphate are available in Mono Lake, and have similar useful shapes and charges, but the third point distinguishes the two: phosphate bonds for a long time, while arsenate does not. Even in a lakeful of arsenate, life chooses phosphate.

To understand phosphate's unique qualities, start where a chemist starts: its tetrahedral (that is, four-sided) shape, as shown in Figure 1.2. When dissolved in water, the molecule tumbles freely, and the four oxygens whirl around the phosphorus center. From a distance, these atoms blur together into a tiny ball of negative charge, joining with the opposite charges on water molecules in an intricate dance.

No other element is used for this purpose by life because no other element has chemical properties like phosphorus. All of the possible chemical options are shown on the periodic table. Take a moment to find the periodic table (Figure 0.1) at the beginning of this book and consider all those boxes, each representing a different element. (As we go through this book, refer back to that table like a map. It works like the map of Middle Earth in Tolkien's *Lord of the Rings*.) At first, it seems like the 90 naturally occurring elements give lots of options to make something else like phosphate. But the options quickly narrow.

First we have to cross off the elements in the first two rows of the periodic table (from #1 hydrogen through #10 neon). These elements are all too small to fit the four bonds to oxygen needed for our tetrahedral arrangement, so they don't fit rule #2. Then we have to cross off the elements in the third row and below (after #37 rubidium). These are big enough, but too rare or too insoluble in their oxygen-bound "-ate" form, so they don't fit rule #1.

Then consider the columns. The number of electrons exposed in the outermost level of each atom increases from left to right in the table. The farthest left column has one electron and the next column has two (as does the block of elements running from #21 scandium to #30 zinc in general); #5 boron's column has three available electrons, #6 carbon's has four, on out to the rightmost row, which has eight electrons. About one electron in the outer shell is needed for each bond to oxygen. So we have to cross off everything left of #5 boron, which has too few electrons to bind four oxygens, and also cross off the rightmost row, which has too many electrons. This rightmost row, in fact, is entirely composed of gases that don't react with anything else, snobbishly refusing to form bonds (and somewhat spitefully named the "noble gases").

When you sit back and look at our once-promising collection of candidates, everything is crossed off except for aluminum, silicon, phosphorus, sulfur, and chloride. (We skipped over a few exceptions, but even those are crossed off with a little more investigation.) Just by asking how life can form a tetrahedron in water, we have reduced the possibilities from 90 naturally occurring elements to five.

But four of these five don't work as well as phosphorus. The oxygen-covered forms of aluminum and silicon are rocks, not metaphorically, but literally. They form strong bonds to each other and like to share their oxygens in long, amorphous linked chains. Sand and glass are sharp-edged silicate materials. These are too solid to allow for the flow a living organism needs to change and respond to its environment. They are as frozen in place as a victim of Medusa and would slice through the flowing structures of life.

On the right end is chlorine, which surrounded by four oxygens is perchlorate. Perchlorate reacts so readily that it is used in rocket fuel. Perchlorate is related to the molecules in chlorine bleach. Like arsenate, it is poison, not food. Some fascinating microbes eat chlorate molecules, but they do not appear to build with them.

That leaves phosphate and sulfate. Like phosphate, sulfate can bond biological molecules and is found doing so *outside* the cell for chemical reasons described later. But phosphate can perform the chemical trick of linking itself when sulfate cannot. If a cell chains two sulfates together in water, half of them will be gone in minutes, while two chained phosphates will last a thousand days. As a chaining molecule, phosphate is eminently transferable. One oxygen can be its left hand and another its right hand. Phosphate can be passed between all sorts of different groups by switching what it's holding with its left hand, and then its right.

Some cell signals are turned "on" by attaching phosphate to the turned-on molecule through one of its oxygen hands. I imagine the phosphate glowing faint green, both because pure phosphorus can glow green, and because it signals that a pathway is "on," a "green light" to the rest of the cell. As the signal propagates, more proteins are attached to phosphate, and the cell begins to fill with glowing green beacons. Eventually a protein is turned on that tells the cell to move something, to make something from DNA, or to do whatever the cell may require. Phosphate is transferring information from one part of the cell to another. Because phosphate is bound to the protein with a medium-strength bond, that bond can be broken with a flick of an atom, allowing the signal to be turned off as quickly as it was turned on.

Phosphate may be even more useful for energy transfer than it is for signaling. Two or three phosphates linked together are held in tension. The medium-strength bonds that hold the phosphates together also hold 7 to 10 negative oxygens next to each other. These oxygens repel the negative oxygens in water, protecting the bond from being broken by water's reactivity, but they also repel each other (if opposite charges attract, like charges repel). When the bond is broken, the oxygens repel apart and accelerate the breaking, releasing energy as they go. Even more important, the released phosphate interacts better with water. Overall, because ATP is more stable broken than joined together, breaking its three phosphates apart releases energy.

Four slightly different triple-phosphate (abbreviated "TP") groups are found in the cell, called ATP, GTP, CTP, or TTP, depending on what molecular "handle" they have attached to them (abbreviated A, G, C, or T). Your energy-transferring processes turn the food you eat into linked phosphate groups, the most important of which is ATP. Your muscles get energy to move by splitting phosphates off ATP. Also, the first thing you do to break down the sugar glucose is put a phosphate on it. Phosphate both sends signals and holds chemical power.

Finally, phosphate's energy helps build the most important of all biomolecules, the long-term information storage molecule DNA and its sister molecule, the short-term information transfer molecule mRNA. There are four types of "TP" molecules, each of which matches one of the four "letters" in DNA.

The four oxygens that surround phosphorus give it a negative charge in water, even when it is linked on its left and right. This is why both DNA and RNA end with "A" for "acid," because a negatively charged phosphate is an acid. An acid is something that has shed a positively charged hydrogen in water, leaving a negative charge behind.

Phosphate may be the only way for nature to make abundant charged chains in water. The negative charges repel and spread the long DNA chain out like a ticker-tape, which is more easily read than a tangled mess. The whole point of DNA is to hold information that is read like an open book, and phosphate keeps the book open.

Steven Benner is a chemist who redesigns DNA's "bases," which is another name for what I have called the "handle" end of the nucleotide, the A, C, T, and G parts of DNA. Benner has chemically welded together new alternative bases with remarkable success, and he has gone on the books as saying that the naturally occurring base structure "is a stupid design" that he can improve. I take his word for it on that.

But the reason Benner redesigns bases is that he couldn't redesign the phosphate at the other end of the molecule. No chemist can redesign the periodic table to find another element that links DNA as well as phosphorus. Years ago Benner started his design program by trying molecules other than phosphate in the DNA that had none of phosphate's negative charge. They would tangle and fold without phosphate's charged self-repulsion. Benner's research bore fruit only when he turned to the other end of the molecule. (Other scientists have introduced positive nitrogen-based charges to the backbone, but nitrogen doesn't work for phosphate's other purposes, because three nitrates together are no more stable than three sulfates together.)

In a cell, phosphate is so important and abundant that it causes the problems of abundance, particularly when it comes to charge. If all of these phosphates are transferring information and energy inside the cell, wouldn't that mean that the inside of the cell should have a negative charge overall? Such a huge charge imbalance would crumple the cell, or promote frequent electrical discharges that would undermine the stability needed for life.

The cell needs something else to balance phosphate's charge. It needs something positive that can stick around well, but not too well. It needs something discreet that gets out of the way when it's not needed. So another element must stand beside phosphate, if phosphate's negative charges are to work inside the cell.

MAGNESIUM: PHOSPHATE'S INVISIBLE PARTNER

When we teach about ATP in high school biology, we teach it wrong, because we have to. By editing it down from "adenosine triphosphate" to "ATP" we make it possible to fit into a high schooler's brain along with the rest of adolescence. Come to think of it, it's a miracle that we get ATP to stick in there, considering. But our terminology, shrunken to three letters, edits out ATP's constant chemical partner.

Negatively charged ATP must be balanced with positive charge. In the cell, positively charged magnesium provides this balance. The chemist in me says we should write it as Mg-ATP, but the contingencies of history make that collection of five letters pretty much unpronounceable in English ("Em-gat-puh"?). We are stuck with "ATP" just like we are stuck with QWERTY keyboards. Whether our labels recognize it or not, magnesium is essential wherever phosphate functions, from ATP to RNA.

Many elements have two positive charges like magnesium, but magnesium's size sets it apart. If we need a positive charge, we need a metal on the left side of the periodic table, so cross off the whole right side of the table. On the left side, water reacts with each element and takes away its outer electrons, so the leftmost column loses its one outer electron and has one positive charge as a result. One positive charge is too weak to effectively balance the negative phosphates, so cross that column off. Three positive charges are too strong—they'd stick but we'd never we able to pry them off to move phosphate around. This excludes the column under aluminum on the right. Between these two extremes are elements that are mostly +2.

Next we consider abundance (rule #1). Since phosphate is abundant in the cell, its balancing agent must also be abundant. Like before, we cross off the bottom half of the table because the bigger elements down there are not abundant enough. Even if the first long row, from scandium to zinc, was abundant enough (which I'm not sure about), all of those elements stick a little too tightly to phosphate as well. They form high-strength phosphate bonds and solid phosphate rocks, not medium-strength phosphate bonds.

We are left with one column: beryllium, magnesium, calcium (the three atoms below them went home during the previous steps). Since atomic size increases as you move down on the periodic table, beryllium is smallest and calcium is largest. All three of these have two positive charges to effectively counterbalance phosphate's negative charge.

But like Goldilocks with her three porridge bowls, we can reduce this list from three options to one. First we have to account for the fact that we need something that sticks to three phosphates in ATP, not just one. Magnesium fits perfectly between the oxygens on two adjacent phosphates in ATP, but calcium is too big. This makes magnesium the optimal elemental partner for phosphate.

What about beryllium? Even though it's small and appears as easy to make as the other small elements, because of a quirk of nuclear physics, not much beryllium is actually made in stars (see Chapter 3). This is actually very good for us, because beryllium is stickier than magnesium, so sticky that it is toxic to life. Beryllium allergy is caused when it sticks to an oxygen-rich pocket on the surface of a cell and provokes an immune response. Because the beryllium never unsticks, it sets off alarms and immune inflammation that wreck the body.

Magnesium's stickiness to phosphate is used by the bacterium that causes tuberculosis. It puts magnesium in a protein toxin that uses the magnesium's positive charge as something like bait. The magnesium attracts the phosphates from your RNA strands and then reacts with them, snapping them into little pieces. Magnesium forms the edge of a knife that tuberculosis uses against RNA.

But the tuberculosis bacterium must be careful with this magnesium knife, because it can cut up the bacteria's RNA, too. The microbe makes another protein that acts as a sheath for the knife and covers the magnesium. Even though magnesium is just one atom among thousands in the protein chain, it is at the very center of how this protein knife works. Chemists are already trying to design a small molecule that can stick to magnesium and blunt tuberculosis's magnesium knife.

Because of the phosphate-magnesium balance, RNA in the cell carried around a positively charged cloud of magnesium ions, a sort of magnesium aura. Magnesium is especially important when RNA must fold up into a compact shape to do some work, like when the ribosome folds up to build new proteins. The ribosome's negative phosphates are knit together with magnesium, stabilizing it with a web of positive and negative charges. Ribosomes in plants without magnesium fall apart, and the plants turn old before their time because they are missing this one element.

Many smaller chains of RNA form compact structures with specific shapes that help specific reactions along. For a century we've known that proteins do this, which we call enzymes. When RNA does an enzyme's job, it's called a *ribozyme*. Scientists who study ribozymes know the value of magnesium. If they forget to add magnesium to their experiments, their ribozymes will fall apart. That's the kind of thing you don't forget twice.

Experiments have systematically stepped through the periodic table and tested how different metals can tie together the phosphate chain of an RNA ribozyme. This reveals that magnesium is indeed the best element for this job. Magnesium, with its two positive charges, worked so well that we can say, given the choice between one magnesium and a hundred singly charged potassium atoms, this RNA would still choose magnesium. Magnesium also binds on a timescale faster than milliseconds. Magnesium fits quickly as well as tightly into the RNA chain of the ribozyme.

Magnesium helps with DNA, too. For example, it is an essential ingredient when enzymes proofread DNA. On Figure 1.1, where DNA rises from Mono Lake, magnesium is found in its favorite place, nestled between the DNA phosphates. Most pictures of DNA don't include this aspect of its structure because the magnesium moves around so much, but if the magnesium wasn't there, the negative phosphates would crumple the cell within seconds. Magnesium and phosphate pair as effectively as steak and red wine.

CHEMICAL RULES AND A *WONDERFUL* DISPUTE

The magnesium-phosphate pairing is only the first of many. In a sense, chemistry is the study of pairings. Chemistry studies the bonds between atoms, and every bond is a pairing with two and only two sides. Even the immature form of chemistry known as alchemy had a place for this idea.

On the biochemical timescale, some pairs are permanent, some are fleeting, and some are repulsive and anti-bonding. Some are very picky about direction, and some don't care which way is up. But all of these bonds come from atoms attracting and repelling through electron imbalances. All are the consequences of the contrast between two physical pieces: miniscule, flighty, negatively charged electrons dancing about; and heavy, solid, positively charged atomic nuclei anchoring the electrons through opposites attracting.

The same physical laws that say that positive and negative charges attract add up to say that, in water, magnesium and phosphate pair especially well. One member may be less famous, like magnesium, but general chemical rules tell us that it is still important. When that pairing chemistry is changed even slightly, like if arsenate pairs more fleetingly with its partners, the consequences can be life-threatening.

Life has a vested interest in collecting phosphate and magnesium, and in rejecting arsenate. A set of protein sensors monitors levels of each and keeps them in balance. The phosphate-binding (arsenate-rejecting) protein from Mono Lake bacterium lets phosphate in. At least two proteins open doors into the cell specifically for magnesium. One particular magnesium transporter opens in response to an ATP-magnesium "key," again showing the complementarity of those two molecules.

Not all elements are welcome. Another protein patrols the interior of the bacterium *E. coli*, looking for arsenate that sneaked inside. When arsenate fits into its arsenate-shaped binding site, the protein changes shape and binds DNA, which activates arsenate-cleaning proteins that block arsenate's oxygens with carbon-hydrogen groups. The intruder is neutralized.

Similar arsenate-cleaning proteins are found in plants and animals, especially those chronically exposed to arsenic. In some cases, the periodic table can explain mysterious chemical effects. For example, *Leishmania* parasites are susceptible to treatments with the element antimony (#51 Sb on the table). But in certain areas of India, such treatments don't work. Some scientists think they know why, with some experiments in mice to back them up: antimony is right below arsenic on the periodic table and the two are therefore chemically similar. In those areas of India, arsenic frequently contaminates drinking water, and the people who live there have upgraded their internal arsenate-cleaning processes. Because antimony is so chemically similar to arsenic, the two are swept up by the same processes, throwing out the antimony that would otherwise destroy *Leishmania*. The law of unintended consequences extends to chemistry and can be understood with a glance at the periodic table's columns.

A flotilla of proteins in the cell respond to specific elements, whether as sensors deep inside the cell or transporters opening doors in the cell membrane for needed elements. In the *E. coli* bacterium, at least five different transporters bring iron inside the cell, which is a huge energetic investment and shows that iron must be pretty important to life. There are also five transporter proteins for zinc, although only two let zinc in and the other three push it out. Copper has two doors opening out; nickel, manganese, and molybdenum each have one door opening in; some magnesium proteins also work for calcium; and one protein ejects both nickel and cobalt. Sodium and potassium also have proteins responsible for them, pushing sodium out and potassium in.

In this way, the proteins in the cell tell us which elements are crucial for life. Why in and not out, or out and not in? Each has its own chemical reason and rule. Many, like phosphate, play irreplaceable and unique roles. Each element has a distinct history, too. Some have been around since the original quickening of life, and some came late to widespread biochemical use. This history was ordered and sequenced by the rules of chemistry. It even has a chemical direction, moving from left to right on certain segments of the periodic table.

All of this says that, whatever life is, it is not entirely random. Neither is it entirely determined. In chemistry, gases are random, solids are set, and liquids are the in-between stage that flows with motion that is random at the atomic level but predictable at the human level. Life itself is more like a liquid than a solid or a gas. Life is carried by a river made of chemistry that is pulled along, not by gravity, but by the predictable rules of chemical stability.

The great science communicator Stephen Jay Gould disagreed with this. Gould's book *Wonderful Life* (1990) describes the evolution of life as a "lottery" with "thousands of improbable stages" (pp. 47, 238). Most strikingly, Gould refers to the "tape of life": "Wind back the tape of life . . . let it play again from an identical starting point, and the chance becomes vanishingly small that anything like human intelligence would grace the replay" (p. 14). Gould was speaking of an event that we will return to in Chapter 9, but his argument has been so successful that it needs to be addressed from the beginning.

Gould documented well the damage done by wrong scientific stories. This made him skeptical of grand narratives. So this book's grand narrative must start from the evidence before it is knit together in a narrative. This story is built from three areas of evidence: rocks (geology), genes (biology), and the chemical rules that tie the two together. To represent these other disciplines, I'd like to introduce Gould to the chemist R. J. P. Williams.

R. J. P. Williams cowrote a book titled *Evolution's Destiny: Co-evolving Chemistry of the Environment and Life* (2012) about how chemistry guided evolution. I owe Williams for much of this story. In fact, Williams has been writing books like this for decades now, based on chemical laws. Only in the past few years has the biology caught up with Williams's chemical predictions, and by and large, it tells the story he expected.

Overall, I think Gould is right when it comes to individual species, but I think he is wrong at the broader levels of ecosystems and planetary evolution. Williams gives reasoning, evidence, and tools that tell a grand narrative tied together by chemistry.

From the proper perspective within this narrative, natural history can indeed be predicted. In the past decade, microbiologists have also challenged Gould by actually running the tape of life repeatedly in the lab, and they too have found predictable patterns.

From this, I conclude that the tape of life is likely more predictable than Gould thought—something less like a tape and more like a river, its liquid flow channeled by the solid banks of chemical laws. We've already seen why this river would flow with phosphorus and magnesium, and not with arsenic. The remaining chapters will show why for the rest of the elements.

In each case we begin with the living, thriving specimens at hand. The chemistry they nimbly perform tells of the underlying rules and pairings of chemistry. Chemists can mimic (as through a glass darkly) some aspects of life in the lab. These experiments tell us the chemical rules, and the best summary of these is in the grid of the periodic table. The periodic table is a map that will guide us through the history of chemistry on this planet.

By paying close attention to biochemistry, a story begins to emerge of order in disorder, of ingenious organisms persevering through ages of time, of poison and food, of sunlight and water, death and life. It is a story written in elements and channeled by chemical rules of energy, flow, and pairing. We enter the story midway along the road of life, and we begin by looking behind us and counting the footprints.

2

PREDICTING THE CHEMISTRY INSIDE A CELL

scraps of paper in a tiny sculpture // sulfur's special role // how to reject gold // cadmium's deadly mimicry // electrons and oxygen are opposites // the chemical power of the metals at the heart of life // the chemical left hand matches the biological right hand // in medicine, bismuth trumps zinc // zinc knits proteins together // the birth of the Accidental Enzyme

THE WORDS INSIDE THE WALL

The process of scientific discovery is something like a walk near Freswick Castle. I assume you've never been there. (Neither have I, but a friend has.) Freswick Castle stands at the end of Scotland's northeast end, at the mouth of the Burn of Freswick in the district of Caithness. As of this writing, it is unlisted in Google Maps, and I had to manually scan the coast to find it. Outside the castle is a simple, unlabeled structure that doubles as a biochemical parable.

The castle itself is narrow and three stories tall, with orange shingles and gray stone, set on an arc of narrow beach between hills to the north and cliffs to the south. The building is approximately the cruciform shape of a shrunken cathedral, with the rightward wing moved to the top of the structure so it resembles a lowercase *f*. If you wander the grounds near Freswick Castle, you will discover a stone wall in the wind-blown waves of yellow-green grass, worn but still standing firm like Hadrian's Wall. From above, it is a period preceding the castle's *f*.

Let's approach this as a scientist, with measurement. From the castle side, this structure resembles the circular stump of a roofless tower, eight feet tall and twice that wide. The stones are ancient sand, compacted and weathered, stained different shades of red from iron deposited millions of years ago, but the mortar is new.

But inspection is not enough—we should go in. Walk around to the other side, and an opening appears, as shown in Figure 2.1. The structure is not a closed circle, but it is a spiral wall open to the sea, and to you. Inside, a small stone bench invites you to sit. A window slit next to the bench is an eye to the outside. Surrounded by a jigsaw of rocks, you can hear the echo of waves all around and watch the blue-gray sky above. If the spiral's opening is a mouth, then you are Jonah in the whale. You are both inside and outside at once.

The architect of this singular structure wants you to discover it this way. (I hope I don't subvert his purposes too much by publishing them here—if you happen to see

FIG. 2.1 A view of the entrance to the sculpture *ekko* by Roger Feldman, Freswick Castle, Scotland.

this in person, please act surprised.) My art professor friend Roger Feldman traveled to Scotland for two summers to repurpose old castle rubble in this manner for others to discover. He calls it *ekko* and Vimeo has a video with more information about it.

Roger wants you to experience his sculpture by entering into it. In the same way, you have to experience science by placing yourself inside it. Imagine approaching a cell as a tiny, mysterious, found structure like *ekko*. You cannot physically walk into it, but you can ask questions with well-planned experiments and can envision what it's like from the results of those experiments. Hundreds of experiments can be summed up with the imaginative exercise of walking into a cell like you're walking into *ekko* and looking around. Inside this imaginary cell, some things will be familiar and some will be alien.

Every one of us stumbles across mysteries in nature. Before Darwin, William Paley told a famous story of discovering a watch in the grass of an English meadow. Paley said finding life is like finding that watch. Here we have the same island and metaphor, but with very different implications.

Like *ekko*, and unlike Paley's watch, a living cell is open to, and changes with, its environment. *Ekko* is not a mechanism but a place, a piece of nature carved out by a wall. It is made from the same stuff that surrounds it, and there is a simple elegance to it so that I can imagine how it may have come together on its own.

At the microscale, the stones that form *ekko* would be a dozen atoms wide and would be tossed by the constant atomic motion like feathers in the breeze. Now imagine if these tiny stones had positive and negative charges on their sides. They would flow together, clicking like magnetic bricks, forming lines and sheets. If one of those sheets came to rest in a curled, not-quite-closed spiral position, that would be *ekko* on the atomic level, built not with hands, but with chemistry. The cell looks like that. Each cell has a wall of self-assembling molecules that form from a simple pattern of charges, for example.

That's all there is to *ekko*, but inside the cell the mystery deepens. There are no signs or instructions outside or inside *ekko*, but deep inside every cell, there are words. These words are written in DNA, which is nothing less than a chemical string of letters that are read in a particular direction. DNA has only four nucleotide letters, written in carbon with serifs of oxygen and nitrogen and connected with phosphate. These carry information depending on their order, making words that form instructions written in a radically foreign language.

Finding a cell is like finding a micro-*ekko*, and finding DNA deep within the cell is like finding some scraps of old paper on the bench inside. The scraps of DNA hold old historical legends. Extending the metaphor, a philologist friend who loves those old languages can translate and contextualize those legends. In the same way, chemical experiments can reconstruct the past. In the rest of this book, we will read the messages from DNA and combine them with messages from rocks. The laws of chemistry tie the two together in a story that is the history of the universe.

A TAPESTRY FROM THREE THREADS

In our investigation of these scraps of paper, it soon becomes obvious that DNA words are not the only type of chemical word-string found within the cell. All life speaks with three kinds of chemical threads: DNA, RNA, and proteins (Figure 2.2). In this sense, all life is trilingual.

Well, maybe bilingual, because DNA and RNA are basically the same language. Each "letter" of DNA and RNA is only different by two atoms distant from the information-carrying bits. Two strands of DNA and RNA made with the same letters will easily align with each other if you mix them in a tube.

On the other hand, the third type of thread is proteins, which form a language completely different in chemistry, shape, and purpose. The ribosomes I mentioned in Chapter 1 use RNA letters as a pattern to put amino acid letters together into protein strings, changing the language from RNA to protein. Biochemists appropriately call this process "translation."

This means that DNA, RNA, and protein are woven together as three threads of information in every cell. In complex organisms, DNA is protected and defended from insult in the cell's nucleus, because it holds the ultimate pattern for everything that follows. When the cell needs a protein, the DNA opens up its library stacks, exposing a

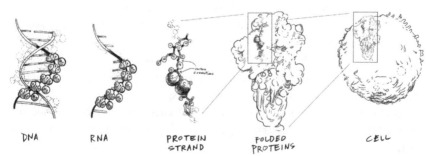

DNA   RNA   PROTEIN   FOLDED   CELL
                          STRAND   PROTEINS

FIG. 2.2 The three threads of information and structure in life: DNA, RNA, and protein. Proteins fold into three-dimensional structures that can carry out chemistry inside, outside, and on the surface of cells.

particular string of letters telling how to make the needed protein. This information is the gene for that protein. The DNA is copied into a short and portable RNA string that moves to the ribosome. If each string is a thread, the ribosome is a cross between a loom and a spinning wheel, because it reads strings of RNA and spins strings of protein.

Once a protein is made, the paradigm shifts from information to action. The string of protein curls up into a three-dimensional glob (the technical word is "globular structure," but one syllable will do). This shape of the protein is able to work in a three-dimensional world of chemistry. This is a thumbnail sketch of how the cell works.

Most biochemistry for the past century has focused on how proteins bring other chemicals together. Proteins break and make things: they chop food molecules into tiny bits, combine those bits with oxygen to make ATP energy, change shape to cause muscle contractions, sense the environment, capture light, and piece together small things to make bigger things, including DNA itself.

Proteins and DNA share most of their elements in common. These four elements comprise more than 96% of your weight. They can be remembered by the nonsense word their chemical symbols spell out: CHON, from carbon, hydrogen, oxygen, and nitrogen. Sugars, fats, antibiotics, and most of biochemistry can be built from CHON. Taken together, they are an elemental protoplasm that can be molded into structures with diverse shapes and charges.

Of these four, carbon is the central building block. It is the smallest element that can form the most bonds (up to four) to the most other elements, including itself. To carbon's right on the periodic table, nitrogen forms three multipurpose bonds and often carries a stable positive charge in water by attracting H+ protons. To nitrogen's right, oxygen forms two bonds and attracts electrons, so it can carry a stable negative charge.

Rounding out the CHON quartet is hydrogen, which can only form one bond, but it is small and abundant, so it is used to cap off open bonds all over the molecule. Hydrogen can also drop off the molecule and hitch a ride on a passing water molecule. In this way, it quickly moves around and reacts with other molecules (as in the case of phosphate vs. arsenate for the Mono Lake phosphate-binding protein).

The central nature of the CHON elements came home to me when I discovered a work of art I didn't expect, in the same way someone might discover *ekko*. In Chapter 1, I mentioned how the ideas of R. J. P. Williams form a chemical counterbalance to the biological arguments of Stephen Jay Gould. We will meet Williams's ideas later in this chapter, but first we will meet his art, done with four colors for the four elements.

A few years ago, after reading one of Williams's books, I was wandering around the campus of Harvard. In the science library building, I looked up to see a mobile of four colored tetrahedra hanging from the atrium. I found the plaque and saw that this was a mobile of the protein thread essential to every living thing. The plaque under it read:

CHAIN OF LIFE

John Robinson and R. J. P. Williams

Looking closer, I saw that the mobile showed the four basic elements in proteins: black for carbon, green for oxygen, blue for nitrogen, and small silver balls for hydrogen capping the open ends. That is all Williams needed to make the backbone of a protein chain.

Together, these four small elements make DNA, RNA, proteins, and more. They provide a huge variety of complicated shapes and all three kinds of charge (positive, negative, and neutral). Life doesn't need much more. Run an image search for "antibiotic structure" to see how creative bacteria can be when developing complex shapes with these four elements. Whenever I need a complicated chemical shape for a biochemistry test, I look at a list of antibiotics. The microbes are far more chemically creative than me.

But there are limits. CHON can make almost any shape, but it is limited to small amounts of charge. As mentioned previously, life needs phosphorus to make a negatively charged tetrahedron (carbon can't hold enough bonds). At least that can happen—a stable, positively charged tetrahedron doesn't happen in life because abundant elements can't make one.

Also, life can't make a cube out of carbon. Carbon's four bonds form 109-degree angles naturally, not the 90-degree angles required for square shapes. Under duress, a microbe can squeeze carbon into a square, but a cube has too many stressed angles. Carbon cubes are explosive, which is in general not a good property for life. Cubes will be made by life, but out of iron and sulfur, not carbon.

Because you can only make so many shapes out of CHON elements, the numbers of letters in the cell's three languages are also limited. There are four letters in DNA, one more in RNA, and then 20 amino-acid letters strung together in proteins. That's all my biochemistry students have to memorize: everything from the smallest microbe to the president of the United States makes proteins out of the same set of 20 amino acids, decoded from DNA using the same code.

Proteins are the point of interaction between life and the chemicals of the world, so they need to be as diverse as the world they encounter. Life can achieve more diversity from 20 amino acid letters than from four or five DNA/RNA letters.

Another aspect of proteins that helps them do chemistry is something they lack: charged backbones. Protein backbones are made from neutral CHON atoms, so they can curl into compact loops and U-turns and other complicated shapes. DNA's negative phosphates all repel, making DNA a straight line that is good for conveying information, while a protein can better adopt the precise shape it needs to move atoms around.

P IS FOR DNA, S IS FOR PROTEINS, AND O IS FOR OXYGEN OUTSIDE

Looking around the imaginary *ekko* cell, if you put on imaginary chemical glasses that allowed you to see different elements in different colors, four colors would dominate your view, one for each of the CHON elements. In the "Chain of Life" sculpture, these were black, green, blue, and silver. After that, you would next see two more colors, for phosphorus (red, maybe) and sulfur (yellow, definitely). These colors would be in two different places.

Much of the phosphorus is in long strings of DNA, carrying information, and in small bits of ATP, carrying energy. On the scale of *ekko*, the proteins inside the cell would be the size of sesame seeds, floating around, changing chemicals from one form to another. These proteins would be speckled with the sixth yellow color, that of sulfur atoms. The phosphorus in DNA and the sulfur in proteins are chemically opposite in one particular sense: in DNA, phosphorus is coated with oxygens and is "oxidized"; in proteins, sulfur is coated with hydrogen or carbon atoms and is "reduced."

Hydrogen and oxygen are on opposite sides of the periodic table and have opposite chemical properties. If acid is the chemical opposite of base, then a reduced molecule is the chemical opposite of an oxidized molecule. A reduced molecule will tend to have positive charge from its hydrogens, while an oxidized molecule will be more negative from its oxygens.

These opposites are important because every cell is reduced inside and oxidized outside. Inside the *ekko* cell's wall you see more hydrogens and electrons, while outside you see more oxygens. Take sulfur as an example: outside the cell is the oxygen-coated "-ate" form of sulfur, but inside the hydrogen-coated form of sulfur predominates. This one chemical observation is crucial, and it forms the beginning of a thread that we will follow through the rest of this book, a thread that organizes both life's insides and its outsides.

Molecules outside cells—in the blood, for example—are often covered with negative-charged sulfates. If you've ever heard the actor-doctors in a TV hospital calling for "more heparin, *stat!*" then you've heard an order for sulfate, or more accurately, a molecule that fills the blood with a carbon-linked net of negative sulfates. Heparin's network of charges repels other molecules and inhibits blood clots from coming together. Scientists have made a molecule that works just like heparin by putting sulfates on the end of nondescript carbon chains, showing that it's all about the sulfate, not the carbon.

When I say that the insides of cells are more reduced, I mean that there are more hydrogens and more electrons to balance the hydrogens' positive charge. The cell needs to move these electrons around, but it's surrounded with water, which blocks electron movement. The only elements available for electron-moving are sulfur atoms, carbon rings, and lesser amounts of certain trace metals. These three things are movable open boxes for electrons, like sedans or SUVs with tiny open seats.

Oxygen and nitrogen-oxygen molecules can also absorb electrons, but their help must be refused. These molecules are so small that once they pick up an electron they become unbalanced and highly reactive. The cell has to use them very carefully. If sulfur atoms and carbon rings are four-seaters, oxygen is like a motorcycle that can pick up an extra electron on its handlebars but then runs the risk of dropping it accidentally. If you drop an electron where it's not supposed to be, it will start a reaction and break things apart. Sulfur carries electrons safely because it is bigger, and it is bigger because it is lower on the periodic table than the CHON crowd, but still small enough to be abundant and available in water.

The cell keeps its sulfur in two electron-moving molecules: a small molecule about the size of ATP called glutathione and a short protein chain called metallothionein. (There's a clue in each name, because *thio* is Greek for sulfur.) Both of these molecules have exposed sulfur atoms that are like sponges for electrons.

The cell uses these sulfur molecules to mop up stray electrons, and sulfur is such a good mop that the cell can send waves of electrons around as a signal. If a stray electron accidentally falls off a transfer molecule, a sulfur atom from glutathione or metallothionein will scoop it up. This sulfur-based cleanup system is even more important than we thought: for decades, scientists missed seeing sulfur-rich bubbles inside yeast cells that store glutathione for a rainy (or electron-rich) day.

Life hoards electrons for three reasons:

1. Life needs a constant electron supply to *feed* into the energy-generating, oxygen-consuming fires of the mitochondria.
2. Life uses the electrons as *glue*, to bond and build the intricate structures of proteins, nucleotides, fats, and sugars. All structures are held together with electron bonds, and building them requires a lot of electrons (and also the positive hydrogens that come along with those negative electrons to balance charge). The center must hold for life, and that center is held together with electrons, so life keeps every electron it can.
3. Life *signals* with electrons. For example, in tadpoles, signaling with electron-imbalanced peroxides leads to growth. More visibly, one kind of male dragonfly turns from yellow to red when it matures by pushing electrons onto a yellow electron-poor pigment molecule. This pigment plus electrons turns red, like a molecular traffic light, and when other dragonflies see the red, they get the signal that this is a full-grown dragonfly. Electrons flip the switch. Many other temporary signals can be sent in all sorts of cells by manipulating electron levels in the cell, using glutathione's sulfurs in your own cells.

In proteins, sulfur can do one other unique trick: it can bond itself. This atomic narcissism means that if a protein has two sulfur-containing amino acids pointed at each other, they can bond each other face to face and form a loop that changes the shape of the protein. This bond is medium strength and easy to reverse by adding hydrogens and electrons. Moving a few tiny electrons on or off sulfurs can change the shape of a huge protein, so that a flicker of electrons can break bonds and move protein mountains.

For example, one important enzyme (pyruvate kinase) that breaks down sugar works differently with electrons/hydrogens on its sulfurs. So does an enzyme that sends proteins to the trash. A third example is how one bacterial protein senses bleach with a sulfur and adopts a new shape when bleach is around, alerting the rest of the cell.

Chemists have a word for electron-moving reactions like these: reduction-oxidation or "redox" reactions. These reactions move tiny negative charges around by transferring negatively charged electrons. A second type is "acid–base" reactions, which move tiny positively charged protons (H^+).

Life can use both redox and acid–base reactions to accomplish its purposes, but it appears to prefer redox reactions for its most important functions, like gaining energy from food and building new structures. In water, redox reactions are easier to control than acid–base reactions. In acid–base reactions, the protons can easily hop on a passing water molecule, while free electrons cannot.

Therefore, under life conditions, redox reactions can accomplish precise chemical changes if the appropriate electron acceptors and donors are placed in the right locations, without requiring big changes in temperature or acidity. Redox reactions can help organize ornate structures without drastic and deadly chemical shifts. Other chemical reactions are too slippery when wet.

If controlling electrons means life, then losing control of those electrons means death. This is where the small, electron-hungry elements can get out of hand. On the periodic table, the farther northeast you go, the more electron-hungry (and the more dangerous) the elements get. Oxygen is the most abundant, farthest northeast element. Fluorine is even worse, but it is not abundant in its gas form, because it's already reacted with something else. Bleach is a form of chlorine, also in that corner. It is similar in its indiscriminate electron-shuffling reactivity, being slightly weaker than fluorine—but you still don't want to drink it.

If too much oxygen seeps inside the cell (the main inside part called the cytoplasm), then the oxygens pull electrons off sulfurs and pass them around pell-mell, scrambling the bonds that the cell has worked so patiently to construct. Oxygen is shunted from inside the cytoplasm and directed to the furnaces of the mitochondria where it is needed. It can be used for transient signaling waves, but even these must be cleaned up by sulfur before they damage the cell.

The result may sound paradoxical, but it is actually fundamental: *oxygen and electrons are biochemical opposites.* Oxygen and electrons must be kept apart like Mentos and diet Coke, lest there be disruptive explosions. Oxygen must reside outside the cytoplasm if electrons are to reside inside. The wall of the cellular membrane defines the inside of the cell as a low-oxygen zone (unless it's for brief signals), and life itself hinges on this dichotomy.

THE MICROBE THAT LEARNED TO REJECT GOLD

Continue to look around inside the *ekko* cell and, in lesser amounts, you'll see metals dissolved and floating around. In fact, you'll see a peculiar pattern of particular metals: lots of manganese ions but only a few coppers. This pattern comes about because of chemistry. The columns of the periodic table predict which metals you'll see (and how much of each). For one thing, you won't see much gold, because the cell works very hard to keep gold out.

At first glance, cold, hard metal seems incompatible with warm, flowing life. But even the nutrition label on your cereal box shows that you need metals like iron and calcium to live. Chemically, metals and carbon-rich life molecules meet through the common medium of water. Metals dissolve by losing electrons and becoming positively charged (like the +2 magnesium ion from Chapter 1). This positive charge interacts well with the negatively charged oxygen in water.

Metals meet life as positively charged, dissolved ions, and in this way, some microbes work with gold like tiny goldsmiths. If you visit the Michigan State Museum, in person or online, you may be able to view an art installation called "The Great Work of the Metal Lover." This artwork looks like an old Frankenstein movie set. A large round flask filled with water is connected to various tubes and wires. Tiny flecks of gold glint in the water. A certain extreme microbe swims in there, challenged by a toxic gold chloride solution without oxygen.

Like in Mono Lake, bacteria answer their environmental challenge with clever chemistry. The bacteria in this artwork press electrons onto the dissolved gold, subtracting away the gold's positive charge to zero. Uncharged gold doesn't repel itself anymore and gloms together, forming micrometer-sized spheres that float in the flask. Some call it the Midas microbe, but I think it's more Rumplestiltskin.

This bacterium's name is Latin for "gold-loving" when it is actually "gold-rejecting." It is not hoarding the gold; it is getting rid of it. In *Lord of the Rings* terms, it is Frodo, not Gollum. The gold stays outside the cell as the peptide finds it and transforms it.

Only a handful of species perform this trick, because they learned to survive gold's toxic shock. Most metals lose one or two electrons when they dissolve in water, but gold loses three, giving it a large +3 charge that sticks to negative charges indiscriminately; +3 gold ions stick to sulfur, nitrogen, and oxygen, rudely shoving other essential metals that are merely +2 out of the way.

Large amounts of pushy gold are not prevalent in most environments, and therefore few microbes have had to evolve molecular shields against its toxicity. These shields are molecules that use gold's own stickiness against it. One microbe makes a sacrificial, reactive mini-protein with negatively charged oxygens that lure the positive gold ion in. Then a reactive carbon (an aldehyde) tosses a few electrons onto the ion, bringing its charge down to zero. No longer compatible with water, the gold ions form spherical micronuggets like in the previous artwork.

Eucalyptus trees have learned a version of this trick. When they grow above a gold deposit, their roots run deep enough to draw up gold atoms, which are of no use to the plant. They get rid of this toxic gold by tying it together in micronuggets, which they deposit in out-of-the-way places in their leaves and twigs. You could pluck off a leaf and analyze it for gold to find a gold mine below. In *Lord of the Rings*, Tolkien described Lothlorien, Galadriel's home, as the "Golden Wood." A properly placed eucalyptus grove could use the same name.

Pressing electrons on gold is easy if you have extra electrons—you can do it in your kitchen with green tea to make gold nanoparticles. This is because green tea has antioxidants in it that reduce the gold's charge. Look at that word: an "anti"-oxidant is an electron, or something that "reduces" other molecules by pushing electrons onto them. The green tea has several electron-rich molecules that act like the aldehyde in the Midas microbe's mini-protein. Antioxidants are electron-rich molecules that give electrons to counterbalance the disruptive effects of other electron-hungry molecules.

THE TROUBLE WITH CADMIUM

So inside our *ekko* cell, we don't see +3 gold, or +3 aluminum, or many other +3 metals for that matter. Aluminum and everything in aluminum's column on the periodic table have three outer electrons, which they lose in water to become +3 ions. Inside a cell, ions with +3 or more charge can cause problems by sticking to life's painstakingly arranged structures like so much stray chewing gum.

The cell keeps itself clean on these elements with special protein pumps that push sticky ions outside the cell. The metals that remain inside are carefully controlled and moved around by other specialized proteins that herd their respective metals like shepherds. But aluminum's column must not be worth the trouble, because very few of those elements can be found inside most cells. They are rejected by the cell's protein machinery, effectively crossing that column off the table.

Plus-three metals are just the beginning of what the fastidious cell has on its "no-fly list." Most metals are summarily ejected from the cell or are never allowed inside in the first place. The tricky part for the cell is when there are chemically similar pairs from the same column, like phosphate and arsenate. They are similar enough to use the same doors and different enough to cause problems.

Cadmium and zinc are another such pair. On the periodic table, cadmium is directly below zinc, making cadmium a bit bigger than zinc but a lot more toxic. (Notice how mercury, another toxic metal, is below cadmium.) Cadmium is so similar to zinc that it is often an impurity substituted into zinc ore. Cadmium can fit into zinc's seat but can't do all of its chemistry.

Life uses sulfur chemistry to detox itself from cadmium. The metallothionein molecule not only sponges up electrons but also patrols for cadmium. Cadmium is sticky (like zinc), especially to sulfur, so metallothionein's sulfurs hold it in place and keep it from reacting with something essential. Sulfurs are flypaper for cadmium.

Inside a few organisms, cadmium does something surprising that serves as a clue to the history of life. In these microbes, cadmium sticks in the place of zinc *and does zinc's chemistry*. For these organisms, cadmium has changed (or is changing) from toxic to useful.

The enzyme in question is one of the oldest zinc protein enzymes, carbonic anhydrase. This enzyme effectively hammers an extra oxygen atom onto CO_2 carbon dioxide, making CO_3^{-2}, the three-oxygen carbonate ion. The added oxygen pins down the gas so it can dissolve better in water. In the normal enzyme, zinc holds water still, so its oxygen can be added to CO_2. The cadmium enzyme uses cadmium to do the same thing.

The first cadmium may have bound instead of zinc by accident, but then the accident became useful. Cadmium in the environment moved inside the cell and was changed from toxin to tool by sticking in the right place.

If this happened for cadmium and zinc, it could have happened for other metals over history. The periodic table would provide a pattern by which new metals would be introduced and old metals removed. New metals may have been toxic at first, like cadmium, but they could become useful with simple, predictable changes.

Many things stick to metallothionein's sulfurs, so they serve as all-purpose cleaners for the cell. They have been caught mopping up extra lead and mercury ions, for example, which are also toxic and sticky. In fact, the metals are toxic *because* they are sticky. They jumble the chemical order of the cell.

Some environments have so many toxic molecules that survival requires extreme ingenuity to keep out or kick out the toxins. Perhaps the most ingenious is a red alga named *Galdieria sulphuraria*. It makes up most of the biomass in sulfurous hot springs, because no other organism in its right mind would take up residence in such a hot, acidic, metal-rich environment. This chameleon-like alga switches between a pale yellow form that eats sugar when there's no light and a green form that "eats" light when there's no sugar.

The secret to *G. sulphuraria*'s success is in its genes. It has detox systems that eject arsenic, aluminum, cadmium, and mercury. It has highly selective protein doors that let metals in, an arsenic-capping enzyme, and an enzyme that neutralizes mercury by pushing electrons onto it. A full 5% of *G. sulphuraria*'s genes are transporters/protein doors, compared to an average level of 2%–3%.

It looks like *G. sulfuraria* copied its genes from its bacterial friends. The genes correspond so closely to other bacterial genes that if they were English words instead, the alga could be convicted for plagiarism. It has cribbed genes from at least a dozen phyla of bacteria. It has even copied from itself to make multiple copies of genes for experimentation. The genome of this alga is clear evidence that genes rapidly move among microbes, accelerating adaptation. Because it is so open to new genes, *G. sulfuraria* is the master of metal rejection and transformation, a microbial wizard-hermit.

THE METALS INSIDE CELLS ARE ALWAYS THE SAME

G. sulfuraria lives in a unique and extreme chemical environment. Even so, I can predict what chemicals will be inside it. Specifically, the concentrations of most metals inside *G. sulfuraria* will be consistent with other microbes and even our own cells. Despite the huge differences in outside environments, inside there will be DNA and the DNA will use CHON + phosphate (+ magnesium). There will be proteins, and the proteins will use CHON + sulfur.

This internal consistency extends to the diverse array of metals inside the cytoplasm. Each different metal will be present at about the same level in all cells, because all metals follow the same chemical rules, and the patterns of chemical reactivity are the same whether you're an alga in a hot spring or a human in a cubicle.

To explain why, let's start with a friend's experiment gone horribly wrong. There is a certain danger when you're in the lab mixing metals with cells—not to you, but to your experiment. One of my colleagues was testing how snail cells respond to zinc. He set up a series of experiments adding more and more zinc. At a certain point, the snail cells froze up and turned to thick jelly. Zinc had jammed the flow of life. Zinc's chemical stickiness to oxygen, nitrogen, and sulfur is useful at lower concentrations, but at higher concentrations zinc adheres to everything, crosslinking the cell as if filling it with glue. Life must flow, and solidified life is no life at all.

Chemically, zinc has only a +2 charge, but it is held in a compact sphere that concentrates the charge, so it tends to stick to negatives more than larger atoms. Most metals turn to glues at high enough concentrations. Calcium likes to stick to oxygens, so it will jam up the negative phosphate oxygens in DNA and phosphate compounds like ATP. Enough calcium and phosphate will form solid, hard-edged calcium phosphate crystals that slice through the flowing oily structures of life. Cells eject calcium to keep this from happening.

Outside, the calcium's stickiness is more useful. Some bacteria build biofilm structures outside the cell by placing DNA as a phosphate-rich scaffold. The DNA attracts calcium and hardens into a shell-like coating.

Other sticky ions will also stick to phosphate. A worm called *C. elegans* that is grown in high levels of +3 aluminum shows ill effects as the aluminum sticks to phosphate. The cell has to choose between aluminum and functional DNA—it can't have both. To stay alive, the cells must cast the aluminum into the outer darkness. Another bacterium detoxifies itself from uranium poisoning by collecting phosphates in one place. The uranium is attracted to the phosphate and sticks to it.

Phosphate's stickiness is a general property and will work with essential metals as well as toxic ones. One microbe keeps copper ions out of the way by packing them in subcellular compartments with sticky phosphates and calciums. Too much phosphate, on the other hand, depletes metals that the cell needs. Yeast with high phosphate levels suffer from iron depletion because the iron sticks to the phosphate.

Because too much stickiness is bad for life, and more stuff means more possibilities for stickiness, there can definitely be too much of a good thing in a cell. Even an essential metal turns toxic when too much accumulates. Because metals with +3 charges are stickier, they reach this point at lower concentrations than those with +2 or +1 charges, and must be ejected more stringently. The periodic table describes ordered, predictable trends in both stickiness and size. Since stickiness correlates with the amount inside the cell, then that amount is also ordered and predictable.

This is best shown with an experiment, and the experiment is best shown with a picture (Figure 2.3). In the lab, chemists can measure how much metals stick to oxygen, nitrogen, and sulfur (the negative non-metals in the cell that stick to positive metals) using small claw-like molecules. The results can be arranged from left to right in periodic table order for the metals that make +2 ions in water. This graph makes a mountain that is an upside-down V. Stickiness increases from left to right, with copper being the peak of the mountain because it is the stickiest.

This "stickiness" graph works as well for proteins as for chemical claws, and in a number of studies, the order of stickiness to proteins is predicted by this graph. The only deviations come if the oxygens/nitrogens/sulfurs are arranged in a shape that one metal particularly likes. This graph works for making catalysts, too, because catalysis involves sticking to something.

The chemical stickiness graph is reflected in a parallel biological graph. The graph on the right in Figure 2.3 shows how much of each metal is floating around inside the cytoplasm of cells, arranged in periodic table order. In bacteria, reptiles, birds, and mammals, the result is similar: concentration *decreases* from left to right, with copper as the bottommost point, meaning its concentration in the cell is the lowest. The biology graph forms a right-side-up V, so that biological concentrations mirror chemical stickiness. The levels of metals in each cell are dictated by chemical binding.

The underlying chemical principle is that life needs to flow. It needs enough of each metal to stick to non-metals a little bit, but not so much that it sticks everywhere and all the time. Life is balanced precisely at the point where there is enough free ion to use its sticky chemistry, but not so much that it cross-links the structures inside the cell. The inherent chemistry of the metal ions forces each cell to have just enough, but not too much, of each metal on hand. In this way, chemical laws of binding shape the concentrations of free metal ions in biology into a V.

(The cell can get around these requirements by using special "chaperone" proteins that specifically bind to the particular metals. Recent work suggests that on average 70% of the metals used in the cell are not bound to chaperones and follow the trends in Figure 2.3.)

In this way, the cell is like a kitchen pantry. For most recipes, you need a lot of flour, but not as much baking soda. As a result, you keep a big bag of flour in the pantry but a little box of baking soda. Weak-binding metal ions are like flour, because the cell needs a lot of them so that they stick in place. Strong-binding metal ions like copper are more like baking soda. A little of a strong-binding metal will

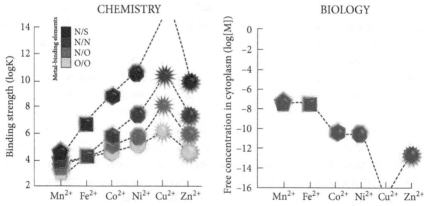

FIG. 2.3 The Irving-Williams series of metals binding. *Left*: binding of metals to molecules containing different elements in a chemistry laboratory. *Right*: concentrations of the same free metals, not bound to anything else, inside biological cells. The two graphs correlate, so that when a point on the left is high, the point on the right is low. Note that the metals are in the order set by the fourth row of the periodic table.

Data from R. J. P. Williams and Frausto da Silva, *The Chemistry of Evolution*, pp. 67, 135.

stick tightly to non-metals, so you don't need as much around, and too much will jam things up.

This metaphor breaks down on the upper end, because buying too much baking soda won't cause your kitchen to freeze up—but if it did, you'd keep a close eye on how much you bought. You could even imagine that having a lot of flour and a little baking soda is a constant trend for all types of global cuisine, because the underlying chemistry dictates that you'd always need more of the first than of the second. Pantries on every continent have big bags of flour and little boxes of baking soda.

Metals that are not on the graph can be understood in the same terms. For example, gold's three positive charges make it so sticky that it would be off the top of the "chemistry" graph. As a result, its concentrations in a living cell would have to be a mirror of that graph, kept so low that it would be off the *bottom* of the "biology" chart. This shows why the cell has to reject gold, and for that matter, any abundant +3 ion.

It is remarkable that the simplest arrangement of elements possible—that is, the order of the periodic table—produces a distinct trend that works for both simple and complex situations, both in chemical test tubes and biological cytoplasms. This universal trend of chemical *strength* is also universal for biological *amounts*.

A CHEMICAL HYPOTHESIS AND A PLANETARY TEST

The chemical trend shown on the chemistry graph is named the Irving-Williams series, after the two chemists who first put it in order. "Williams" is R. J. P. Williams, who first noticed that the biological concentrations are the inverse of his own chemical series. Beyond his apparent skills in mobile-making, Williams has published several books with coauthors about how this type of chemical observation can help explain past and present biology.

The Irving-Williams series shows how simplicity lies behind apparent complexity. It was first used by chemists to predict how metals react and bind to small molecules. The beauty of it is that it also works with large molecules like proteins in complex systems like life, and over billions of years of natural history. The series is a clear trend that works in many contexts and fits the universal shape of the periodic table. Physics has many laws of nature, from gravity to electromagnetism. The Irving-Williams series is a chemical law of nature.

It also has been a law of nature for a long time. The physical laws haven't changed much since the universe began—and physicists have checked. In particular, the fine structure constant in 7-billion-year-old light from 7-billion-year-old interstellar methanol is the same as today's constant. This means that yesteryear's methanol forms bonds in the same way as today's methanol. The Earth is less than 5 billion years old, and the chemistry behind the Irving-Williams series has been constant throughout time.

We can therefore apply the Irving-Williams series to the geochemistry of old rocks and the biochemistry of old life. Chemical bonding also tells us how certain elements are bound tightly into rocks while others float in the air. It tells us how energy is stored in the chemical bonds of fat and sugar and in atmospheric oxygen for later use.

It also tells us something fascinating about our planet: it was not always this way. When the Earth first formed, it was a very different chemical environment. How and why did it change? Williams published a hypothesis, based on chemistry, that the change was not only dramatic: it was *inevitable*. Chemistry determined the first conditions on Earth, and then changed it over billions of years to what we see today. Life changed as its environment changed, and chemistry tells how. Williams and geologist Ros Rickaby captured this in the title of their 2012 book: *Evolution's Destiny: Co-evolving Chemistry of the Environment and Life.*

Williams is a chemist through and through. When he talks about evolution, he doesn't talk about species, but groups species into broad groups that do the same chemistry, which he calls "chemotypes." Animals are one chemotype because they take in oxygen and food, then combine the two for energy. Plants are a different chemotype, because they take in carbon dioxide and sunlight, then make sugar used as food for animals. Single-celled organisms that take in oxygen are another chemotype. Those that breathe hydrogen are yet another.

Williams takes the long view. He sees the random variation among species as an engine that drives how chemotypes are fully realized as a group. This is a very different emphasis from Stephen Jay Gould's *Wonderful Life* (1990), which emphasizes the random walk of speciation for individuals. These two descriptions may very well be compatible, Gould's view from up close at the organism level, and Williams's view from afar, across eons of time, at the chemical/chemotype level.

I first encountered Williams's ideas in the middle of the first decade of the 2000s, just as DNA sequencing technology matured to the level that we could routinely read whole genomes of all sorts of species. I remember being intrigued but not convinced that this wasn't just a case of a chemist seeing chemistry everywhere around him. I also knew that the perfect test of Williams's theories was emerging. Science was just beginning to read the genes of organisms great and small and to look for evidence of their sequence of development as written in the DNA language. Those genes could be grouped together and read for signs of which metals were used when. Elements that were available long ago should be used by old genes, and elements that only became available late should be used by new genes.

I'll save you the trouble of skipping ahead to where the data are described. R. J. P. Williams was right. The chemical trends he identified are reflected in the genomic data, with a few exceptions, and he predicted it all from chemical rules like the Irving-Williams series. This means that, over 4 billion years, chemistry shaped biology. It also contradicts *Wonderful Life*. Stephen Jay Gould's "tape of life" would look much the same upon replay, because its events, when examined from the proper perspective, were shaped by universal and predictable chemical laws.

The process of evolution must be predictable—because R. J. P. Williams predicted it. As to why I can say this and what it means, that requires nothing less than telling the story of the natural history of the universe from the perspective of the chemist.

BREATHING DEPENDS ON A METAL

Based on its stickiness, size, and chemical properties, each of these metals reacts differently. Let's return to our micro-*ekko* model of life and look around to see what these metals are used for. The elements needed for life can be sorted into three categories, which are shown on the "Origami Periodic Table" (Figure 0.1) in the front of this book.

1. *Building*: Most of the cell is built from the big six elements for life used as *building* blocks: the CHON elements, phosphorus, and sulfur. These were the six "colors" first seen when looking around micro-*ekko*. Calcium as used in bone can be considered a part-time member of this team.
2. *Balance*: A second group of seven elements floats around in the water, unattached, like tiny glinting snowflakes. These are used because they don't form long-lived bonds, so they can be pumped in or out to *balance* (or unbalance) the cell's overall charge. This group includes sodium, potassium, calcium, magnesium, chloride, and the oxidized elements phosphate and sulfate. The singly charged metal ions of sodium and potassium and the doubly charged ions of calcium and magnesium are therefore used in large quantities.
3. *Chemistry*: The rest of the elements found in the cell are present in smaller amounts and follow the constant concentrations dictated by the Irving-Williams series. These are the metals located in the middle of the periodic table, the region of transition metals. Most of these have two positive charges, but they have slightly different shapes, sizes, and reactivities, and some have special abilities, which can end up making them useful for very different things. Each of these has the ability to *catalyze* different reactions and must be kept around like ingredients in a pantry for when the cell needs to cook something up.

A metal from group 3 (the group of catalysts) is usually found stuck to protein, where it could be doing one of three things: it could help carry out a reaction; it could tie a protein together; or it could just be randomly stuck, along for the ride. In a word, a bound metal may be catalytic, structural, or accidental. If there's one thing metals can do, it is catalyze reactions. A 2010 study of all cofactors (the assistant molecules or ions that help enzymes do their jobs) found that the huge majority of reaction-activating cofactors are metals, not CHON structures.

Metals can activate all sorts of reactions. Chemists use big chunks of metal as catalysts to make reactions go faster. It's like the catalytic converter in your car, in which a gas or liquid flows over a metal surface that attracts molecules. An inert surface can sometimes get chemistry going, but metals can also give or take electrons to jump-start a reaction. Metals shuffle the bonds between other atoms, accelerating reactions that may take years to completion in seconds.

Your body has *little* chunks of metal, sometimes atom-sized ions, and sometimes networks of a dozen or so metal atoms. This isn't enough to set off the metal detector at the airport, but it is enough to make your innards work right.

The metals are placed exactly where they need to be by the large, unwieldy looking CHON structures of proteins. It really can be all about the metal: chemists have made a porous rock-like structure that will change glucose sugar to fructose sugar (just like your body's enzymes). This structure holds tin and titanium atoms in the right place, and those metals catalyze the reaction of glucose to fructose. A rock structure that holds the right metal in the right place mimics one of your metabolic enzymes.

So the chemistry in rocks and the chemistry in the body overlap. From the chemist's perspective, something as vital as life's blood can be seen as a complicated solution of dissolved metal. In red blood, red iron, and red rocks, red is the color of iron bound to oxygen (caused by different structures that result in similar colors). In rust, the oxygen has combined with iron haphazardly into a fragile, brittle mess. In rocks, it is held in a solid crystal lattice. In blood, the oxygen is placed onto the iron precisely by the large CHON structure called hemoglobin.

If hemoglobin can be called a catalyst, it is not for a chemical change, but a *positional* one. Hemoglobin "catalyzes" the movement of oxygen, from the lungs into the body. In the lungs, hemoglobin picks up oxygen, then releases it in the capillaries. If you are anemic and don't have enough iron in the hemoglobin as a result, then your blood won't carry enough oxygen, and your brain and muscles start to waver from lack of O_2.

The job of carrying oxygen is given to Fe (iron) and not Cu (copper) or another metal because of the chemistry of metal stickiness, as shown by the Irving-Williams series. On the chemistry graph in Figure 2.3, Fe (iron) is on the left end, with medium-weak strength. This means that Fe (iron) can stick to oxygen in the lungs but let go of it in the capillaries. If the bond was too weak, as with Mn (manganese), the oxygen would not attach to the metal in the lungs. If the bond was too strong, as with Cu (copper) or Zn (zinc), the metal would never let go in the capillaries. (Some deep-sea and/or spineless organisms carry their oxygen with copper at cold temperatures and high pressures because they need a stickier interaction and can get by with less oxygen.)

BLOCKING METALS, BLOCKING FUNCTIONS

This is why CO (carbon monoxide) is so dangerous. This colorless, odorless gas, formed from incomplete burning of carbon in car exhaust, is a bonded pair of atoms that looks very much like O_2. Carbon monoxide mimics oxygen, seeping into hemoglobin and sticking to the iron like oxygen does—except that carbon monoxide sticks *more tightly* to iron than O_2. The iron atom is blocked and can transport no more oxygen. Carbon monoxide renders the iron in your blood ineffective. By blocking this one element, your cells will asphyxiate, even as your lungs scramble to take in more oxygen. The oxygen cannot move where it needs to go.

Hemoglobin is not merely a passive carrier. The power of the iron inside hemoglobin can be redirected so that the transport protein becomes an enzyme that catalyzes a true chemical transformation. One group of chemists uses dissolved hemoglobin to make plastic polymers, because it's easy to make chemical links with the iron, even with all that CHON protein around it. Your body uses the iron in hemoglobin to make nitric oxide for signaling by moving around electrons on oxygen and nitrogen. When moving oxygen around, hemoglobin acts like a train car, but the same molecule can cook up some nitric oxide and act more like a kitchen. Hemoglobin is versatile because iron is versatile.

Hemoglobin is the household name for metal-binding proteins, but it has lots of company. About one-third of proteins have metals bound to them, and usually the metals are essential to holding the protein together, to catalyzing a reaction, or sometimes both. Copper is used in plants to move the electrons that harvest solar energy. Zinc holds together proteins called zinc fingers that grab DNA and turn genes on and off. Because copper and zinc stick tightly to sulfurs and nitrogens (as shown in the chemistry graph), a protein with the proper arrangement of sulfurs and nitrogens will gain the power of these metals.

Medicines like Pepto-Bismol will target sulfur-metal bonds like these. The stomach ulcer bacterium, *Heliobacter pylori*, has a long tail riddled with sulfurs. This sulfur tail carries a few zinc atoms along wherever the protein goes, so that it can then lend newly folded proteins a zinc atom, keeping the proteins active and the ulcer bacterium infective. The zinc can be displaced from this protein tail by a metal that biochemists and bacteria rarely encounter: the triply charged metal ion of bismuth, the active ingredient in Pepto-Bismol. Pepto-Bismol may destroy ulcer-causing bacteria by sticking to sulfur, displacing zinc, and generally wreaking havoc with this protein-making protein. Before the Industrial Age, bismuth was not in the stomach environment, so *H. pylori* never has seen it before, and has no defense against it (yet). We humans purified a sticky, new metal like bismuth and used it against bacteria, turning a rare element's natural stickiness to sulfur into a useful weapon.

Some scientists have even hypothesized that carcinogens work via Irving-Williams-style stickiness. In particular, nickel and chromium—a metal with *six* positive charges!—may cause some forms of cancer with their extreme stickiness when they displace iron in crucial growth-signaling pathways. If the metal stays stuck, the pathway stays on, the cell keeps growing, and a tumor is born.

SWITCHING METALS, SWITCHING FUNCTIONS

Put these three observations together:

1. It's easy to move metals around and stick them to proteins.
2. Metals can be crucial to the structure and catalytic activity of proteins.
3. Chemists prove their points by building new chemicals (at heart, we're kids with tiny LEGOs).

You can see how chemists would want to switch metals around from protein to protein in order to build new metal-protein complexes. It's not quite as dramatic as the image of the mad scientist switching the head of his lab assistant with that of a duck, but it has the advantage of actually being possible. When the metal changes, so does the enzyme.

The enzyme carbonic anhydrase that can use cadmium as well as zinc to add an oxygen to carbon dioxide was our first example of metal-switching chemistry. If another metal replaces the zinc in this enzyme, some bizarre chemistry can result. If manganese is put in there instead of zinc, the enzyme can make a weird triangular epoxy group on a target molecule. If rhodium is put in, it moves hydrogens around instead. Changing the metal changes the character of the enzyme. Another reason a cell must pump strange metals out is to avoid creating new enzymes.

One frugal beetle has figured out that by switching metals you can use the same enzyme to make two different things. Normally this beetle's enzyme uses cobalt or manganese to join together 5-carbon chains into 10-carbon chains, which it can use to defend itself as a larva. But when it needs a 15-carbon chain as a life cycle hormone when it grows up, its cell will use the same enzyme with a one-atom change: magnesium instead of cobalt or manganese. The chemistry of a single magnesium ion is sufficient to change the output of the enzyme and make a growth signal instead of a chemical defense.

Likewise, a bacterium named *Citrobacter freundii* has an essential enzyme that moves phosphate from ATP to something else. If manganese binds in magnesium's place, the enzyme changes to form two bonds to the same phosphate, making a little phosphate circle. If the bacterium needs to make phosphate circles, it just ships in a little manganese and an old enzyme performs new chemistry.

Even the same metal can result in different chemical results if that metal is held differently by the enzyme. In one family of enzymes, some members of the family add oxygen-hydrogen groups to a target molecule, while others subtract carbon-hydrogen groups. They do these very different chemical reactions by holding metals with different grips, like a pitcher preparing to throw a fastball rather than a curveball. The metals are powerful enough to do different things and are directed by their context.

Sometimes changing the metal on an enzyme will change its target molecule. One DNA-cutting enzyme cuts DNA in different places with magnesium, manganese, or cobalt than with calcium. If cadmium, zinc, or nickel binds, it won't cut at all. The metal is the tiny rudder in charge of a huge protein ship.

All of this adds up to mean that different elements make different reactions possible, and more metals imply more possibilities for chemical diversity. Bacteria that grow in soils with different levels of potassium, calcium, and selenium will build different types of molecules. The more potassium and calcium are in the soil, in fact, the more diverse are the shapes that those bacteria can build. Some of these unusual chemical shapes may be useful as new drugs or natural pesticides, made possible by metals.

Chemistry shaped biology as chemical changes paved the way for biological changes. All metals were not available at all times, and some useful metals were locked away, stuck in geological dead ends when the planet first formed. On the day when one of these metals became available for the first time, new chemical reactions could catalyze new life and new complexity.

DESIGNING METALS INTO PROTEINS

Science is about *connecting* facts as well as collecting them. So if this gallery of examples shows that metals are very important to their proteins, we should be able to reshape and redesign proteins to use the power of metals. If we put a new metal into the hands of a protein, it might do something amazing.

My own lab experience is in a similar area of protein design. Proteins are incredibly complex molecules. If given the chance, they will surprise you. My first design project, as a nervous, new faculty member, actually was a failure in that we put in a year of work and the protein didn't do what we predicted it would do. (I'm sure the proteins were building our collective character.) We changed our hypothesis and then found success.

When designing a protein, you must remember that proteins are like birds. You don't control the protein after it flies away into the test tube. The complexity of the protein is greater than your imagination or your hypothesis. This has happened for other labs—and sometimes the surprise is a good one that shows the power of metal chemistry.

The simplest thing to do with a metal is to just stick it onto a protein. This is the opposite of shooting an arrow at a target. Instead, you design a protein target, and then you set it loose into a metal-rich solution to find its metal-ion "arrow." If you design three or four nitrogen/sulfur/oxygen atoms into the right places on the protein, that will often hold a metal tight.

Many have designed metal binding sites into proteins. In what may be the most colorful result, scientists put sulfurs and nitrogens in a pattern that recreated a special double-copper binding site. This double-copper arrangement is colored a bright purple both in the natural and artificial proteins. The bright purple color in the test tube shows that the design worked.

In 2010 a team of scientists designed a protein so that zinc would bring four copies of the protein together, serving as a zinc zipper between the four protein chains. They arranged half of a zinc binding on each protein chain, so that the zinc would bridge the two half-sites, pulling two protein chains around itself and making a full site. Zinc pulled four protein subunits together, and copper did, too, because of its stickiness, as shown in the Irving-Williams series.

Some scientists used this zinc zipper design to string proteins end on end in long, thin "nanotubes." The proteins link up to form repeating zigzag patterns that weave together into a bigger tube. This group has gone on to weave arrays of protein linked together like chain mail with zinc.

Another group focused on finding new protein scaffolds to hold the zinc rather than building bigger structures. They organized a comprehensive search of all known protein structures to find proteins that had the best shapes for supporting a claw of nitrogens to bind zinc. They designed two zinc half-sites into this new protein, and zinc made it zip together—but then they noticed something unusual. This protein had an extraordinary trick up its sleeve.

THE ACCIDENTAL ENZYME AND
THE DELIBERATE METALS

It started with a mistake. When the structure of the protein-zinc-protein complex was determined and all the atoms were put into place, they didn't fit together perfectly. Three nitrogens bound the zinc as expected, but the protein had moved around (as proteins will do) and the fourth designed nitrogen didn't stick. In its place was a jagged funnel to the surface. The zinc was sticky enough that it bound to three nitrogens instead of four, leaving a crack in the protein. This particular crevice was bigger than the zinc, and about as big as a normal enzyme's active site (Figure 2.4). It was big enough that a small molecule could easily fit in, and in some pictures of the protein a stray molecule was caught fitting in.

The scientists realized that the shape of the crack was similar to that of the zinc protein carbonic anhydrase mentioned earlier in this chapter. Maybe, if a molecule with a fragile bond in the right place flew into the crevice, then the zinc would push electrons around and break the bond. Maybe, by making an imperfect zinc zipper, the researchers had accidentally made an enzyme.

The lab set in motion. Molecules with fragile bonds were mixed with the designed zinc protein and they were quickly chewed apart. No one had set out to create an

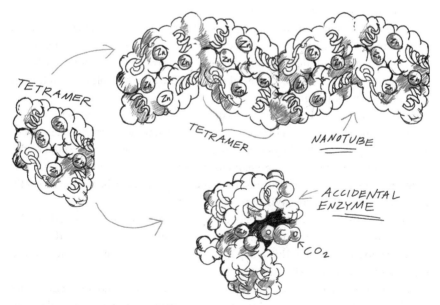

FIG. 2.4 Designed zinc proteins. First zinc held together four proteins (*on the left*); then this design was used to make a nanotube (*top right*) and, with another scaffold, an accidental enzyme (*bottom right*).

enzyme, but even so, they had, just by holding zinc next to a crack in a protein. (Now, a related laboratory has used this same idea to make another type of enzyme called a beta-lactamase from a zinc-binding protein.)

This was not the first time a new enzyme was built around zinc. In 2007, a different laboratory made a part-random protein with a quartet of sulfur atoms placed to bind zinc. Several of these random proteins were able to form new chemical bonds, so the pool of random proteins created what biochemists call a ligase. This is about the power of a random pool of proteins, but also the "ligase" power inherent to zinc. It is possible that zinc is good at ligating chemicals in the same way that magnesium is good at bringing phosphates together (in DNA polymerase, for example).

Thankfully for researchers' egos, this works even when it's not an accident. One lab made a long, thin protein with a three-nitrogen claw at one end that binds zinc, and a three-sulfur claw at the other end that binds mercury. The mercury sticks so strongly to the sulfurs that it holds the whole protein together. The zinc, on the other hand, can add oxygens onto carbon dioxide like the carbonic anhydrase enzyme—another metal-powered enzyme, this one a little less accidental.

Another lab randomly mutated the shapes in the cores of simple, four-column proteins, millions of times over. If the "heme" form of iron found in hemoglobin is mixed with these randomized proteins, about half of them will take it into their random cores. Some of these random heme-binding proteins could catalyze reactions. For example, they could use their heme iron to move electrons on and off the reactive peroxide form of oxygen.

This is another type of accidental enzyme, showing that you can make a simple enzyme without really knowing what you're doing, so long as you can try enough possibilities and use iron's natural catalytic power. The same random "bind a metal and let it do the work for you" strategy works well for zinc and copper, because of their inherent stickiness and chemical power.

So even though in a cell there aren't as many metal atoms as CHON, sulfur, or phosphorus, these metals are disproportionately important. Their concentrations are set by the Irving-Williams series, which is ordered by the periodic table. The periodic table's columns contain other patterns, such as toxic mimics of essential nutrients, as in zinc/cadmium and phosphate/arsenate. These patterns determined what life needed and what it rejected, all the way back to the beginning.

Therefore, the story of chemical life must start, as chemists do, with the periodic table itself. If life is built from the periodic table, then what is the periodic table built from? Like physics itself (and, to tell the truth, this *is* physics itself), the answer to this question would make Plato proud. The rows and columns of the periodic table are built from Platonic ideals, from the abstract combinations and logical consistencies of math.

3

UNFOLDING THE PERIODIC TABLE

the map inside your chemistry book is built from odd numbers // the music of the electron spheres, given a twist // the periodic table unfolded in stars // the Big Six elements in rocks // spinning out a new moon // fizzy water changed the Earth // oxygen tilts the table

THE PERIODIC TABLE FOLDS LIKE A MAP

Our starting point is not hidden, nor is it far off. It is not an extreme place like Mono Lake or Freswick Castle, but it is a central concept expressed on a single page. The periodic table is the center of chemistry, and therefore of this book.

You can spot it at a distance from its vaguely cathedral-like shape. You can see the chemical symbols that it contains on magnets and T-shirts and restaurant signs. Its regular columns are not quite symmetric, but that is because it has been twisted out of its natural shape by the contingencies of history. Rearrange it just a little and a simple mathematical pattern appears.

To see this pattern, imagine that the periodic table is made out of beads on an abacus, arranged in the familiar U shape. Then push all the beads to the left:

Row 1 = H-He
Row 2 = Li-Be-B-C-N-O-F-Ne
Row 3 = Na-Mg-Al-Si-P-S-Cl-Ar
Row 4 = K-Ca-Sc-Ti-V-Cr-Mn-Fe-Co-Ni-Cu-Zn-Ga-Ge-As-Se-Br-Kr
Row 5 = Rb-Sr-Y-Zr-Nb-Mo-Tc-Ru-Rh-Pd-Ag-Cd-In-Sn-Sb-Te-I-Xe

By row, there are 2, 8, 8, 18, and 18 elements. The pattern continues in the rows below, but it is obscured by the fact that on most tables 14 elements have been moved out of the sixth and seventh rows. On the table here I have put them where they belong. These rows have 32 elements each.

This can be simplified even more. The rows increase, first by 2, then by 6 more (2 + 6 = 8), then by 10 more (2 + 6 + 10 = 18), then by 14 (2 + 6 + 10 + 18 = 32). The series 2, 6, 10, 14 is the doubles of counting up by odd numbers: 1, 3, 5, 7. Put another way, each row is equal to $2n + 1$ with n = integers from 0. So the periodic table is built by counting and adding odd numbers.

These patterns come out with some origami. There are lines on the figure in the front of the book that show where you can fold to bring blocks together, folding it as

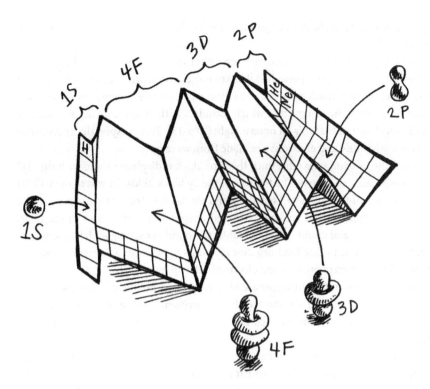

FIG. 3.1 How to fold the periodic table. The dotted lines on the periodic table in Figure 0.1 show how it folds into blocks of 2 (on the far left), 6 (on the far right), 10 (the transition metals on the middle right), and 14 (the rare-earth metals on the middle left).

in Figure 3.1. You can fold away the 14 block, then the 10 block, then the 6 block, so that only two columns remain.

The beauty of the periodic table is that the columns and rows are put in chemical order. For example, all of the elements in the first column will easily lose one electron and usually form single bonds. The atoms used by life are marked in Figure 3.1, so that you can see that life uses the hydrogen that gives the H to CHON, the sodium of table salt, and the potassium (K) of bananas. Potassium is like sodium is like lithium is like hydrogen in terms of the fundamental chemistry of how electrons move. The major difference is that the elements farther down are bigger.

If the leftmost column is made of jovial, easily reacting elements, the rightmost column is made of elements that keep to themselves. Helium, neon, argon, krypton, and xenon may have the best names of all, but they don't do much chemistry. These elements don't lose electrons or form bonds at all, and they tend to be aloof gases that are impossible to use in life, so none of them is marked. There's no such thing as a neon vitamin.

These patterns continue across the table, so that the columns form a rhythm. Primo Levi said that the periodic table is poetry, because column by column, it rhymes. To me, it has the regular pattern of a good crossword puzzle, which veteran solvers call "solid fill."

Eric Scerri writes that this orderly repetition is almost musical: "The elements within any column of the periodic table are not exact recurrences of each other. In this respect, their periodicity is not unlike the musical scale, in which one returns to a note denoted by the same letter, which sounds like the same note but is definitely not identical to it in that it is an octave higher" (p. 19). This is especially appropriate because both music and electrons are made from waves.

Refold the origami map table until the "2p" block is displayed along with the "1s" block, eight elements across. The top row of the 2p block holds the rest of the CHON elements, four of the six most essential elements for life. The two remaining major elements, phosphorus and sulfur, are buried deeper, in the third row of the table. As we dig deeper and count up on the table, the numbers get bigger because we are adding protons one by one (and negative electrons to balance each proton's positive charge). All these extra protons and electrons take up more space.

Since the atoms in the third row are bigger, they can hold more than four bonds, and for the first time we see the tetrahedral oxygen "-ate" molecules, phosphate and sulfate. The seven elements used by life for balancing charge are located in this block, too.

One more fold to go and we'll have everything we need for life. Unfold and refold so the 3d block of 10 is exposed, which is called the "transition metal" block. Life needs only small amounts (milligrams) of iron (Fe), zinc, copper, and the others. Also, it appears that life needs only one example from each column. This even holds true for molybdenum (Mo) and tungsten (W) because lifeforms will tend to use *either* molybdenum or tungsten, not both. Usually the top example, being the most abundant, is the easiest to find and the one life uses. There's no need to search for silver if copper (above it) will do the same needed chemistry for cheap.

In the Irving-Williams series (Figure 2.3 in Chapter 2), the "stickiness" of these metals increases from left to right, peaking with copper and zinc. The three leftmost columns on this block aren't used because they would be off the chart on the low end, and are not sticky enough to be useful to life. Therefore we shouldn't expect to find a scandium or titanium-based organism any time soon, and vanadium usage is rare. Those elements simply cannot stick or make other chemicals change, so life tosses them aside and keeps to the right-hand side of the transition metal block.

All the messiness of life is somehow circumscribed by these tidy columns and rows. As C. P. Snow describes understanding the periodic table, "For the first time I saw a medley of haphazard facts fall in line and order. All the jumbles and recipes and hotchpotch of the inorganic chemistry of my boyhood seemed to fit themselves into the scheme before my eyes, as though one were standing beside a jungle and it suddenly transformed itself into a Dutch garden" (p. xviii). Chemical history, too, is transformed from a jungle to a garden, because everything is made from 100 or so elements arranged in rows of doubled numbers, counting 1, 3, 5, and 7.

BALLOON ANIMALS AND QUANTUM MECHANICS

The numbers are simple, but the rules seem awfully arbitrary. Why are 1, 3, 5, and 7 such important numbers, and why multiply them by 2? Why is this elegance hidden beneath the structure of the periodic table? To answer these questions, we have to build an atom from scratch.

The periodic table counts up by atomic number, which is the number of protons in that atom's nucleus. On the nanoscale, protons are big, heavy, and immobile, and they have a positive charge. Each positive proton balances its charge with a negative electron, and electrons are small, light, and mobile. In an atom, the electrons buzz around the nucleus like children at a birthday party, while the protons sit back and watch bemusedly, like the tired parents. But the children are attracted to the parents (after all, they have cake), so they never get too far away.

To build an atom, you have to think like an electron. A physical description of how something moves is its "mechanics." Consider what happens as we add protons and their balancing electrons, one by one. This is how helium and hydrogen differ— helium has one more proton and one more electron. (Neutrons also help to hold the nucleus together, but they are uncharged.) From the protons' perspective, only the electrons are moving.

But they don't move like the objects we're used to. The physics equations that work for cars and hockey pucks break down when applied to electrons. Our new language and new equations for how electrons move can be borrowed from the physics professors who study waves. Electrons move like waves in a water tank or ripples in a river.

Anything that is as small as an electron can cohere as a particle or spread out as a wave, depending on context. In the early twentieth century, Einstein found that light *waves* can behave like *particles*, and soon afterward, Schrodinger calculated where electron *particles* would (probably) be by applying the equations for three-dimensional *waves* to them. Both light and electrons can behave like both particles and waves.

In Figure 3.2, notice that the ends of the waves always meet. This is because standing waves can't exist in halfway states. When you pluck a guitar string, it vibrates in a wave, but *it can only vibrate in whole numbers* because the beginning and end of it are fixed in place. Electrons are tied down by their protons like a guitar string is tied down to a guitar.

This results in a surprising simplicity. Electron waves tied down to protons must be *quantized* to simple numbers like 1, 2, 3. There's no wave "1 and a half" between 1 and 2 on the diagram in Figure 3.2. One way to think of this is to imagine the electron as a circular string, flat on the page in two dimensions. If this is a wave that is plucked, the wave must match itself coming around the circle. This means that the wave can only exist in quantized intervals, and if you call the first one "1," then the second one is "2," with twice as many wiggles in the wave, and so on.

To describe this "quantized" movement, we need a "quantum mechanics." Don't worry if you never got this far in physics. Many chemists forgot it after the test. All

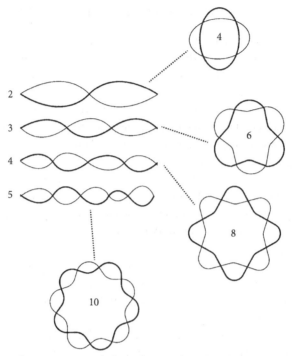

FIG. 3.2 Electrons form waves using whole numbers. Different kinds of waves (one-dimensional lines and two-dimensional circles) with increasing energy can form waves only as whole numbers because the ends of the waves must match.

you need for chemical quantum mechanics is an imagination for what a wave would look like in three dimensions, because that is exactly what an electron looks like.

Imagine a perfectly still pool of water, still enough to reflect like a mirror. You hold a pebble in each hand, then drop both at once. The ripples expand in perfect circles, in two dimensions, across your reflection on the flat face of the water. Where the two ripples meet, they conflict. One ripple might make the other flat, or they may add to make a bigger ripple. If something makes this kind of regular, interfering pattern when it meets, it is a wave. The surface of water does this, light does this, and electrons do, too.

Waves in three dimensions are not as easy to visualize. Instead of being circles in a flat surface, bounded by the shore, three-dimensional electron waves ripple out from the protons as spheres in space, bounded by the pull of the protons' positive charge. An old theory imagined electrons as orbiting the nucleus like the planets orbit the sun, but electrons act like waves, not planets, and there's nothing keeping them from falling in toward the proton. Low-energy electrons spend most of their time near the nucleus itself, but because of the weird rules of quantum mechanics, they are never completely tied down. They whiz around in a sphere, pulled back in by the protons' positive charge. Electrons are anchored peripatetics.

Hydrogen atoms are the simplest atoms, with one proton and one electron each. Their electrons move around in a simple sphere. When another proton and electron are added to make helium, the two electrons pair up in a bigger sphere. One of the weird things about elections is that two can fit together if they have what physicists call opposite spins. No more than two can fit, so three electrons is a crowd. All future electrons must be fit into higher levels. For historical reasons, this shape is named the *s* orbital (for "sharp," but I will forever associate it with "sphere"). As the first energy level, it is called "1s," and it is already full with helium's two electrons.

Moving up to the second energy level of electrons, we also move up to the next level of the periodic table, adding another proton and electron to make #3 lithium. We're now on the second floor of energy levels, and the first thing on this floor is another sphere, a second s orbital called 2s. Lithium gets one electron, and then beryllium next to it gets another with opposite spin, and the 2s orbital is filled.

So far so good, but this second floor has more room on it before we move on to level three. The higher energy of the second level allows the electron wave to twist, just once, around itself. If the s orbitals are round, inflated balloons, these new orbitals are balloons twisted once, like in a balloon animal, to make an elongated dumbbell shape, seen above carbon-oxygen-nitrogen on the origami periodic table.

These new shapes are called *p* orbitals (because alphabetical order would be too easy). The electrons still pair up two by two, but three of these elongated 2p orbitals fit together. One goes up and down, a second one left and right, and a third one in and out, like three dumbbell-shaped balloons tied together into an asterisk shape.

Three orbitals in three directions means *three pairs* of electrons, so six electrons fit in the 2p level. The next six elements after #4 beryllium fill these spaces one by one: #5 boron has one p electron, #6 carbon has two, #7 nitrogen has three, all the way down to #10 neon. Now the second level is full and we can move up (in energy) and down (on the periodic table) to the third energy level.

On the third level we build up again, with two 3s electrons, then six 3p electrons; #11 sodium gets one 3s electron, and #12 magnesium gets a second, both spheres. Then #13 aluminum through #18 argon fill up with six 3p electrons, like before, oriented up, sideways, and in and out.

The third level has a third twist possibility as well. This one is called the *d* orbital and looks like a more complicated balloon animal with four twisted lobes (shown in Figure 0.1), making fascinating shapes that no balloon could ever achieve. The d orbitals have five possibilities, so the electrons can make twisted shapes in two more ways than the p orbitals, which could make shapes in two more ways than s orbitals. This pattern holds true as far as the eye can see or the computer can calculate. Five d orbitals times two electrons per orbital equals 10 elements to fill up the d orbital block.

(One side note: for complex reasons, 4s is slightly more stable than 3d, and so the 4s levels fill up first. On our tour through the table, #19 potassium and #20 calcium fill up the 4s orbitals first, and then #21 scandium through #30 zinc fill up the 3d possibilities. This is why the 3d orbital is on the fourth row of the table. These orbital stabilities are illustrated by a more complex but still predictable "diagonal" pattern taught in chemistry courses.)

On we go through the table: #31 gallium through #36 krypton add six 4p electrons, and then levels continue to fill up until we get to a point where a *fourth* twist of orbitals appears, which is called 4f ("f" for "fourth"?). S has one orbital, p has three, d has five, and—there's no shocking twist here—f has seven orbitals, times two electrons, making 14 possibilities in an "f block" 14 elements wide.

Math predicts what an 18-element-wide 5g block would look like, and even a 22-element wide 6h block, and so on, because the electrons follow the same patterns of waves that we can see in a water park's wave pool. The shapes of these, more twisted with each step, have been displayed on a site called The Orbitron, although even they gave up on representing h orbitals at last check. The shapes are bizarre, but the orbitals themselves continue to follow the predictable rule of two additional possibilities for each new level, added on top of the previous possibilities.

We run out of elements long before we run out of imagination or calculation. In the region past #26 iron, there are so many protons that the atoms start to creak and complain, becoming less and less stable. Added protons and electrons follow the same rules, but if the universe hasn't made enough of a particular heavy element, you have to BYOP—bring your own protons—by smashing them together.

Past about #100 is the realm where you can't find those elements in a chemistry lab anymore, much less in a lifeform. Thankfully, the first 100 elements, and possibly even just the CHON elements by themselves, can combine in so many different ways that chemists still have plenty to study.

AN UNFOLDING UNIVERSE STARTING
AT SQUARE ONE

These rules, as simple as a children's game, built the periodic table. Square one was hydrogen. It is the easiest element to form, just one proton and one electron attracted by opposite charge. Once temperature dropped enough, protons and electrons made in the Big Bang coalesced into hydrogen. About 90% of the atoms in the universe are still hydrogen atoms.

In this beginning, all the matter from all the stars in all the galaxies was confined to an infinitely dense point, smaller than a single electron. All the energy was confined as well, making it so hot that electrons could not settle down on protons, so hot that light could not escape the pull of matter, so hot that the four fundamental forces were melted into each other. The universe was a Jackson Pollock painting wrapped into itself a billion-fold.

Only after the universe expanded enough to cool down did hydrogen condense. Some helium was made when two protons happened to collide, but then the expansion continued, the universe cooled, and the elements were frozen in place. The universe was hydrogen and helium and not much else, not even stars.

We can trace the path of the periodic table as it was expanded from hydrogen and helium to iron and then to uranium by the machinations of stars. One force gathered and another force dispersed. Gravity is weak, but it patiently gathered hydrogen and helium. The process accelerated as matter accumulated. The core of the newborn star grew more and more dense as gravitational energy turned into a thin glow of light. The pressure inside the core increased to unbearable levels: first the electrons were squeezed away from the protons, and then the protons and neutrons themselves were squeezed together into new configurations.

This is nuclear fusion, instigated by intense gravity and pressure from mountains upon mountains of material. Two protons stick together to make helium from two hydrogens, and then another hydrogen can make #3 lithium, or another 2-proton helium can make #4 beryllium. Adding one by one or two by two can fill out the table of elements, like climbing a ladder with rungs made of elements.

A plot of how much the universe has right now of each element starts high at hydrogen and helium, and tapers off as the elements get bigger (Figure 3.3). This is the result of 13.7 billion years of hydrogen and helium smashing in the hearts of stars.

Life can only use an element if it can count on finding enough of it to eat. On the graph of abundance (Figure 3.3), I've drawn a border (after R. J. P. Williams) that shows which elements are abundant enough to be used by life. Below that line, an element must be hoarded if life is to use it. Molybdenum (Mo) is the only element below that line in current widespread use—and we will see in Chapter 7 that it is indeed

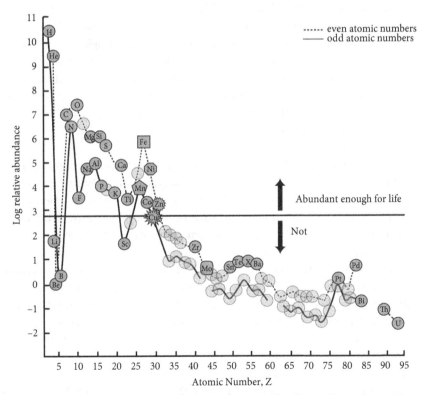

FIG. 3.3 Relative abundance of the elements in the universe, ordered by number in the periodic table. Only elements with a log relative abundance greater than about three are abundant enough to be used for life.

Data from R. J. P. Williams and Frausto da Silva, *The Chemistry of Evolution*, p. 3.

worth saving. Other elements below this line, like selenium and iodine, are used in sporadic cases, but these are decorations for life, not building blocks.

One reason titanium, scandium, and the other residents of the left end of the transition block elements are not chosen by life is because they aren't sticky enough. Another reason is that they aren't abundant enough, because scandium is below the line and titanium comes close. There's a lot more iron and nickel in the universe, so life selected those elements as catalysts (and even those only in milligram amounts).

While smaller elements are more abundant, there are some exceptions, shown as valleys on this graph. These are scarce because the laws of physics make those atoms unstable. Some interesting patterns come out of this. For example, the physical laws of protons and neutrons assembling say that elements with even numbers of protons are more stable than those with odd numbers. The deck is stacked with even elements.

Even though they are smaller than the CHON elements, #3 lithium, #4 beryllium, and #5 boron are not nearly as common, because they fall apart in the extreme heat and pressure of the star's core. Often they are made not from stellar fusion but as the fragments from bigger elements falling apart. It is like we are climbing a ladder to reach the sixth through eighth rungs, which would be the heavier elements on the periodic table, but the third, fourth, and fifth rungs are thin or missing. The missing rungs pose a serious problem for making a chemically rich universe.

The heavier, more stable elements must be made with complex machinations. Three #2 heliums can come together to make a #6 carbon that is especially stable, and then #7 nitrogen and #8 oxygen can be made from this seed. But three things collide much less often than two. In bumper cars, it's easy to run into one other car, but to run into two other cars at once takes particularly strategic driving. Three-atom collisions are so rare that we may despair of ever making carbon at all. The ladder remains broken.

The only way to climb the ladder, even given billions of years, is if carbon's rung could be lowered, so that we could jump high and grab its rung. Remarkably, this is exactly what we see. Carbon's rung is lowered by a special, "resonant" energy level that stabilizes six protons in a carbon nucleus, lowering its energy. A three-way helium collision can just reach the carbon rung with the tips of its fingers.

No one thought to look for such a weird energy level until Fred Hoyle saw the problem of the broken rungs on the ladder. Hoyle predicted that the nucleus of carbon must have a resonant energy level and even calculated its energy as 7.6 megaelectron-volts. Other scientists looked for it in the lab, and there it was. A universe without this resonant energy level for carbon would be chemically dull and lifeless. For life to exist consistent with our physical laws, carbon's energy level is forced to adopt this value. In other words, it is *constrained*.

There are other constraints on how the mass and energy numbers have to be arranged so that the universe would have enough of the three big CHON elements to survive. Most of these are from the physicists rather than the chemists, but they are so prevalent that many physicists require complicated explanations—as complicated as an infinity of universes, which, interestingly, *could* emerge from the current Big

Bang calculations—to explain why our universe is so precisely calibrated for complex CHON structures and therefore complex life.

Looking in on these physics equations as a chemist, I am surprised in two different ways. First, the number of fundamental physical constants needed is surprisingly small. For example, the night sky still echoes with light from the Big Bang. The shape of this light can be described with just six numbers. In 2015, four years of data from the Planck satellite were processed and released, and only minor tweaks to those six numbers were needed to explain all that data. Those six numbers work very, very well.

Yet these physical constants can be surprisingly sensitive to change. It's as if there were only six knobs on the dashboard of the universe, but if one of them were turned by a few percent, the universe wouldn't work anymore.

These physical constants result in chemical particles, which are likewise constrained. One constraint is hidden in the ordinary/extraordinary fact that protons and electrons match so well in charge despite their different mass. This balance of charge and imbalance of mass gives shape and function to the entire periodic table.

If you change the mass of the proton by a few percent, then the universe doesn't work anymore: either stars never form, or they don't ignite, or they don't explode right, all of which are crucial steps in getting a full periodic table out of just hydrogen and helium.

Other calculations suggest that if the light quark mass were more than 2% different, the ladder would be messed up, and carbon and oxygen would not be stable enough to be made in stars on the scale needed for life. A universe in which carbon and oxygen are stable but beryllium and boron are not is mathematically and physically special.

In his book *Weird Life*, David Toomey discusses the best mathematical alternative to our universe, which is a universe missing one of the four fundamental forces of physics (what physicists call "weak interactions"). In this universe, the Big Bang would make hydrogen with an extra neutron, supernovas would have to happen differently, and any element larger than iron would be present in only tiny amounts. Given how important heavy elements turn out to be later in this book, I'm skeptical that this scheme could produce a universe as rich as ours—and this is the best current alternative.

Let's return to our planet and fill out the deck of cards we have to play with for Earth's chemistry. Another valley of conspicuous absence in Figure 3.3 is at #9 fluorine. Again, this looks good for life. Of the elements in that next-to-last column, which all accept an electron readily and form single bonds with alacrity, fluorine is at the top. Its position at the extreme northeast corner suggests that it should be very small and exert a highly reactive pull on elections. But in Figure 3.3 there is more chlorine than fluorine, and chlorine plays a major role in life's processes, while fluorine does not. Oxygen remains as the undisputed dominant chemical in the reactive northeast corner of the table.

The most noticeable peak on the right side of the abundance graph is for iron, nickel, and manganese. Even though it has 26 protons, iron is on the top 10 list for most abundant elements, because iron's nucleus is the most stable nucleus possible. We get elements bigger than iron because neutrons fly around with such alacrity that elements up to zirconium can be made in star cores by chance collisions.

At this point, all these elements are still wrapped up in the heart of a star. To get them out, another stellar process must take over, this time not in the birth of a star, but in its death. Eventually, stars run out of energy and die. Right now in our sun, the outward pressure of mass and energy released by fusion in the core is balanced by the inward pressure of gravity, so that tomorrow's rising sun will be the same size as today's. But when enough stable iron accumulates in the sun's core, no more fusion can happen and no more energy is released. The sun's fusion fuel is spent.

The star collapses, pressing its matter and energy together. If the star is massive enough, this collapsed form gathers so much energy that it explodes in a supernova, shedding its outer layers all over the galactic vicinity. This is bad for the star, but good for us. An unexploded star kept all its elements to itself, while an exploded star has spread all those out for future stars and planets to use. This process also creates more heavy elements.

There are still some peaks to explain with more bizarre processes. For example, #79 gold is more abundant than it should be. One theory is two dead neutron stars (the heavy cores left behind after a supernova) collide with such energy that smaller nuclei are smashed into gold, in a refining fire that builds. One collision like this could make ten moons' worth of gold that could be passed around from nebula to nebula until it accretes into a rocky planet like ours and becomes a ring around your finger.

Such complicated dances of contraction and expansion, of fusion and explosion and collision, take time. The first generation of stars to form in the first few billions of years after the Big Bang must not have filled out the periodic table very much. It must have taken a few stellar generations to make the gold and uranium we find under our feet—about 13.7 billion years, in fact. In light of building up a universe from hydrogen, 13.7 billion years doesn't seem like such a long time after all. (In 2015, Peter Behroozi and Molly S. Peeples estimated that the universe has the potential to form 10 times more planets than it has already formed, so cosmically speaking, we are still in the first inning of the game.)

The periodic table not only *folds* like a road map but actually *is* a map of the rough order in which elements were made. This map shows the path of the universe's development over time, starting from hydrogen and building up to gold. It is also a map of *when* they can form: the light elements with low atomic numbers (like carbon, nitrogen, and oxygen) form early, while the heavy ones (like gold and uranium) form late, if they ever form at all. The periodic table shows why it's easier to find iron than gold—to a first approximation, it is because iron has 26 protons, while gold has 79.

Once the elements are made, they can join together according to chemical rules. With a little more unfolding, we can see *why* certain atoms pair up, and *when* they pair up as well. These are the chemical rules that lead to R. J. P. Williams's inevitable chemical order, and the periodic table becomes a map for Earth's development through time.

ELEMENTS PAIRING TWO BY TWO (OR MORE)

Our planet formed according to a set of physical and chemical rules. First, gravity formed the Sun and the Earth from a formless disc of matter. On an arm of the Milky Way, a cloud of dust spun into a medium-sized star, and after a few hundred million years of galactic pinball, a series of eight large planets and countless smaller asteroids and planetoids remained. The collisions and accretions heated and stirred the mix of 83 long-lasting elements, and they began to interact with each other.

After 9 billion years of star formation, enough large elements had spread around the galaxy to make this mixture chemically diverse. Electrons were shared, chemical bonds formed, and chemistry began. The elements combined following the same statistics as a deck of shuffled cards. More abundant elements paired up more often. Strong pairings lasted longer. These pairings became the planet around us.

Just as a biologist classifies birds by size and color, a chemist classifies bonds by strength. At the end of every physical chemistry textbook, long tables list numbers for pairwise combinations of elements. These come from centuries of investigation of chemists mixing things together (usually just how much the temperature changes and how much gas is produced, but, yes, there is the occasional explosion and singed eyebrows). From these universal numbers, you can calculate the most stable configuration of chemicals when mixed and heated—in other words, you can predict the future of a reaction, or infer the chemistry that would happen on a new planet as it cools.

These numbers corroborate the evidence at hand, or rather, underfoot. The evidence lockers for which elements pair together best can be found in the mineral exhibits lined up in glass cases at any natural history museum. My personal favorite is somewhat old-fashioned, but classic: the Redpath Museum at McGill University in Montreal.

A stately, blocky, columned structure sitting at the back of McGill's central lawn, the Redpath Museum is Canada's oldest. Like Montreal itself, it feels European. Like Freswick Castle, it seems taller than it is wide. It is small, but its space is absolutely maximized. Inside is a dark, wood-paneled lobby. Up the stairs to the right is a bright, airy hall huge enough for dinosaur and giant sloth skeletons, surrounded by an oval gallery and topped with a sky-blue ceiling. Between the stairs and the hall, hundreds of minerals are crowded together in low glass cases.

The Redpath Museum doesn't have the biggest or most expensive specimens, but it does have the best-organized and best-labeled ones, from the perspective of this chemist. In too many museums, the mineral is labeled with only name and place (and, if you're lucky, donor, which doesn't exactly convey useful science). Such labels don't tell you much beyond historical accidents and abstract nomenclature.

The impression from such arrangements is that geology is like biology, a list of species that can be grouped by similarity and usefulness. These names seem like contingent templates imposed upon nature by human minds. After a few dozen examples, they numb the mind with seeming randomness.

But the Redpath Museum does it right by making room for one more line, which may at first also seem like a jumble, but wait—it actually is a *code*. (Unlike Dan Brown's museum codes, this one works.) It's the chemical formula for the mineral in question, describing which and how many elements are in each rock.

Each glass case is a different chemical category. If you look for repeating patterns, you might see that some of the square glass cases correspond to squares in the periodic table. Order emerges from the chemical codes, and we can see a number of intriguing patterns. These are written in the elements in the rocks and connected to the periodic table.

First, when you scan the glass cases, you will see six elements repeatedly: the aforementioned oxygen, iron, calcium, and magnesium, and the as-yet-unmentioned aluminum and silicon. The geologist Robert Hazen calls these the "Big Six" elements of geology. After the three sets found in life—six for building (CHON+P+S), seven for balance, and eight trace metals—Hazen's Big Six are the elements found in most rocks. The sets overlap significantly because life ultimately comes from rocks, and to rocks it must return.

The Big Six may predict future events. Three of these elements (Na, sodium; Ca, calcium; and Si, silicon) will increase in groundwater a few months before certain types of earthquakes. When rocks start to stress and crack open, these elements dissolve in water and are harbingers of greater cracks to come. Silicon is released into the water because much of the heft of the average rock is made of Si (silicon) and O (oxygen). Oxygen makes up almost half of the Earth's mass, and silicon is close behind with more than one-quarter. Aluminum comes in a distant third, with less than one-tenth of the mass.

Oxygen plus silicon makes a common chemical group called a silicate. Glass cases full of silicates stretch the width of the Redpath Museum. Look in the back of the physical chemistry textbook, and you will see that the heat of formation of silicon oxide is one of the largest numbers from combining the Big Six. Physical chemistry predicts that silicates are prominent.

The atoms in silicate groups link together in intricate crystal shapes, some packed tight and some more open, some with regular angled holes like a trellis, some in rings, some in sheets, and some in chains. Carbon builds biological structures, but silicon builds *geological* structures that are diverse and beautiful, ranging from the pure white angles of quartz to a rainbow of minerals, such as topaz, garnet, and beryl.

The other four members of the Big Six can also be found in the silicate case at the Redpath Musuem. Silicate is like phosphate in that it has a large negative charge that must be balanced by positive charges. Every naturally occurring metal can provide such a balancing charge, each with a slightly different shape that fits into the silica network differently. Aluminum, iron, magnesium, and calcium provide this charge most often, completing the Big Six.

Sometimes the negative charge comes from elements without oxygen. If oxygen or sulfur provides the negative charge, the rock is an oxide or a sulfide. If a member of the column on the periodic table under fluorine does so, the rock is a halide. All of these come from the northeast corner of the periodic table (which is the corner where elements most attract negative electrons).

If this seems like a lot of rocks, well, yes, it is. Our planet has the most different rocks in the solar system. A branch of the Redpath Museum erected on the moon, or Mars, or any nearby planet would be only one-third as big. Robert Hazen has assembled a geological theory called mineral evolution, that the changes the planet Earth has undergone over the past 5 billion years have brought the number of possible minerals from a few hundred species (yes, geologists use that word, too) to more than 4,000 today. The other planets are in states of arrested development, and they have only about 1,000 minerals. Mineral evolution is not exactly the same thing as biological evolution, but they both involve a change over time, whether in rocks or in life, or, if we're lucky, in both at the same time.

WHEN METEORITES, MERCURY, AND THE MOON STOPPED EVOLVING

Long ago the Earth started to develop in geological complexity when other parts of the solar system did not. Scientists like Hazen have to work without a time machine (so far) but make do with the next best thing: meteorites. Meteorites are remnants of the original planetary disc that are caught in Earth's gravity after eons of dormancy. Outer space is a good preservative, so a meteorite that dates back to the formation of the solar system has not changed since then.

The Redpath Museum has a few meteorites, although you can easily miss them. They're just tiny black rocks, some speckled, a few of bulbous metal, but none is very colorful. Geology was a lot simpler 5 billion years ago. In his paper describing mineral evolution, Hazen says the earliest meteorites have only 60 different minerals, about 20 of these in grains too small to see. The planetary disk and intense solar radiation worked like a blender on high speed, smoothly mixing the elements into hard, dark chunks of rock, rich in Hazen's Big Six elements. The other 77 elements were occasional impurities smoothly dispersed throughout.

The slightly cooled meteorites were still in a pinball machine of other rocks, allowing for chemical mixing via repeated impact. Carbon made tiny diamonds and graphite sheets because its four bonds lock it into a sturdy crystal. Magnesium was there with iron and silicate, iron was with sulfur, and nitrogen was there with silicon and titanium (chemically, nitrogen and titanium work surprisingly well together, making a super-hard gold-colored coating for drill bits today).

Two major features formed to break up the smooth texture. Calcium and aluminum stuck to each other and dissolved in residual water, gathering together as whitish, glassy drops. Also, in big meteorites, heavy iron and nickel sank down into a dense core with traces of copper and zinc. In this way the number of minerals in the region of the newly accumulating Earth increased to 250.

Just when you think all meteorites are the same, one comes along with purple salt in it. A group of seven boys playing basketball saw a meteorite fall near Monahans, Texas, in 1998. The rock was brought to Johnson Space Center and was broken open within 48 hours, an unusually fast turnaround. Inside were purple crystals of halite (essentially the same sodium chloride as is in your salt shaker), with water molecules tangled up inside. The purple color probably came from cosmic rays jostling a few potassium nuclei. These crystals prove that water was already changing the chemistry of the cooling rocks.

The elements in these early meteorites tend to have more carbon and water in them, emphasizing the light elements that form light, volatile gases. These elements were disproportionately lost from the hot, molten Earth early on. Hazen thinks the Earth originally had more than one hundred times of these volatile, gassy elements back then. Like a balloon hissing out gas, hydrogen, helium, the carbon gases, and water slipped from the grasp of gravity and were lost to space. Then a surprise strike stripped away even more light elements.

Across the solar system, gravity consolidated small rocks into large spinning planets. Relatively late in the process, 4.5 billion years ago, as the rain of rocks was lessening, a planetoid the size of Mars came out of nowhere and slammed into the proto-Earth, shown in Figure 3.4. (According to this hypothesis, it's still under your feet somewhere.) This boiled and roiled the planet, ejecting gases to the vacuum of space, and the newborn Earth split in two, with a smaller bit spinning out and twirling up into our moon. Today the moon is still spinning away, an inch and a half every year.

FIG. 3.4 Illustration of the Impact Hypothesis. This is the currently favored theory for the creation of the moon. The details of this impact, such as the angle and size of the impacting body, are still under active debate.

The moon was spun out of the upper parts of the Earth, but Earth's deep iron-nickel core remained intact. The moon's core is much lighter than it should be, and the Earth's core likewise heavier. (In Chapter 4, the extra iron in the Earth will protect life with an invisible shield.) The moon's scarred surface shows that it has taken many bullets for us by attracting wayward asteroids and interrupting meteor-induced extinctions.

As the moon spun away after impact, the Earth's rotation slowed and stabilized. The impact released so much heat that it dried and degassed our planet. If the collision had not removed so much gas from the Earth's surface, our atmosphere might be thick and opaque like Venus's. Any nascent oceans were boiled off by the impact, so the impact theory also helps explain why the moon is bone dry. As for where all Earth's oceans come from, we're still working on that, but it may have come from below: a lot of moisture could have fizzed out from the Earth's larger, heavier molten interior.

When we talk about the probability of life forming, we usually look down at the Earth below us and out at the placement of Earth the right distance from the Sun. But we should also look up at the serene moon above. We are lucky to have that lantern in the sky, maybe even luckier than we suspect now. Time will tell as we discover more about planets and moons in the rest of the galaxy.

The collision that caused the moon remelted and mixed the rest of the Earth. Then the Earth and the moon cooled from the outside in. The solid crust, in contact with cold outer space, was rich in magnesium and iron, which balanced their positive charges with negative silicate. Magnesium in particular makes strong bonds and fits right into a silicate network with no space wasted (similar to how it fit with phosphates in Chapter 1), making the first rock in Earth's foundations out of four of the Big Six elements: Mg, Si, O, and Fe.

The other two of the Big Six, aluminum and calcium, stay melted at this point because of their positions on the periodic table. You can imagine all the abundant elements gathering on each side of a gym like the beginning of a junior high school dance, positives on one side and negatives on the other. Aluminum, silicon, calcium, iron, and magnesium on one side looked nervously at the other side of the gym, where negative oxygen and sulfur milled about.

The chemical tension was palpable. The boldest elements needed to step out and start the dance. Plus-three aluminum and plus-four silicon had more charge in a smaller space than plus-two iron and magnesium. They snapped up all the oxygen because it was more negative than the sulfur, and didn't leave much for anyone else.

Plus-two calcium and plus-one potassium and sodium didn't have a chance against the denser charges. These were the wallflowers in the dance. Calcium is plus-two but is also lower on the periodic table than magnesium, making its charge more spread out and weaker than magnesium's. Therefore, when the crust of the early Earth and moon hardened, the rightmost compounds (silicate, iron, and magnesium) solidified, while the leftmost/lower ones (aluminum, calcium, potassium, and sodium) stayed liquid.

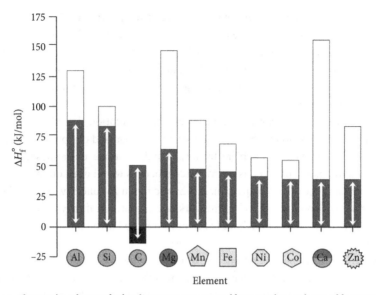

FIG. 3.5 Elements' preference for binding oxygen versus sulfur according to heats of formation, calculated as (heat of formation with sulfur) − (heat of formation with oxygen) and shown as the gray bar highlighted with arrows. The elements on the left have a greater preference for oxygen. The total heights of the bars correspond to the absolute heats of formation with oxygen.
Data from R. J. P. Williams and Frausto da Silva, *The Chemistry of Evolution*, p. 11, Table 1.4.

Figure 3.5 shows this. Notice that it's not the strength of the bond to oxygen that matters (the white bar), but the strength of this bond relative to a bond to sulfur (the gray bar and white arrow). Since they have the biggest arrows, aluminum and silicon end up with oxygen, and we find them at the oxygen-rich "-ate" forms, while iron and magnesium are more often found by themselves or with sulfur. The "aluminate" and silicate that formed were like phosphate down to their tetrahedral structure: so many oxygens were clustered around that these are little balls of negative charge.

(These numbers are taken from the general tables in the back of a physical chemistry textbook, by the way, showing how these current tables contain the chemistry that guided the past.)

Silicate, iron, and magnesium solidified together in different ratios. This first ratio making the first rock was probably dunite, a green rock so rich in dense magnesium and iron that it promptly sank. (Such magnesium-rich rocks are dense but fragile, and they could not hold up the continents to come later.) After sinking into the hot depths of the planet, the dunite would melt again, and the cycle of repeating melting and cooling formed a current that stirred the liquid mantle. This brought together new combinations of atoms and made new minerals like basalt, which has some lighter elements mixed into the iron and magnesium. Basalt floats just enough to rise to the exterior of the planet, plus it contains hidden strength that dunite lacks. As basalt formed, Earth's face dried up.

Rocks like basalt and dunite (with extra magnesium and iron) are called "mafic" rocks, a made-up word (combining "magnesium" with "ferric," from the Latin word for iron). Mafic rocks are heavier, darker, and older, weighed down by the magnesium and iron in them. Rocks that have less magnesium and iron in them tend to have more silicate and aluminate, and are lighter in both mass and color. These are called "felsic" rocks because they are like feldspars and have more silicon in them. Over time, silicate content increases: the sinking dunite is less than half silicate, the mildly floating mafic basalt is half silicate, and the floating felsic rocks to be made next are two-thirds silicate.

Felsic rocks are like white quartz, and mafic like black basalt, hinting at a yin-yang relationship between the two. My fourth son, Benjamin, was born a month after his cousin James, so they are almost the same age and share one-quarter of their genes, but put them side by side and the contrast comes out. James is like mafic rock, thick and strong with dark eyes and hair, and also older. Benjamin is younger, small and lanky, with bright blue eyes and a shock of hair so blond it's nearly white—he is the felsic rock. (*Frozen* almost works, because Anna looks mafic and Elsa looks felsic, but their ages need to be reversed for this analogy.)

The coldness of space gave the Earth and moon a thin and brittle carapace of basalt, but this was no match for the forces roiling within. The light elements that make up water and volatile gases permeated the planet's chemistry, making magma like fizzy water rather than flat. As water and other volatile gases trapped inside boiled and fumed, they expanded and burst through cracks in the crust as volcanoes and lava flows. This steam-driven motion added about 250 more minerals to the early Earth, bringing the total to 500.

This happened on the moon, too. The moon's large dark "seas" are lava flows pushed out by escaping gas and colored dark from iron. The lighter-colored moon rocks are felsic, with more calcium and aluminum.

Then the moon's geological development stopped as rock formation ceased. The moon was too small: it didn't hold enough heat to keep its insides liquid, and it didn't have enough volatile gases to keep pushing out its lava. The crust froze together and the volcanoes stopped. Mercury and the asteroid Vesta are so small and dry that they are frozen together in this point of time.

If not for the larger size and wetter composition of Earth, our planet would look like those, a gray swirl of light and dark rocks, pitted with asteroid craters and devoid of life or even movement. Mercury and the moon are time machines to an alternate past in which Earth did not have the chemicals that it needed to continue its mineral evolution.

OXYGEN'S CORNER OF THE MAP

But better things were in store for Earth and its solar system siblings Mars and Venus. Earth continued to move, grow, and develop until a small museum could be opened in Montreal with glass cases of minerals, including old rocks like what would be found on the moon, but with about 3,500 possible newer minerals that could not.

If you're in the main hall at the Redpath Museum, looking up at the blue domed ceiling, turn around. You'll see two graphs on either side of the opening you just walked through, chemistry/physics on one side and biology on the other (like in the previous chapter). On the left is a chart of biology through time, the species spreading out in an evolutionary tree, branching up from the past to the present. On the right are charts of physical parameters through the same time: heat flow, sunlight, Earth's rotation, meteorite impacts, and glaciations. Between these meteorological columns there is one that is undeniably chemical: oxygen in air (and, because this is Quebec, the undeniably classier French translation "Oxygène libre"). Here, a sky-blue line grows from a trickle one-third of the way up the column to a wide river at the top, representing today's atmosphere of 20% oxygen, a chemical river that entailed life. The chemical oxygen column on the right grows in sync with the biological tree of life on the left.

Oxygen has a special place on the wall of the Redpath Museum because it has a special place on the periodic table. It is the most common element in the northeast corner of the table, and all elements in this corner pull most on electrons. As the most compact and abundant electron-loving element, it was the most reactive element in the early universe. We just saw how it reacted with silicon and aluminum, leaving nothing free/*libéré*.

The rest of the periodic table also tells the chemist what kind of bonds can form. Slow, strong bonds form from the top row of CHON elements, because their small size allows them to get close together. As we move southwest, away from oxygen's corner, the atoms become bigger and their bonds become longer, faster, and easier to break.

These elements form bonds moderate in strength and speed: phosphorus, sulfur, and the plus-two metals from magnesium and calcium to iron and zinc. A phosphate group forms a medium bond that is strong enough to stick in the jumble of a cell, but also weak enough to break when its job is done. Iron sticks to oxygen with a medium bond that can grab it in the lungs and let go in the capillaries. Of the group of many elements, calcium and zinc also form medium bonds and will have unique roles in the story to come.

Farther southwest from oxygen's corner, we encounter bigger elements that make longer, weaker, faster bonds, especially plus-one sodium and potassium. It is chemically impossible to build with sodium in water—water dissolves all sodium bonds. These elements are only useful for balancing the overall charge in a cell, although this will become extremely useful in Chapter 9.

Put together these two general ideas: the northeast corner is the "stickiest" and longest bonding, while the bottom, bigger elements are less abundant. The periodic table wasn't an equal-opportunity palette of 100 elements from which any kind of chemistry could be assembled. Each kind of bond had only a few options. From these, life chose the six CHON+P+S elements, the seven ions for balance, and the eight for trace metals. Life's version of the periodic table only has about 20 elements (which are shaded on the origami periodic table).

The blue oxygen line on the Redpath Museum's chart is our first hint that many of the chemical changes over time are connected to oxygen. A tree of mineral complexity could be added next to the biology chart's tree of species, and it, too, would expand over time, following the same trend as the oxygen line. In this chapter, mineral evolution was driven by hydrogenated oxygen (water!), and later it will be driven by oxygen itself.

However, at this point in the story, oxygen is so reactive that it has already reacted. On the molten, well-mixed Earth, the oxygen was swallowed up by a cage of aluminum and silicon. There was none left for the atmosphere. The story of life appears over before it starts.

Yet there *was* just enough movement, change, and flow to keep the system changing and diversifying. The Earth was not a dead lake, but a river that would eventually hold life. The change had to happen without oxygen at first, but we know the end of the story, and we know that the oxygen will be emancipated.

The key philosophical elements are flow and consistency. Life flows between consistent bounds. It would take a special planet to walk the tightrope strung between the two by maintaining both flow and consistency for billions of years. With those present, the Earth was able to change and react, and after another billion years or so, solar power would finally unlock the oxygen lying dormant in its rocks.

4

THE TRIPLE-POINT PLANET

Earth's silent siblings // eating acetylene on Titan // three shields to guard the Earth // an arrow through the sky // summing it up with energy rate density // 4 billion years of liquid balance // the Earth cracks, moves, and evolves // the limited palette of the first ocean

THE PLANETARY PATHS NOT TAKEN

Let's move to a vantage point a little quieter: the surface of the moon. It is so still that Neil Armstrong's footprints remain undisturbed. The only reason the US flag there appears to "fly" is that a wire holds it up.

The moon and Mercury stayed still as Mars, Venus, and Earth moved on down the road of geological development. The moon is a "steady" environment, a word whose Middle English roots are appropriately tangled with the word for "sterile." Nothing moves on the moon, but in its sky Mars, Venus, and Earth move in their orbits, just as they moved on in complexity 4 billion years ago.

Out of the whole solar system, Mars and Venus are the most like Earth in size, position, and composition. Mars is smaller, but Venus could be Earth's twin in size. If Earth and Venus were separated at birth, then something happened to obscure the family resemblance: liquid water brought life. To chemists, liquid is the third phase of matter, between solid and gas, and its presence made all the difference.

Mars gleams a bright blood red even to the naked eye, while Venus is choked with thick yellow bands of clouds. Mars is cold enough to have carbon dioxide snow, while Venus is hot enough to melt tin and boil water. Earth's blue oceans and green continents provide a bright, primary contrast. These three siblings have drastically different fortunes.

At first, they looked the same, colored with black mafic basalt and glowing red magma. The original planets were all so hot that their atmospheres were driven off into space. The oceans and the air came from within. Steam condensed into oceans on each planet's cool basalt surface.

Oceans changed the planet. Water is a transformative chemical, small yet highly charged, seeping into the smallest cracks, dissolving what it can and carrying those things long distances. Venus, Earth, and Mars do not look like the moon because they have been washed in water.

Mars is dry now, but the Curiosity rover left no doubt that the red planet was first blue with water. Mars has rounded pebbles, conglomerated sand-rocks, and alluvial fans from river deltas. Once it had enough water to cover the planet with an ocean hundreds of feet deep. Today only a few transient, briny trickles darken its surface. The oceans of Mars have been either lost or hidden in the red rocks, perhaps enough to support a future colony of explorers who could roast rocks to harvest steam.

Earth has very old rounded pebbles about the same age, showing that our oceans also formed early. Hazen's "Mineral Evolution" essay describes rounded pebbles of fool's gold pyrite and a uranium rock that will react quickly with even a hint of oxygen. Because all the oxygen was locked up in rocks, these pebbles persisted long enough to tumble through water, and eroded rather than corroded.

No one knows exactly what tipped Venus off the course toward life at this point. One recent theory casts Venus as Icarus, a little too close to the sun: the sun boiled its oceans a little too much, and the crust took just a little too long to cool. Heat coming in exceeded heat going out. The steamy atmosphere locked in more heat, which created more steam, which locked in more heat. Soon the oceans of Venus became the clouds of Venus.

Water as steam is more vulnerable than water in an ocean. Solar radiation shatters its bonds, reshuffling H_2O steam into H_2 (hydrogen) and O_2 (oxygen). Hydrogen is so small that it slips from the planets' grip. Gravity holds heavy gas molecules close to the surface, but hydrogen is the lightest molecule possible. If oxygen is a family car, then hydrogen is a moped. This hydrogen moped is so light that solar radiation kicks it up to rocketship velocities, and it flies to outer space. The oxygen left behind quickly reacts with something else as oxygen likes to do, and then the planet has lost its water for good.

To continue the freeway analogy, carbon dioxide is like a fully loaded tractor trailer, being about as heavy as a gas molecule can get. It only stays airborne because its structure is perfectly balanced. It takes more than sunlight to break its double bonds. CO_2 is therefore a homebody molecule, keeping to itself, not reacting much, and sticking with its planet. Venus's atmosphere is now about 95% carbon dioxide, which retains heat and makes the surface of Venus a toxic, choked desert.

Venus shows that too much gas on the surface of a planet could doom its prospects for Earth-like conditions and the liquid flow of water. The four gas giants, from Jupiter to Neptune, are swollen spheres of hydrogen and helium. These gas giants have so much of the first two elements that they lost the chance for interesting chemistry with any of the next hundred. Gas giants are chemically simple, with few of the elements we need for simple life, much less complex life. (Their icy moons are another story in a later section.)

There is some hope yet for these planets, because they contain realms very foreign to our everyday experience, like supercritical fluids of hydrogen at the high pressures deep in Jupiter's core. Whatever life could happen would drastically differ from us, and it would be hidden from the sun's energy as well. But could the complex array of reactions needed for carbon chemistry and water chemistry happen there? Looking for life on a gas giant is like hanging your World Series hopes on a high-risk, high-reward baseball prospect. It usually doesn't work out.

THE MISSING OCEANS OF MARS

Mars didn't have the same problem as Venus—it had its own problems. We know more about Mars than Venus because probes sent to Mars aren't crushed into electronic slush, and the Martian atmosphere is helpfully transparent to our telescopes and probes. A series of rovers, culminating in the Curiosity mission, has given us a close view of Martian geology and has offered clues as to where all that water went.

In accordance with ancient Roman theology, it turns out that Venus has depth to her beautiful color, while the striking red color of Mars is superficial, even skin-deep. When Curiosity dug into Mars, only the first few scoops were red, with dull gray dirt underneath. But the boring gray color is exciting to those hoping to find life.

The color difference reveals a chemical difference storing energy that life could use. The red surface is iron oxide, as in Georgia red clay, covering the surface in a thin layer like the paint on a barn. This layer is chemically boring because the iron clings so tightly to the oxygen. When Curiosity used its electronic nose to smell the different molecules in the gray dirt underneath, it smelled like clay, with clear overtones of the Big Six elements—both the Big Six for rocks from Chapter 3 *and* the Big Six for life from Chapter 1 (CHON, S, and P). The silicate rocks were shaped by life-friendly mild temperatures and pH values.

Curiosity also found two forms of sulfur, both hydrogen-coated sulfide and oxygen-coated sulfate. Bacteria here on Earth can feed on the energy difference by switching one form to another. This chemical pair becomes energy if oxygens and hydrogens are moved off and onto the sulfur. We might scoop up some of this gray Martian clay and grow microbes in it, without oxygen or sunlight—except that one crucial piece is missing. This party would have to be BYOL (bring your own liquid).

Mars is missing the liquid phase of matter. The planet could pass through an airport TSA checkpoint with its liquids in a quart-sized bag (relatively speaking). The red planet has snow, but no rain. Carbon dioxide falls as snow in winter, and summer sun turns it directly to gas when the summer comes. In chemical terms, on Mars snow does not melt; it sublimates.

Like on Venus, Mars's water was boiled to steam and split by sunlight. The leftover oxygen reacted with iron in the Mars rocks, giving them their iron-red patina, like the copper-green patina on the Statue of Liberty.

The atmosphere on Mars is mostly missing, too. Its problem may have been location: the asteroid belt is next door. Mars didn't have a big moon to protect it, and strayed too close to the freeway of speeding rocks. A constant rain of meteors buffeted the Martian surface with cosmic shotgun blasts that blew away its atmosphere. Scientists are struggling to explain how Mars stayed wet for even a few hundred thousand years, while Earth has stayed wet for many times that long.

This same problem occurred in other parts of the solar system. Jupiter attracts and accelerates meteors, so the rocky moons surrounding Jupiter have atmospheres blown off. Their atmospheres are so thin that Mars air seems positively dense.

On Venus, the oceans boiled off. On Mars, they were blown off. Two different *modi operandi*, two different planets, one result: no oceans. Without flowing oceans, both of these sister planets look uninhabitable.

But farther out, this sweepstakes has more recent entries. In particular, one other big rock in our solar system has retained a thick atmosphere and liquid oceans on its surface, despite all the physical forces conspiring to remove them. If the three sister planets are primary colors, red, yellow, and blue, this planet-ish rock is a secondary color. It is the mysterious orange moon of Saturn named Titan.

TITAN LIFE?

Titan orbits Saturn, so from its surface, the sun is smaller than a dime. Not that you'd be able to see it anyway, because a dark orange haze obscures everything. Titan is just a little smaller than Mars, but it is abnormally big for its family. It is 20 times the size of all the other moons of Saturn combined, and about the same size as Jupiter's four big moons. Unlike those, it has retained an atmosphere, and beneath that, it has exotic oceans as liquid as our seven seas.

In my opinion, Titan is the best candidate in our solar system for the presence of extraterrestrial life because it has the three phases of matter flowing into each other. In fact, we *may* already have evidence in hand that life exists there, although from this far away, it's hard to tell.

Titan is both like and unlike Earth. The most striking likenesses can be seen in pictures taken by the Huygens probe in 2005. Huygens was launched as part of the Cassini mission to explore Saturn. Actually, it was more dropped than launched. As it fell through Titan's atmosphere, the clouds parted and it sent back the first pictures of the moon's surface. An online video titled "Huygens: Titan Descent Movie" shows what Huygens saw.

The watching scientists described it this way in *Scientific American*: "the pictures caused jubilation and puzzlement in equal measure. None of us expected the landscape to look so Earthlike. As Huygens parachuted down, its aerial pictures showed branching river channels cut by rain-fed streams. It landed on the damp, pebble-covered site of a recent flash flood. What was alien about Titan was its eerie familiarity" (p. 36).

The landscape was familiar because it was a solid surface scoured by a flowing liquid. On Earth, water scours soil, but on Titan, liquid methane scours solid ice. Titan has methane weather, whereas we have rain and snow. We have a water cycle, but Titan has a methane cycle.

On Titan, sand-like grains heap up in dunes like a subzero Sahara (no one knows what they are). Some kind of lava flows darken the landscape (no one knows what they are, either). Hard to see, but perhaps most amazing, are the lakes and rivers of liquid methane pooling and flowing over the surface. The pictures could have been taken from your last lakeshore vacation—if you had an orange filter on the whole time and only took pictures of rocks.

But the differences between Titan and Earth are just as significant. First, it is very, very cold. Titan's temperature is closer to absolute zero than it is to the freezing point of water. The outer crust of the planet is a thick sheet of miles upon miles of ice, with a liquid ocean far underneath. Those methane lakes are as still as mirrors, unbroken by waves. Methane rains in torrents stronger than any earthly monsoon, and then leaves the surface dry in decade-long droughts until the next flash flood. Anything living here would have to be very hardy and very small.

Titan's orange haze comes from methane. The sun can break hydrogen off methane gas like it can off water gas. The broken methanes stick together, making long carbon chains and even rings. (When severe smog is orange, I wonder if it's from hydrocarbons.) On Titan, these carbon chains are so big that they sink to the surface.

There the chains can react and release energy like gasoline. Gasoline is a carbon chain that we react with oxygen, but there's no oxygen available on Titan, just inert nitrogen and hydrogen. Hydrogen turns the chains back to single-carbon methane, which releases its energy. In this way, the small hydrocarbons carry a trickle of the sun's energy down through the haze to Titan's surface, just enough for a microbe or two. It's the same principle behind the flame on an acetylene torch, but with hydrogen instead of oxygen, it would burn one-quarter as bright (Figure 4.1).

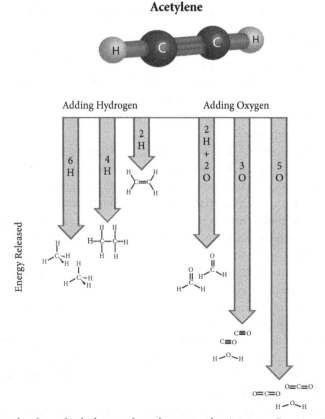

FIG. 4.1 Acetylene burned in hydrogen releases less energy than in oxygen. Energies calculated from heats of formation under standard conditions.

So life on Titan could move in the methane lakes and eat acetylene and other hydrocarbons as food. We can't see microbes from this distance, but we should be able to see their food in the atmosphere. In particular, scientists looked for the small hydrocarbons acetylene and ethene on Titan, as well as the hydrogen needed to burn them. But they couldn't find either—simulations have come up empty. The tastiest molecules for life are conspicuously absent.

This amounts to suspicion but not conviction. Other physical processes that we don't understand could remove these molecules. (For example, cyanide storms scour the surface, and who knows if that could deplete the hydrogen?) If these molecules remain unaccounted for, the future may see this as the first hard evidence of extraterrestrial life.

How would a methane-based organism transmit information and grow and move and compete? What would its CHON structures look like—would it even have N or O? Would it be able to bind and use metals?

Automated probes won't answer all these questions. What would answer them is a biochemist at Titan's surface, assuming the nontrivial problems of long-term human survival in a frozen, dimly sunlit realm can be solved. The biggest problem is getting the biochemist back, because there is no feasible basis for a return trip. Is there a biochemist out there willing to make the biochemical discovery of the century—on a suicide mission? I myself am looking at the floor and pointedly avoiding eye contact. I love biochemistry, but I love other things (and other people) more.

If Titan is the best we can do in this solar system, then our options are decidedly limited. Titan's life must be simple and slow, because the energy released from hydrogen is scant and the temperatures are extremely cold. The extreme weather patterns would buffet anything larger than a cell and bash it against the solid-ice shoals of the planet. Simple life has demonstrated the ability to adapt to harsh conditions on this planet, so I think it's worth a look, but I don't expect much.

Liquid flows exist at a few other places between here and Neptune, and all of these are candidates for simple life. However, none of these has as much gas-liquid-solid contact as Titan, nor are there obvious sources of the plentiful energy needed to support complexity—or for that matter, the faint energy needed to support simplicity.

Jupiter's big family of moons holds tantalizing liquid flows. Europa and Ganymede both have liquid oceans hidden in darkness under miles of ice. Likewise for Saturn's moon Enceladus, which is warmed by underwater vents like those suspected in the origins of Earth-life. Jupiter itself has supercritical helium and hydrogen deep inside. After the disappointingly dry rocky planets near Earth, the gas-giant moons seem like veritable oceans of possibility.

But, chemically, I doubt that life there extends beyond the level of the black mold that grows in my lab's cold room. These underwater oceans only contain two phases of matter for flow and exchange: solid, liquid, and gas together are difficult to find. Also difficult to find is energy itself. Without energetic power and chemical diversity, I am skeptical that we'll find much biological diversity. (Still, what I would give to run tests on extraterrestrial mold!)

The closest match to Europa's oceans on Earth is Lake Vostok in Antarctica. Millions of years ago, this lake froze over, and its liquid depths have been sequestered ever since. A group of Russian scientists carefully drilled into it and read the genes they found there. The good news is that there's plenty of life down there. The bad news is that it is a diverse menagerie of *simplicity*: 94% of the genes they found were from the simplest form of life, without any compartments inside. A few multicellular beasts may lurk deep in the lake, feeding off geothermal energy like the tube worms at deep-sea vents, but they remain unseen. It would likely be the same with these other oceans, or with Titan. Finding microbes on Titan would be enough for me.

If even simple life, or (please let it be so) complex life can be found built from another part of the periodic table, I promise to write a different book about how it works. But first things first. Until we have an alternate biochemistry to study, the Earth-bound water-based system I see around me fills both my imagination and the pages of this book.

There may be other ways to arrange a solar system to get sunlit, liquid water. One intriguing option is if a planet like Titan orbits a cooler, bigger type of star called a red dwarf. Red dwarfs are cooler, so the planet would have to be very close to catch enough heat for liquid water. That closeness has its own drawbacks, because gravity could lock the planet in place, making weather even more extreme than Titan's. But the same effect could heat the planet and keep the oceans liquid. We don't know enough to be sure.

EARTH'S INVISIBLE IRON SHIELD

Venus, Mars, and Titan show what can go wrong. Earth's success was also due to what went *right*. On Earth, the three phases of matter were preserved by five shields: two physical and three chemical.

Two physical shields work through gravity. The moon is a silver shield against meteor strikes. Jupiter, as well, shields Earth by corralling asteroids into a well-defined belt comfortably distant from our planet. Both of these are visible to all, but we have three other invisible shields. These are direct chemical results of three elements in three phases: iron in the solid Earth, oxygen in the gaseous air, and calcium in the liquid sea.

Evidence for the iron shield can be seen near the poles, flickering an unearthly red, green, and blue in the winter sky. Although I've lived in Seattle for more than 15 years, I have yet to see the aurora borealis with my own eyes (most likely because I grew up in Florida and get cold quickly). For people like me, there are videos from the vantage point of the International Space Station. On these, the light courses through the upper atmosphere like a cold, living fire.

These curtains of light happen when solar energy strikes oxygen and nitrogen atoms and scatters their electrons. This blunts the force of the sunrays so that the molecule-shattering energy is softened into colored light. The energy is channeled toward the polar regions by an electromagnetic force field.

That magnetism comes from a metal sphere of iron and nickel hidden at the Earth's core. After the collision that formed the moon, all the extra, dense iron and nickel sank to the center of the Earth and solidified. The Earth's solid crust held in the heat so that the interior remained liquid. A flowing band was sandwiched between the solid core and the solid surface, allowing the core to rotate and create a magnetic field. This is a different way in which flow creates the conditions needed for life. Earth's extra-large spinning core makes for an extra-strong field, extending far out into space, collecting and deflecting charged ions (Figure 4.2).

The periodic table set iron's density, abundance, and solidity, which created this shield. Mars, Venus, and Titan currently have no magnetic field, which lets the solar wind destroy gas molecules like water. (Mars had one but lost it.) The gas giant planets have magnetic fields from the flow of gas, but only Earth has enough flow inside itself to be both rocky and magnetic for billions of years.

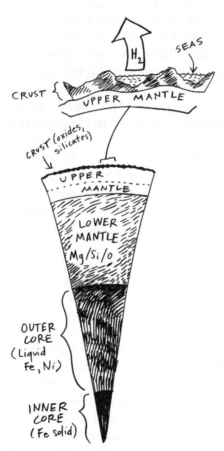

FIG. 4.2 Different chemical phases inside the Earth and on its surface. The solid-liquid-solid pattern inside the Earth allows the iron core to spin, generating a protective invisible shield against radiation. At the surface, solid, liquid, and gas coexist as hydrogen escapes to space over time.

Other kinds of radiation may inhibit life in other corners of the universe. Gamma-ray bursts course through most galaxies with enough energy to snap apart the chemical bonds that hold life together. These are so prevalent that some scientists have calculated that only 10% of galaxies could hope to hold a living thing together long enough for it to persist and evolve—only on the outskirts and at a relatively late time (where and when we happen to be, in fact). Large parts of the universe are sterilized by invisible forces. If it weren't for Earth's invisible shields, we wouldn't be around to know it.

SHIELDS OF AIR AND WATER

Two other shields come from chemistry in gaseous air and liquid sea. The air shield is the ozone cycle, which comes from oxygen, so it wasn't around at the very beginning. It unfolded as photosynthesis began to release oxygen into the atmosphere about 2.5 billion years ago.

The ozone cycle is $1 + 2 = 3$ with oxygen atoms: 1 O atom + an O_2 oxygen molecule = O_3 ozone. A high-energy photon of sunlight can crack the normal O_2 bond in half, and the single O atoms that result can stick to another oxygen, making ozone. Ozone is an odd, gangly molecule that absorbs a band of high-energy ultraviolet radiation that is particularly abundant in sunlight. Ozone catches dangerous UV energy and shatters, propelling O atoms around the atmosphere. This exuberant process transforms dangerous, bond-splitting UV radiation into harmless atomic wiggling. Through ozone, oxygen provides molecular sunscreen to the Earth.

Contrast the ozone cycle with the atmospheric cycles that predominate on other planets: Venus has a sulfuric acid cycle from its interior sulfur dioxide volcanoes, while Mars has a peroxide cycle from solar radiation shearing the oxygens off carbon dioxide. The Martian atmosphere has a little water in it, and even a trace of ozone built from these oxygen atoms, but its reactions are uncontrolled and out of kilter, making dangerous peroxide H_2O_2 molecules. Their chemical cycles are destructive, but ours is protective.

The ozone cycle is connected to the oceans because when sunlight splits water vapor into hydrogen and oxygen, hydrogen exits into outer space and oxygen is left behind. For a brief moment, this can form ozone. Liquid water in contact with sunlit air makes a fleeting ozone umbrella.

Water is also involved with the final shield, which happens when H_2O (water) and CO_2 (carbon dioxide) mash together into H_2O-CO_2 (carbonic acid). I tried to use this reaction in a lecture demo once. I put a chemical that would turn blue in acid in some non-acidic water. When I blew on the water, the carbon dioxide in my breath made carbonic acid and turned the solution a deep inky blue. At least, it did when I practiced it beforehand. In class, I couldn't get it to work, despite a few minutes of huffing and puffing. Demos always work great until the class is watching.

Perhaps it is because no one is watching, but this reaction works fine at the air-ocean interface. Over time, carbon dioxide in air turns oceans acidic. The acid helped chemical mixing by breaking down rock and creating new clays in the ancient oceans. But the real benefit came about when this acid combined with other elements.

High amounts of CO_2 (carbon dioxide) on the early Earth stifled the young planet in an opaque, hot blanket of gases with the potential to turn the oceans to the pH of lemon juice. But the oceans didn't acidify that much, and the CO_2 blanket was transformed, because the chain of chemical dominoes was not yet finished. Calcium and magnesium are part of the Big Six of geological elements, so rainfall brought large amounts of these two ions to the ocean. The negative charge on carbonic acid attracts the positive charge on calcium and magnesium. Iron, calcium, and magnesium fit tightly with carbonate and form a network of bonds. A lattice forms and a solid rock is born. Depending on which metals are around, it forms limestone, dolomite, or calcite.

This is the same reaction that made the tufa in Mono Lake. The salt and phosphate and most other minerals stayed in the water, while calcium carbonate crashed out of solution and precipitated as solid limestone, crystallizing into tufa towers. Whole mountain ranges of carbonate minerals like Italy's Dolomites may come from ancient CO_2 brought down to Earth by an open liquid ocean. Through these reactions, the ocean serves as a liquid shield. Long ago, calcium-rich oceans protected a planet from excess CO_2 by turning it to rock. The Earth's CO_2 fell to reasonable levels in a few million years. No CO_2 was removed from Venus's atmosphere, possibly because it was too close to the sun and too hot for liquid oceans. Some interesting carbonate-laden Martian meteorites suggest that Mars may have gotten partway down this road, but it never finished. As a result, the CO_2 stayed in the air, and to this day, both Mars and Venus retain atmospheres chemically similar to car exhaust.

AN ARROW THROUGH THE SKY

Molecule by molecule, these reactions changed the atmosphere over time. In the gas phase, each of the four CHON elements has its own story:

H: H_2 (hydrogen) *decreased*, drifting up and away into space;
C: CO_2 *decreased*, pulled down and packed away by the water-calcium shield;
N: N_2 (nitrogen) was released from the mantle over time and *increased*; and
O: O_2 (oxygen) *increased*.

On the periodic table, each of these is the gaseous form of a particular element arranged from left to right. Over time, in the air, the elements on the left decreased and the elements on the right increased. An arrow can be drawn on the table from left to right, representing how the atmosphere evolved away from H_2 and CO_2 and toward O_2 (oxygen). Its flow was like a river, pulled not by gravity but by the rules of chemistry on a planet with water oceans, pulling hydrogen up into space, carbon down into rock, and oxygen out into the air.

Over 4 billion years, the composition of the atmosphere on Earth moved toward the right on the table. This orderly change is summed up as the increased *oxidation* of the atmosphere. A number of processes worked together to shift the atmosphere in this direction—including the biological influence of photosynthesis, which will start in Chapter 7—but physical processes pushed the atmosphere in this direction as soon as there was an atmosphere to push.

The motor pushing this directional change across the periodic table from left to right is the way heat ("thermo") moves ("dynamics") in the universe, or to use the technical and faintly intimidating term, "thermodynamics." It all depends on the fact that heat spreads out naturally.

Explosions explode and water splatters because matter and energy spread out if given a chance. Every piece of matter and packet of radiation in the universe must, on the whole, follow the same rule and ultimately spread out. The scientists who work in the borderland between chemistry and physics have quantified this universal tendency in thermodynamic terms as a state of increased *entropy*.

Hydrogen leaking into outer space is spreading out. Evaporating water is spreading out from the compact liquid state to the gas state. These all make intuitive sense, but spreading out even explains how the gas carbon dioxide condenses into carbonate mountains, because it's not just matter that spreads, but energy as well. If enough matter and energy spread out in one place, then a connected system somewhere else can *condense.*

The best medium for matter and energy to move around in is liquid. The water-calcium shield requires liquid flow so that the elements can attract and the energy can spread out. Chemists do most of their reactions in flasks that hold flowing, moving liquid, allowing reactions to flow and move to completion. The driving force of entropy needs the liquid phase to drive an ordered change on a reasonable timescale. The reactions of life are no exception to this rule. All life thirsts for water to grease the biochemical gears. Without water, life is, at best, suspended animation—and at worst, death.

This rule of "change requires something to spread out" even works on the scale of the planet. In general, with the exception of hydrogen going up and meteorites coming down, the Earth is a closed system with respect to *matter,* and it does not exchange atoms with the rest of the universe. But it does exchange *energy* with the rest of the universe as it takes in sunlight and radiates heat in return. Earth is colder than the sun, so energy enters as high-energy sunlight but leaves as lower-energy heat. The key is that the exiting radiation is *more spread out.* (One paper proposes that special antennas could capture this infrared heat energy emitted by the Earth, and spread it out still further as a source of renewable energy!)

This can be felt in a closed car on a sunny day. High-energy light comes in through the windows and pushes the air molecules around, raising their temperature. The atoms cannot spread, but the energy spreads through them. You feel the spread-out energy when you open the door and feel the wave of hot air. In a sense, life is sitting in that car, between incoming light and outgoing heat, taking in energy and ultimately spreading it out as heat. The difference is that life can carefully redirect this current of flowing energy to specific places and for specific purposes.

This little trick allows us to gather as the rest of the universe scatters. Even though the total amount of matter on this planet is pretty much constant, that matter can take different forms and use energy currents driven by the sun (and, to a lesser extent, by heat leaving the Earth's interior through cracks in the crust). Matter converts from form to form through the chemistry of bonds forming and breaking. Complex cycles can form in which elements like carbon or nitrogen are recycled among organisms in seemingly infinite loops.

This looks like a perpetual motion machine, but it is not forever. It only lasts as long as the sun. Good thing our star has a few billion years left.

THE TREES OF TITAN ARE ALSO INSIDE YOU

Behind every seemingly stable structure embedded in a changing universe, whether massive star or tiny human, is a flow of energy passing through that structure like a river. This flow spreads out (and keeps the laws of thermodynamics) as the structure stays the same. If the flow is the energy of random motion, we call it heat. If the flow is matter in a liquid or gas state, we call it a fluid. Either way, the flows follow rules that maximize efficiency. As they flow, they build similar structures across the universe that look just like trees.

When the Huygens probe parachuted down to Titan's surface, the surface it surveyed was covered by branching structures immediately recognizable as river basins (Figure 4.3). The flows need to be strong enough to power life, yet gentle enough that life's intricate structures persist. Most likely, the sunlight is too faint and the winds too harsh to support life on Titan, but even in the absence of life, branching river basins form.

The tree-like shape of a river basin allows it to take the paths of least resistance down to the river mouth. Tiny rivulets upstream are repeated and combined into larger flows downstream. The engineer Adrian Bejan has developed a theory for how this flow is built up from its smallest components in almost any material. This efficient complexity happens naturally as water flows down a hill.

Bejan's math shows that this branching pattern minimizes resistance to flow from any area (or volume) to a single point. So if, for example, you need a flow that connects a single root with many branches spread throughout a sunlit space, the simplest solution is to build branches that bifurcate into smaller branches—a tree.

If you need a flow that connects a single esophagus with many aveoli to exchange air in and out of an organism, the simplest solution is a tree-like pattern in the lungs. A tree of arterial blood branches into tiny capillaries, and it is connected to a second, superimposed tree of venous blood that brings the blood back. If the two trees are superimposed in a particular way, then they better retain heat. Very different animals have blood vessels arranged in this particular double-tree geometry (with variations depending on whether they are warm- or cold-blooded). Bejan even applies his theory to explain how rolling loops form in boiling spaghetti, and how tiny, evenly

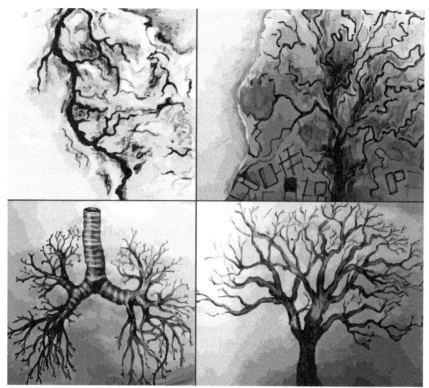

FIG. 4.3 Similar tree-like branching structures in different contexts. Branching structures carry liquid methane in rivers on Titan (upper left), liquid water in rivers on Earth (upper right), liquid sap and nutrients through trees (lower right), and air through the lungs (lower left). Although the Earth river picture appears to flow downward, it is actually sketched from a close-up of a river delta flowing upward to the ocean on top; this emphasizes that the branching patterns look the same for both basins and deltas.

spaced vents appear in cooked rice. (I suspect that most of his grant funding comes from other applications.)

Bejan's trees may look fractal, but he insists that they are "constructal" instead. Fractal patterns are built from the top down and proceed infinitely, while Bejan's complex constructal patterns are built bottom up from finite blocks in a constrained volume. Bejan is an engineer who made a theory that draws branches like an artist sketching a tree—and his "art" works in many diverse media.

Bejan's constructal theory is more about the road than the destination; that is, it is more about optimizing a flow system than a statement of whether the system is at the absolute optimum. It is about minimizing imperfections rather than maximizing perfections. It correlates with the flowing, imperfect systems of life that seem nonetheless optimized for their purposes.

Fractal complexity is descriptive and correlative, while Bejan's constructal theory is a predictable mechanism to approach a specific goal: the most efficient thermodynamic spreading out of a system, achieved most quickly. The branching pattern is the most efficient way for flows to reach their spread-out equilibrium state, which is why it is seen in so many places, from a computer's cooling system to a street map of traffic flows.

We will see efficient tree-like branches of flow in every stage of life's evolution. The complex branching trees of biology and the spontaneously branching rivers of geology have thermodynamic efficiency in common.

RIVERS OF ENERGY ADDED UP WITH ENERGY RATE DENSITY

Not only can the shape of energy flows be drawn by Bejan's tree-making theory, but the density of the flows may be predictable as well, according to the ideas of astrophysicist Eric Chaisson. If the physical chemists are right and thermodynamics are truly universal, if they can quantify spreading out with entropy and flows with flow rates, and if we live in a comprehensive, lawful universe, then this carries a suggestive conclusion: a single, universal number might quantify the river of energy passing through all stable structures, everywhere, at all times. Chaisson thinks he has developed this number, which he calls *energy rate density*.

Energy rate density (ERD) is the ratio of watts to kilograms. As such, the ERD for a system measures the river of energy that is spread out as it flows through a system. If the river flows more quickly and more energy is processed, then the ERD increases, too.

Chaisson's hypothesis is that an increased ERD goes along with increased structural complexity to process that energy. As things become more complex, their ERD increases. As the universe ages, objects with higher ERD accumulate. These objects will be more complex and energy-hungry than the objects that came before. In Chaisson's view, the laws of thermodynamics increase ERD and increase complexity over time.

ERD clearly increases for Chaisson's speciality, stars. A newborn star gives out 0.0001 watts/kg, our 5-billion-year-old sun has grown in brightness and lost mass so it is at 0.0002 watts/kg, and a subgiant star that is about double our sun's age would be 0.0004 watts/kg. (These numbers are small because stars are extremely massive.) Once a subgiant star becomes a red giant, its ERD increases to 0.012 watts/kg, and then grows into a final stage that processes 0.2 watts/kg.

The steps forward in ERD coincide with more structure, with more onion-like layers of more complex elements in the star. The structural complexity and the energy throughput are related. As more and more entropy is generated by the system, matter and energy are spread out from the system, but the system itself becomes more efficient at generating that entropy and becomes more complex as a result.

Chaisson thinks that ERD increases on all scales, for objects ranging from galaxies to stars to microbes to humans to societies. Take galaxies, for example. The Milky Way is older than our sun and has a lower ERD (0.00005 watts/kg, if you're curious), because much of a galaxy's mass never becomes an energy-emitting star. Newer galaxies give out less energy per unit mass than older galaxies and also have more nebulous structures.

Chaisson applies his ERD to Earth's geosphere. Before life formed 4 billion years ago, the physical and chemical systems spread solar energy into heat energy on the infrared spectrum, resulting in an ERD of 0.0075 watts/kg. The energy river put out by the sun fed the energy-emitting processes on Earth and flowed through the planet, creating local structure and causing more disorder elsewhere in the universe.

The same pattern of increasing ERD over time is seen in biological processes. Photosynthesizing microbes and plants take solar energy and process it with a value of about 0.1–1.0 watts/kg. Animals eat the plant energy and disperse it with a value of about 4 watts/kg. Human society processes animal, plant, and solar energy with a value of about 50 watts/kg. Chaisson sees in this trend not just an interesting progression but a driving force of increased energetic efficiency and complex structure over time.

A few caveats apply. Chaisson's numbers are only accurate within a multiple of 10 and require thinking about the average, typical (resting) cases, rather than the extremes. If ERD works, it works on a broad scale only—but this book is concerned with the broad scale. At crucial points later, we will return to Chaisson's concept and apply it to microbes, plants, animals, and even societies. I think it works surprisingly well.

Chaisson's hypothesis is consistent with the laws of thermodynamics. Some systems change so that they spread out more energy. The spreading out over time can increase structural complexity into endless forms most beautiful as the system itself *gathers* matter, complexity, and structure. This system also *spreads* increasing amounts of energy into the rest of the universe.

These apparently opposed forces of gathering and scattering cause all the change in the world. Every liquid river is a mass of gathered atoms that can flow and scatter. Water can dissolve mountains and can evaporate into the air. It is the most prevalent representative on Earth of the liquid phase. As such, water is necessary for protracted cycles of change on Earth.

The first arrow of change, the "arrow through the sky," followed the thermodynamic rules of scattering and gathering. Hydrogen flowed up to space and the carbon dioxide flowed down to rock, both of them scattering as they went, flowing away as oxygen slowly, slowly increased. As the universe spread out, an arrow ran through the upper part of the periodic table, from hydrogen to oxygen. When a sun shines on flowing rivers and oceans, oxygen must increase and hydrogen must decrease.

THE ISLAND THAT CONTAINS OCEANS

Because oceans are liquid, and liquids are by definition an in-between chemical state, they are also by definition chemically special. Liquids are the most chemically useful phase of matter but also the hardest to achieve. In a lab, you can make anything (even helium) solid by cooling it and pressing on it, and you can make anything (even gold) a gas by heating it and decreasing the pressure. But to make a liquid, you have to walk a line of intermediate temperatures and pressures, so that the molecules are adjacent but still flowing.

Chemists show the lines between solid, liquid, and gas with phase diagrams. These graphs are maps in which pressure increases from west to east and temperature increases from south to north (or vice versa). Solids are always in the southwest corner, gases are always on the northeast, and liquids are in-between. These show what will happen when everything is well mixed and has had time to equilibrate at a certain temperature and pressure. Figure 4.4 shows phase diagrams for the solar system, with each planet or moon's surface conditions labeled.

Each map is a sideways Y shape, and the branch point of the Y is a unique place called the triple point. At this temperature and pressure, all three phases of matter coexist. To have a complete cycle of weather phenomena, with freezing, raining, and evaporation, you need to be at or just east of the triple point. This is exactly where Earth is today on the water map, nestled in the liquid region but able to access the steam and ice regions with small changes in temperature.

Mars is too far west to have liquid oceans, and Venus is too far north. (Note that Mars is close enough to the liquid that salty water might exist momentarily on its surface, fitting with recent observations.) Titan and Triton, Neptune's largest moon, are

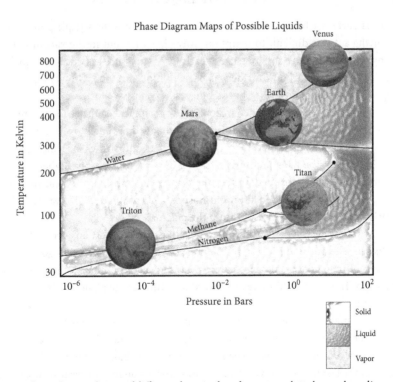

FIG. 4.4 The surface conditions of different places in the solar system plotted on a phase diagram of temperature versus pressure. Note how the Earth sits in the liquid water region and Titan sits in the liquid methane region, but other places fall outside these regions and do not maintain surface oceans.

so far into the solid region that any surface water is sure to be ice. Deep within Titan's ice crust, pressure increases and pushes the conditions north on the map, probably enough to make a deep and dark subterranean ocean.

Methane forms looser bonds than water, so its Y shape is farther south. For methane, Titan lies east of the fork, on the island of the map that implies methane oceans and weather cycles. Venus, Earth, and Mars are all well north of methane's freezing line, so there is no possibility of methane oceans this close to the sun.

Finally, a liquid nitrogen map shows a Y shape that is farthest south of all, because nitrogen has the weakest interactions with itself. Nitrogen can only be solid or liquid when it is very, very cold. Saturn's moon Triton is near nitrogen's lines on this map, but it is far to the west of the triple point. Some nitrogen may snow on this planetoid, but nitrogen oceans are impossible. Pluto is in the same place on this graph, and it shows evidence of moving nitrogen glaciers, but this surprising motion is still a far cry from the constantly flowing chemical reactor that is an exposed, liquid ocean.

Our solar system only has two places where flowing gas meets sunlit liquid: Earth with water and Titan with methane. Exposed liquid oceans are rare because only a narrow range of chemical conditions will produce them.

BILLIONS OF LIQUID YEARS

Earth is currently balanced on the liquid island east of water's triple point, but this balancing act was difficult to maintain. Any planet left out in the vacuum of space changes in temperature and pressure over time. Air pressure leaked away as hydrogen left the Earth and carbon dioxide was packed into limestone. Gravity and sunlight helped compensate for the loss, but stability can be elusive—just ask Venus and Mars.

The surprise in this case, when we look at the geological evidence for past temperatures and pressures, is how *little* things have changed on Earth over very long time spans. The planet was heated by countless impacts and strikes, of which the moon strike was only the latest and greatest. Pressures were probably higher, too. After the atmosphere was stripped by the moon-forming collision, the heat from inside the Earth quickly resupplied a thick atmosphere of carbon dioxide, water, and other volatile molecules that boiled out from the magma.

How thick was the early atmosphere? To illustrate the best answer we have now, I'd like to tell you about the figure you don't see in this chapter. I was going to make a graph of temperature and pressure over time since the moon impact 4.5 billion years ago. The line would slope downward over time as the planet cooled and leaked air to space.

I never made this chart because, on that scale, it would have made a boring, flat line. Over billions of years, our planet's temperature has not changed much. The fluctuations we all experience between seasons are *larger* than the average change we've had over the last few billion years. Fluctuations occur on million-year scales, icing up the planet and even threatening life itself, but on a billion-year, absolute temperature scale what stands out is the constancy. Three and a half billion years ago the average temperature at the ocean surface was about 35°C. Today it's about 20°C. To be sure, these are averages, but the oceans have never boiled away or frozen. For one-third of the age of the universe, Earth has had an ocean warm enough to swim in, somewhere.

On the scale of Figure 4.4, the 4-billion-year-old Earth would be a few millimeters north of today's Earth—which is a good thing, because another centimeter or so south and oceans would freeze. Earth has never dipped below that line in billions of years, despite being caught between the freezing vacuum of space and the boiling heat of the sun. Venus and Mars very likely were once on the island with us, but Venus moved northeast and Mars moved west, both losing their oceans.

The planet flirted with freezing over a few times, tipping into an ice-covered snowball at least twice, but both times it recovered and warmed up again. Earth has not left the liquid island on the phase diagram map for billions of years.

Most of the geological evidence for this is abstract isotope ratios and rock compositions. Closer to normal experience, you can see fossilized raindrops, in a paper titled "Air density 2.7 billion years ago limited to less than twice modern levels by fossil raindrop imprints." The size and splatter of raindrops back then is the same as today, showing that the atmosphere they fell through was about as dense as what you're breathing now, but with more methane and carbon dioxide, and less oxygen.

The pressure and temperature of the Earth has been so constant that it has created its own scientific puzzle: the "Faint Young Sun Paradox." Carl Sagan and colleagues pointed out that the equations of stellar evolution show that our sun heats up over time (as accounted for by Eric Chaisson's energy rate density). Four billion years ago the sun gave out only two-thirds the light and heat that it does today. Under that faint sunlight, the oceans should have frozen completely, but the oceans' liquid flow is clear from the rocks. So what warmed the oceans?

The complex interaction of different gases in the atmosphere may have trapped heat more efficiently. There are several promising leads involving many gases that were more abundant in the deep past: H_2 (hydrogen), CH_4 (methane), even N_2 (nitrogen). Hydrogen and methane are food for life, so that as life expanded and the sun warmed, the gas blanket over the Earth was naturally and literally eaten away.

Stability in temperature matters a great deal to life itself. Living things are easily a little too hot or a little too cold. As stable as the Earth's surface temperature has been, the places or times when it has been most stable have also been where and when life has thrived most.

Temperature instability coincides with extinction. The Permian extinction was the worst ever, wiping out entirely half of the "family" groups of species on the Earth. This extinction was so bad because it was accompanied by a large-scale heating of the oceans, among other factors. But even during this event, when the oceans were so hot that complex life was hanging on by its collective fingernails, the ocean surface remained below 40°C. To life, this was an existential threat. On the chemical scale, this was a blip of less than 10% on the absolute scale of temperature. It's a good thing temperatures have been so stable.

Additional stability makes life thrive, as shown by study of a certain class of plants in Australia. These are bushes with hard leaves, called sclerophylls, which make clear fossils. Australian geography provides a nice contrast: in the southwest, diverse sclerophylls abound, while the southeast hosts fewer species. In 2013, a broad study of thousands of sclerophyll fossils compared the southeast to the southwest to find out why species flourish in one corner but not the other. The southwest region was good for sclerophylls not because it was particularly dry or particularly hot, but because it was particularly *stable*.

Life is sensitive to even small fluctuations in temperature. In fact, temperature determines which type of soil bacteria dominates, so even a small temperature shift would change the soil microbial ecosystem by large percentages, with unknown effects. We should take care to maintain the temperature stability of the Earth. It *has* been the defining physical characteristic of the planet for the past 4 billion years, after all.

This stability may extend to factors beyond temperature. The "habitable zone," as currently defined by astronomers, is the region around a star where liquid water can condense. Some have proposed that this concept be expanded to add a geothermal habitable zone, where the planet has enough gravitational pull from nearby bodies that its liquid mantle is mixed just enough but not too much. A planet outside this zone would end up like Jupiter's volcano-ridden moon Io, in which Jupiter's gravity yanks out great tides of magma eruptions that obliterate the chemical bonds needed for stable life.

As telescopes have searched the sky for other planets, we have developed a reasonably big sample. We guess that about 10% of sun-like stars have an Earth-like planet around them, and find that in general, planets only up to 1.6 times the mass of Earth can turn out rocky and Earth-like, so that gas giants are much more common.

Compared to other solar systems, ours is unusually arranged. Most solar systems have at least one Jupiter-sized behemoth of a planet nudged up next to the star, closer in than even Mercury. Our solar system is empty in that area, exactly where most of the matter in the planet-forming disk of dust should have been located, bearing "scant resemblance" to other systems, to quote a paper on this. Where did all those potential planets go?

Our doughnut-shaped solar system may have come from a dance of two gas giants. Our system's pattern can be recreated in a computer if Jupiter swings close to Earth's current orbit, barging through the newly forming planetoids close to the sun. Jupiter swept the area clean with gravitational resonance, sending most of the material plunging into the sun. Then Saturn's beckoning gravity pulled Jupiter out again to where it sits today, conveniently corralling the asteroid belt and intercepting the occasional comet.

According to this story, Jupiter gave our planet a clean slate for life, free from interplanetary interference and constant apocalyptic collisions. Even more important, without Jupiter's cleaning up, our planet would have too much atmosphere, and would suffocate under a thick layer of boiling gas, like Venus.

Overall, the clean and safe arrangement of Earth's immediate environment seems rare at best. The good news is that there are so many stars in the universe (thousands times more than the number of grains of sand on Earth's beaches) that even "rare" means "a lot" in absolute terms. There's room for rarity and paradox in that abundance of possibilities. But right now this universe is inherently lonely. If complex life exists elsewhere, it is probably too far away for communication and definitely too far away for travel, even at the speed of light.

To quote Arthur C. Clarke, "Two possibilities exist: either we are alone in the Universe or we are not. Both are equally terrifying" (quoted by Kaku, p. 295). An undeniable thrill also lies in the question—terror and vertigo are characteristics of the best roller-coaster rides, after all. The thrill and terror are contained in the hard, scientific search for alien planets.

Our own existence is remarkably special and unusual, while it also requires certain universal medians and mediocrities. The Earth seems to be in at least the top 10% of a number of categories (e.g., gamma rays, planet placement, solar system spacing, sun size and brightness, etc.), but how many categories and just how high we are in each remains to be seen.

Perhaps we are not at a special point, but on a special *line*—a shoreline between too much order and too much chaos, a liquid island near the triple point. Life must exist where repeating cycles and flows provide the energy that keeps everything from freezing up, without so much energy that it blows things apart. Caleb Scharf calls this equipoise the "cosmo-chaotic principle" (p. 74). The most remarkable part to me is how *long* we've been sitting at this shoreline despite a host of forces that would pull us off it. Our biological intricacy rests on the foundation of eons of constant temperature, pressure, and flow, held in chemical defiance of the hungry emptiness of space.

THOUGH THE EARTH GIVE WAY AND THE MOUNTAINS FALL INTO THE HEART OF THE SEA

Oceans of water provide flow at the surface of the Earth. There are more liquid oceans underneath, as an unseen basement ocean of flowing magma mixes the elements. The water ocean and the rock ocean connect through faults and fissures, intermingling as continents flow together and apart.

To see this deep connection, start with another scientific mystery. Why are the oceans not like Mono Lake or the Dead Sea? Mono Lake has many inflows but no outflows, so it collects everything that cannot evaporate and escape. Likewise, many rivers flow into the ocean but none flow out. After all these billions of years, the ocean's concentrations of mineral ions should be higher than Mono Lake's, and ocean water should be the hardest of hard waters.

The calcium shield has subtracted some calcium and magnesium from the ocean, but it can't account for the huge discrepancy. Metals are missing from the ocean that don't stick to carbonate. Rather, the metals are removed, along with water, into the Earth's interior, like someone left the drain open at the bottom of the ocean. That "someone" is the process of plate tectonics.

Plate tectonics began with the hypothesis that the Earth once fit together like a gigantic jigsaw puzzle and then broke apart, pushing Africa away from South America and creating the Atlantic Ocean between. New rocks form at a gigantic seam in the Earth where upwelling magma presses and spreads. Pressure from this seam pushes huge plates of rock around. The ocean crust is only about 10 miles deep. We are float-ing like houseboats on the magma ocean underneath.

Where plates meet, one dives down under and the other is pushed up. The mega-range of mountains running down the west coasts of both Americas, from the Cascades to the Andes, is pushed up by the Pacific Ocean crust diving down.

For an interactive map of this and other geological features of the planet, I recom-mend the Smithsonian website *This Dynamic Planet*. It shows the outward-pushing divergent seams in the middle of the two oceans in red lines the color of seeping magma. Viewed like this, the Earth has seams of red like a baseball, buried deep underwater.

This interactive map also shows plate motion with white arrows, so you can follow the arrows to the yellow lines where the plates dive back down (called convergent faults). Tons of rock are plunging and melting along these convergent faults, making hot spots for earthquakes and volcanoes. Plate tectonics explains geology like evolu-tion explains biology.

Most geologists think plate tectonics started 3 to 4 billion years ago, when the newborn oceans pressed down on the plates, cracking through to the magma in the middle. (One theory suggests that a large asteroid impact got the continents moving.) At the convergent faults, water mixes with magma. As the Earth melts and subducts into the mantle, ocean water is carried along for the ride, emerging millions of years later at the mid-ocean rift. Water disappears under the West Coast mountains and reappears in the middle of the Pacific, like Pac-Man tunneling from one side of the screen to another.

This mixing turns the Earth inside out and changes its chemistry. One theory says that we have so much nitrogen in our atmosphere because it was released from the mantle via the motion of plate tectonics. Mars and Venus have no plate tectonics and no nitrogen. Other gases like CO_2 and hydrogen sulfide and subterranean minerals saturate the water at Earth's seams, providing a constant input of heat and chemicals for life to use.

It's a good thing that the water spends so much time inside the mantle. There, water is protected from the sun's shattering radiation that would otherwise turn it into hydrogen and oxygen. Water also forms a complex melted substance with rock that lubricates the movement of the continents and the resulting mixing of the elements.

Plate tectonics works because the kinds of rock that formed later in Earth's his-tory are less dense and more buoyant than those that formed first. Recall that basalt is "mafic," weighed down and darkened by the colored metals iron and magnesium, while the "felsic" rocks are lighter in both color and density, having more silicon and oxygen. Basalt floats, but it is like a heavy boat floating low in the magma, with most of its mass submerged. The crust needs to be lighter and float higher for the motion of plate tectonics to take hold, so it needs rocks that are lighter, more felsic, and less mafic.

Complex mixtures of mafic rock melt so that the lighter elements aluminum, sili-con, sodium, and potassium are liberated first. These are lighter on average and more diffuse in density, so they tend to rise through the molten mantle. When they collect at the surface, light gray granite forms.

If you look closely at granite, you will see four colors: black, dark gray, light gray, and grains of white. The white flecks are quartz-like nodules made of little else but silicon and oxygen. Quartz is a pure chemical that fits together tightly, making for tough grains that persist after the rest of the granite erodes away. These grains end up tossed about on the shorelines like so many grains of sand, because they *are* grains of sand. These are relatively light and fluffy for rocks, and they float to the surface like cream in a churn.

Deep magma produces granite too slowly to make a continent. Water steps in to help at this point. Water interacts well with the light, oxygen-rich silicate and aluminate (does oxygen have an affinity for more oxygen?), helping it melt at lower temperatures, then pulling it together and forming large blocks of mineral. Because water is integral to its formation, granite is not found much on other planets.

Water transformed wet sediment and basalt into granite, which rose into great mountain ridges and pulled up other rocks attached to it as well. They keep floating up and up, collecting more granite and forcing denser plates back down into the ocean.

The nearby moon pulled the ocean back and forth in huge tides. Both the water cycle on the surface of the Earth and the mixing of ocean with magma inside the Earth were like the cycles of a washing machine, scrubbing the Earth and suspending dirt and chemicals in the liquid ocean. Stable chemical combinations, like calcium with carbonate, solidified and sank to the bottom. The mixing of the entire Earth served to allow many such combinations to be found. Many ores are "hydrothermal," meaning they were formed by super-hot salty water rising through rock, carrying metals and sulfurs and other atoms along with it that otherwise would never have come together.

Using water chemistry as a guide, geologists trace ancient paths of water like a treasure map, because along the way are rich deposits of copper, zinc, and lead metals combined with sulfur. Hydrothermal actions formed gold and uranium deposits as well. X doesn't mark the spot—H_2O does.

In "Mineral Evolution" (2008), Hazen and coworkers estimated that the processes described in this chapter—granite formation, plate tectonics, and carbonate deposition—added about 1,000 new mineral species to our rocks. Most involve the action of water, so that means liquid oceans tripled the number of different minerals on this planet from 500 to 1,500.

The white angles of quartz, the soft green of jadeite, the glittering fool's gold of iron (sulfide) pyrite, and the glittering gold of true gold—all of these were brought about by the chemical action of water on a hot, flowing planet that was big enough to keep its heat when others were losing theirs.

THE SULFUROUS CRADLE OF LIFE

Because they are connected at the seams, as the rocks changed, the ocean changed, too. Ocean metal concentrations reached a rough balance. Tides and acid rains brought metals in, while the plate tectonics conveyor belt and solid rock precipitation brought metals out. All these metals were positively charged, because metals are located in the southwest corner of the periodic table. The ocean needed negative charges from the right side of the table to cancel out the positives.

The first balances come from the halide column on the far east side of the table (chlorine, fluorine, and bromine). However, halides have only one negative charge each in contrast to metals' two, three, or more positive charges. There are simply not enough halides to cancel the positive charges in the ocean.

The oxygen-bound "-ate" forms of carbon and phosphorus help, too, but they are not enough either. Carbon is spread thin through all three phases of matter, moving up to the air as CO_2 and down to the rocks as solid carbonates. Phosphate and silicate prefer to be in solid rock than liquid ocean, while nitrogen is so light it spends most of its time as a gas. And, of course, oxygen itself is unavailable and locked up.

If you're keeping score at home, you'll see that we have just crossed off all of the abundant possibilities for negative charge from the northeast corner of the table except one: sulfur. Sulfur dissolves well in water in all its forms, only forms a little gas as hydrogen sulfide, and, most important, has three chemical forms (sulfide when paired with hydrogen, sulfite when paired with three oxygens, and sulfate with four oxygens). These three forms of sulfur have similar stabilities and can be converted back and forth with just a little chemistry, forming geological and biological sulfur cycles.

The hydrogen-bound form of sulfur, hydrogen sulfide, is H_2S, which looks almost exactly like H_2O water. Like water, it can lose an H^+ to form the negative SH^- ion, which dissolves very well in water and balances the positive charge of all those metals. The two oxygen-bound forms of sulfur are also negative.

Today, sulfur in its various forms is the second-most abundant negatively charged element in the ocean, following only chlorine (see Figure 4.5). Back then, it was probably about as abundant, but the distribution of its three forms was tilted toward its hydrogen-rich sulfide form.

Sulfur and oxygen have a friendly similarity. Sulfur is able to form two bonds and pull electrons around like oxygen, but it is not quite as strong as its upstairs neighbor on the periodic table, since it is farther away from the northeast corner (which makes it bigger). Sulfur and oxygen work well together, dissolving and interacting freely.

In the ocean, hydrogen sulfide is like water, and sulfur is like "oxygen lite." If the ocean is a vast collection of oxygen with hydrogen carried along (H_2O), then from the periodic table you would predict that the box one step south, sulfur with hydrogen on it, would be the next-best thing—and you would be right.

Parts of the early oceans were rich with sulfur in the versatile form of hydrogen sulfide, providing negative charge and carrying hydrogen and electrons. At the seams in the Earth, electron-rich magma inside met the more oxygen-rich ocean outside. This chemical difference created the potential for new chemical reactions. This geological energy difference could have led to a biochemical energy difference that could fuel life.

Sulfur paired with metals in the order set by the Irving-Williams series, with the six metals from manganese to zinc in periodic table order. On the left side, manganese often pairs with oxygen in rocks, but the four elements on the right pair with sulfur and form sulfur ores. (Iron is the one in the middle of the table that can be found as both oxide and sulfide.) This is the same pattern seen in Chapter 2's chemistry-on-the-left, biology-on-the-right graphs (Figure 2.3), so we could add a third "geology" graph to those, showing manganese oxide on the left, iron oxide and sulfide next to it, and four metal sulfides on the right. The periodic table orders biology, chemistry, and geology.

THE GREAT CHAIN OF BEING (DISSOLVED)

Now all this information can be put together into one picture to set the stage for life. In the sulfur-rich areas (including the seams in the ocean floor), the metals on the right side of the Irving-Williams series paired with sulfur and hid away in rocks. The stickiest one, copper, was completely packed away. Zinc, next door, was a little less sticky and has a more complicated story. The four on the left side, manganese, iron, nickel, and cobalt, are less sticky to both sulfur and oxygen, making them available in the early ocean.

Chemical rules dictate how much of each metal will be dissolved. This concentration for each element, shown on Figure 4.5, shapes how that element is used in life. Here's how it works:

1. The highest-concentration elements in the top rung of Figure 4.5 dissolve so well in water that they don't do much else. These elements are like the kids who spend all of their time in the pool each summer floating around. They cannot build permanent structures, because they flit away to hang out with water instead. Building with these elements is building with sand (or with table salt) underwater. Their charges can be used to balance charges inside and outside cells, so they, along with phosphorus, are the elements of *balance*.
2. The second level of elements are still abundant in water but dissolve less well. Here are the CHON elements that make tight bonds good for *building*. Phosphate is on the bottom borderline of this level, and it can be used for building DNA. Calcium is on the top borderline, and its structures are semi-permanent (sometimes it will join its friends on the level above in the pool). In Chapter 10, calcium will adopt a special role that both balances and builds, and because it is the only element on this borderline, it is the only element that can fill this role.
3. Down at the third level of Figure 4.5 are elements that don't get along very well with water but are still abundant enough that, if life scratches and saves, it can build something small with it, maybe just an atom or two. This is where most of the eight trace metals used for biochemical catalysis can be found, not as big structures, but as occasional atoms contributing to big structures. Many of these concentrations changed over time, becoming more or less available as oxygen increased. This level is used for *biochemistry* and *catalysis*.
4. In the bottom quarter of Figure 4.5, the elements either aren't abundant enough or can't dissolve enough. These cannot be used in life at all. The only exceptions to this rule are cobalt (Co) and tungsten (W). We humans only use cobalt in a few places and tungsten not at all. To us, tungsten is poison.

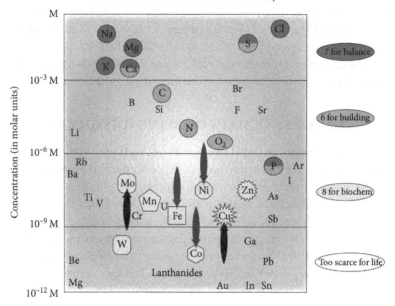

Element Concentrations in the Ocean Today

FIG. 4.5 Element concentrations in the ocean today. Elements are shaded according to their category and arranged according to their concentrations in modern ocean water. Arrows show if element concentrations have increased or decreased since life began. Note how the elements that maintain balance are high concentration, those that are used for building are medium concentration, and those that are used for biochemical catalysis are low concentration (but still present for life).

Concentrations from R. J. P. Williams and Frausto da Silva, *Bringing Chemistry to Life: From Matter to Man.* 1999. Oxford University Press, p. 274, Figure 9.10.

This figure looks like the medieval concept of the Great Chain of Being, in which everything was arranged in an ascendant hierarchy (Darwin himself employed its cousin "the scale of nature" at times). Figure 4.5 is more limited than that old idea. This concerns how much of each element dissolves in water, not the place of everything in the universe.

That said, the picture it gives is of a universe ordered, not by arbitrary rules, but by the chemical rules of how elements dissolve. There are many chemical "thou shalt nots" for life: life can't build sodium chains in water, and life can't make a niobium enzyme. This differs from Gould's free-for-all in which life can choose any number of possible directions. Instead, life's options were and are limited by the chemistry of water. A small set of possible elements could be used for balance, building, and biochemistry—so few that each category can be counted on one's fingers.

The reason that life uses only about two dozen chemical elements out of the 90 naturally occurring options is not that it randomly picked which ones to use. It was forced to use the chemical cards it was dealt by the early ocean. Carbon-based life in water should use these elements in these proportions. That is what chemistry predicts (à la R. J. P. Williams), and that is what our biological experiments see.

Elements dissolved in a liquid can flow, change, move, and form structure. Such movements are impossible in solid or in gas. With this cast of players and set of potential chemical roles, the stage is set for the planet to change forever, so the next chapter starts in the liquid oceans.

5

SEVEN CHEMICAL CLUES TO FIRST LIFE

the chemistry and geology near Fort Casey, WA // reading the rocks and genes // 1. interstellar chemistry // 2. catalytic compartments // 3. serpentine white smokers under the sea // the old iron crystals inside you // 4. cooking life with black smokers // 5. the lifelike hot spring in Russia // 6. phosphate and the first RNA // 7. when messier experiments work better

NEW WHITE STALACTITES AND AN OLD GREEN BOULDER

I was struck by two colorful examples of the hidden chemical structure of the world while I was supposed to be on vacation on Whidbey Island in Washington State, at a place called Camp Casey.

Camp Casey is one of three forts the US Army built at the mouth of the Puget Sound. Each sat on an island and looked toward the others, forming a "Triangle of Fire" across the water route toward Seattle. Now, a century after obsolescence, you can visit the eastern apex of the Triangle, which stands as Fort Casey State Park.

Children play in the industrial labyrinth of dark rooms, concrete steps, and watch towers, as parents worry about the lack of railings and abrupt drops. (It helped me learn as a parent to "let go a little.") The beach breeze and wide-open grounds are perfect for flying kites, a campground sits on the beach, and a ferry across the Sound leaves every hour from next door. It's a nice place.

My fondness for the area comes because I've spent a lot of time there. My university owns the old parade grounds and barracks just north of the fort. Faculty can stay in the old officers' quarters a short walk away. Last time I was there, I was trying not to work but couldn't help myself from doing a little geology. I saw two strikingly different rocks made of chemicals that may have bridged the gap between the two worlds of the flowing and the fixed—that is, the quick and the dead.

The first rock grows out of the concrete fort. It is a white, rippled rock that appears to drip from the walls and flow from the ceilings as teardrop stalactites. It is shaped by water. This water flows through the concrete, dissolving calcium and moving it to the surface. When the calcium in the water meets carbon dioxide in the air, calcium carbonate forms and freezes in a slick white mass. This is the same chemistry that formed the dolomite mountains and absorbed the CO_2 blanket, except here it is seeded by an abandoned concrete maze.

The second, contrasting rock sits on the beach north of the fort. This rock may have pushed non-life forward into life—a transition as abrupt as CO_2 gas turning to carbonate solid, but much harder to imagine. Against the gray sandy pebbles and sun-bleached driftwood, a six-foot-wide rock, mottled green and brown, sits half-buried in the sand. It's the most beautiful rock on the beach, curved as if it is flowing, because it once flowed. This rock looks serpentine and, if I'm right, it is made of a mineral called serpentine.

The serpentine boulder sits where ocean becomes land, at high tide surrounded by waves, at low tide dry and inviting. It is what the Celtics would call a "thin place." A rock like this one may have sculpted the first life on this planet with an unusual chemical reaction that made fuel from water.

Unusual chemical reactions are clues to how chemistry turned into biochemistry. These chemical clues spring from the unique properties of different parts of the periodic table. The structure of the table gives them order and rules out certain possibilities. If the origin of life chemistry can be worked out, even this most miraculous of transitions would be governed by the periodic table's columns and rows.

In this chapter are seven chemical clues. (Serpentinite chemistry is the third.) Each clue is related to a chemistry experiment done in a lab. Some clues contradict each other, but others reinforce. At the very least, all describe important aspects of how life works right now. Other scientists working in this field spend a lot of time talking about clues. Is there any other way to approach a mystery?

ROCKS AND PROTEIN CLOCKS IN SYNCHRONY

Before the clues, there is the question of timeline: When did life begin? This is a bit of a surprise in its own right. I would have thought that, given all the different molecules that have come together in any living thing, this assembly should have taken a long time. Instead, most evidence implies that life formed on this planet as quickly as possible, if not sooner.

Living processes are even harder to pin down in rocks, but various lines of evidence (including unnatural imbalances of neutrons) can only be explained by life 3.5, 3.6, or even 3.8 billion years ago. A study of phosphorus in rocks 3.5 to 3.2 billion years old finds that life was mature enough to use phosphorus in a widespread, well-defined phosphorus cycle. Evidence for life immediately follows the evidence for oceans. An energy-diverting, growing, replicating chemistry followed the presence of liquid water in a geological blink of an eye.

Proteins can be dated like rocks. After collecting genetic information from thousands of organisms, scientists can compare notes on protein similarities and differences, estimating how old the first proteins were and which proteins came first. One of the most prolific of these scientists is Gustavo Caetano-Anollés, who built a "universal molecular clock" that converges with the earliest putative fossils, projecting that the first proteins were built 3.8 billion years ago. These agree that life followed liquid quickly.

Protein calculations may even help to resolve what the temperatures were in life's rock nursery. Biochemist Eric Gaucher and his colleagues looked at an essential protein called *elongation factor* that works with the ribosome to make proteins. A massive comparison of elongation factor sequences allowed Gaucher to deduce that the great-great- . . . -great-granddaddies of elongation factors were heat-stable. The oldest appears to be adapted to an environment that is around 80°C.

These experiments suggest that life's oldest local environments were very hot but still liquid. The planet could have hosted life even when it was young and toasty. As it turns out, three of the clues that follow require hot environments.

Gaucher's studies also shed light on how protein shapes have changed over time. He resurrected two types of ancient enzymes called thioredoxins and beta-lactamases to investigate how the chemical reactions facilitated by a protein may have changed over time. These ancestral proteins are still more heat-stable and acid-resistant, supporting the idea that life formed in a hot, harsh local environment. They are also about as efficient as modern enzymes, suggesting that they quickly optimized their rates of reaction. The path of evolution that the beta-lactamases took narrowed predictably from general uses to specific.

So what could have happened to take the first step/giant leap, turning geochemistry into biochemistry so quickly? The first clue is found, not on the Earth, but between the stars.

CLUE #1: BIOLOGICAL ACIDS FROM HEAVEN

Space is not as empty as it once seemed. True, there are no men on the moon or canals on Mars, but there are oceans on Titan (methane) and three other moons (under ice). Water can be detected many places from its intense infrared color. This color is especially rich in comets and can be found in nebulae and dark lunar craters, even on the dark side of Mercury. Much of this water predated even the sun.

In the past decade we have seen the colors of other elements as well, frozen in bonded arrangements eerily similar to those used by life. It is often repeated that we are stardust because our elements were made by stellar fusion and supernovas. The new finding is that we are also from dust between stars. Nebulae and other space clouds hold material bathed in UV light that scrambles the electrons and bonds. This chemical process makes the vast expanses of space teem with complex arrangements of elements, forming the bonds and shapes that could be assembled like so many jigsaw pieces on the young Earth's surface.

Some of these fragments may be puzzle pieces for the nucleic acids that make up DNA and RNA. Nucleic acids have always been a major puzzle to scientists because they are such complex molecules, having three parts:

1. *D/R* for *deoxyribose/ribose* sugar;
2. *N* for the *n*ucleobase;
3. *A* for the phosphate *a*cid.

All three pieces have been seen, separately, in interstellar space. The most complex of these is the *N* (nucleobase) piece. Space telescopes have found huge amounts of cosmic carbon circles, known as polycyclic aromatic hydrocarbons (PAHs). These hexagons of carbon are stable and even form in the soot on an outdoor grill. The carbon circles can be found intermingled with nitrogen, looking like the first draft of a nucleobase.

The *D/R* sugar piece of the nucleic acid puzzle is seen in space as glyceraldehyde, which is half the size of the ribose sugar in DNA. Glyceraldehyde is a reactive molecule that your body puts together to form other larger sugars with a few steps.

The third piece of the puzzle, the acidic phosphate that makes the *A*, may seem to be the easiest to obtain on our planet, since phosphate is found in rocks. The problem is not with phosphorus's presence but its chemical availability. Like silicon and aluminum, phosphate is *too* stable in rocks, and water only dissolves a tiny bit.

One possible solution to this has fallen into our laps from outer space. Meteorites have a mineral called schreibersite, made of iron and nickel combined with an oxygen-depleted form of phosphate. A team of scientists from Florida and Washington found evidence of it in 3.5-billion-year old rocks. These meteorites could have delivered phosphorus to the Earth at the right time.

Meteorites also carry the amino acids that make up proteins. As with the PAH carbon circles in interstellar clouds, the sheer abundance of amino acids in meteorites is remarkable. One meteorite was found with more than 80 different amino acids in it, and amino acids are common in other meteorites as well. The question is not so much "how did life get enough amino acids?" as "how did it *choose* which amino acids to use?" Life only uses 20 amino acids, while functional proteins can be built from just five or six different amino acids. (One study found that a subset of 12 amino acids was *better* than our set of 20 for building proteins.)

Chemistry says that the early Earth could have cooked up its own amino acids rather than "ordering out" from meteorites. Famous experiments from the 1950s by Stanley Miller and Harold Urey, in which simple gases were mixed in a chamber and shocked with electric sparks to simulate lightning, are a bit dated but have undergone a recent renaissance. Miller and Urey found a mixture of amino acids in their flasks, but after the first few years, progress seemed to slow. A lot of effort was put forward, a lot of grad students tinkered with glass flasks and electric discharges, and a lot of tar was made.

Even so, several recent studies have looked closer and have found that these experiments may have worked better than we thought. When an old batch of vials from 1958 experiments by Urey and Miller including sulfur in the mix were scrutinized with twenty-first-century chemical instruments, two dozen amino acids were found, including one of the two sulfur-containing amino acids sitting in your body right now. New treasures were brought out of the old storerooms.

Another group of researchers noticed that the common analysis protocol involves harsh nitric acid that destroys weak molecules. But remember the calcium-ocean shield in the ancient ocean: carbonate would absorb protons and soften the punch of strong acids. These researchers added carbonate to make their mix more historically accurate, and found more amino acids. The same researchers added iron sulfide, a known chemical component of the early Earth, and found that it protected the molecules in these experiments as well.

The general rule is that the more complicated the experiment, the better, so long as the complications make the experiment look more like the early Earth. With nucleic acids, sugars, phosphate, and amino acids seeding the early Earth, all of the pieces were on the gameboard, but were shattered and scattered. They needed something to bring them together. The next two clues show that there were at least two ways in which different chemicals may have done just that, automatically.

CLUE #2: TINY BUBBLES IMPROVE REACTIONS

Some meteorites do something amazing—they blow bubbles as they dissolve. David Deamer and Louis Allamandola led a team of researchers who observed this. When they mixed an extract from the meteorite in water, all sorts of carbon-rich oily molecules grouped together, like oil tends to do. These were special, hollow drops called vesicles, like bubbles blown in water with water in the middle.

They also found that a mix of chemicals (simulating a comet) recreated the same behavior when illuminated with UV light in temperatures as cold as interstellar space. The small molecules linked into the same sort of carbon circles seen in Clue #1. As much as 1% of a comet is made of carbon circles that form bubbles.

The more meteorites we analyze, the more interesting chemicals we find. Another meteorite, found near Sutter's Mill, contains a rich array of carbon chains, including the fatty acids used by life and strings of carbon-oxygen called polyethers.

The bubbles called vesicles are especially useful because they have water inside, which means they provide a boundary that could hold reactions in one place. Any reactions stuck inside this bubble would be protected from outside interference and would also be able to keep the products of its own reactions, reaping what it sowed. Bejan's tree-making theory works in a bounded, finite space. Once the bubbles set a boundary, Bejan's theory predicts that heat and matter would flow in complex, tree-like patterns within and around that boundary. Branching complexity could grow from flow.

Bubbles form in nature in many different ways. If a molecule has a pattern of water-loving and water-avoiding parts, it can form a bubble. Bubbly sea foam is made of proteins and sugars from the life in the sea. Ocean bacteria communicate and transfer material by blowing water-filled vesicle bubbles.

Better yet, tiny bubbles provide fertile ground for natural experimentation (Figure 5.1). Several reactions that don't work in one big flask work better divided into many portions in small vesicles. Even if most reactions fall, if one reaction succeeds, its success is protected by the wall of carbon around it. The reactions that win the lottery get to keep their winnings.

The process works *better* than a simple random lottery. One group of researchers took all the machinery that makes proteins from RNA, about 80 components in all, and mixed them together with vesicles. If the molecules came together randomly, no

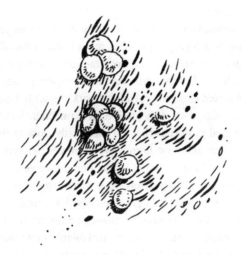

Tiny Bubbles Improve Reactions.

FIG. 5.1 Tiny bubbles such as vesicles provide compartments that protect and enhance reactions.

vesicle had a chance to contain all 80 components. If random, this experiment would have been as impossible as winning the lottery with a single ticket. I don't think they expected it to work.

But work it did, and random it was not. Some vesicles came together and made functional proteins that glowed green like fireflies. The weak interactions among the 80 components must have been strong enough to gather a complete set of machinery in some bubbles, like a child collecting a set of Pokemon cards. This means that the molecules can, in a sense, stack the deck so that even 80-part systems can be wrapped in a single, randomly formed bubble.

This also means that calculations of how improbable life would be that are based on total randomness make big, impressive numbers that mean absolutely nothing. All of these calculations are at heart the same: a lot of time is taken to delineate the many components of a biological system, and then the calculator assumes they assemble randomly. This results in multiplying tiny fractions together and implies that the odds of life forming from chemistry are astronomical.

Whether with protein sequences or elements or supposedly irreducibly complex assemblies, all of these leave out the fact that *we are not dealing with isolated numbers here*. We are dealing with chemicals that attract each other through chemical rules.

Self-assembly defies simple probability calculations. You have to drop your calculator and run the experiment.

So, if you're having trouble in the lab, try putting the reaction in a bubble. One group of scientists was trying to get RNA to replicate in the lab, but many parasitic RNA strands interfered with the desired reaction, sabotaging the results. The scientists made small droplets that protected RNA copying from interference, and the reaction worked better. Complex chemistry works better if it can get its own room.

Oily bubbles even reproduce themselves. A team led by Jack Szostak made simple models of cells called protocells that are chemically close to the bubbles formed by the meteorite extract, and have found that these vesicles give a passable imitation of life. For example, shearing forces in the liquid tore vesicles into strings of new baby vesicles, a process called pearling. A gentler chemical process based on sulfur's electron-grabbing properties and light also caused pearling.

Some bacteria reproduce themselves through a similar mechanical process called "blebbing." These bacteria make so many extra membranes that the chemical forces of oil-in-water attraction automatically gather the extra membrane into a daughter cell.

Through these reactions, carbon chains form protocell walls in the same way that soap forms bubbles. It is also possible that the carbon was helped along its way by the chemistry of the local rocks. One type of clay mineral helps form Szostak's bubbles and also makes long RNA chains. The bubbles even take along small bits of the mineral that could continue the catalysis.

It doesn't even require bubbles if the rocks have bubble-shaped indentations and are exposed to a flow of energy. In the right conditions, heat flowing across an open pore can string RNA chains together that are dozens of units long.

If the right rocks were in the right places with enough flowing energy, could that energy be caught and spread out in a repeating cycle of reactions? The next three clues are three different rocks found in three different places that might have worked this way to capture rivers of energy and catalyze the formation of life.

CLUE #3: HYDROGEN FROM THE GREEN ROCK

It was the color that first made me curious about the green boulder on the beach by Fort Casey, but its texture drew me closer. It was smooth in texture and sinuously curved, like a twisted piece of glossy snakeskin. Dark green was veined with light green and patches of brown. A friend later told me that it was probably island greenstone (although I think I could have made up that name on my own).

Most rocks that match that description are mafic: dense and dark with heavy magnesium and iron. If the green boulder is indeed mafic, and I'm hoping it's olivine, even a form called serpentine—if so, then maybe a few billion years ago this very rock fueled early life.

The key reaction catalyzed by the rock is called serpentinization, because it makes serpentinite. It's rare but can be found in the shadow of the Golden Gate Bridge and on a particular Washington beach (if this amateur geologist guessed right). Serpentinite is made when the mass of the deep ocean presses hot water into olivine.

At such a green boulder on top of a hydrothermal vent, water presses in so tightly that it pushes the iron out of the olivine, leaving hydrated, pure magnesium silicate behind. Serpentinite feels smooth and looks twisted because it has experienced such enormous pressures. The exiled iron shrugs its atomic shoulders and swiftly combines with oxygen, but in all the shuffle it drops a few electrons. (Iron is particularly good at moving electrons around, a skill necessary for this process.)

The magnesium and iron grab oxygens from H_2O water, leaving two hydrogen protons behind. Hydrogen ions lack electrons, so they glean the electrons that iron has dropped, and a molecule of H_2 (hydrogen gas) is born from the rock.

Hydrogen is biochemical mortar. The H-H bond has electrons that make new bonds in new molecules and provide energy for life's energy-intensive processes. Hydrogen provides fuel for life as well as for future cars. Some scientists think we can run serpentinization in the lab to make hydrogen fuel from water and cheap rock.

Mike Russell, Bill Martin, and Nicholas Lane worked together to explain how this clue may work best at a particular place deep underwater. Serpentinization produces heat that moves reactions forward, and the rock helps balance the pH of the water. These rocks form a natural chemical flow or gradient, as warm, high-pH, basic waters circulate through a cold, low-pH, acidic environment. The waters are gently mixed and gently heated, like a hot plate set to medium, not too hot or cold. Reactions happen well at these temperatures that would take millennia in colder places, as the temperature accelerates evolution.

Carbon dioxide bubbling up from the Earth's interior immediately encounters calcium or magnesium outside the vents, forming calcium carbonate spires reminiscent of Mono Lake and the white stalactites at Fort Casey—different chemistry, different place, same product.

These spires have the same nooks and crannies as the Mono Lake spires, and grow in patterns like the "magic garden" sets sold in science toy stores (Figure 5.2). Many of the pores in the rocks at these vents are the same size and shape as cells, so they would hold reactions in separate rooms like the oil in Clue #2. Newly formed spires are soft gels that allow molecules to migrate and react in near-infinite permutations. Did these chambers serve as incubators for early life, molding reactions and holding them apart from each other like a Petri dish holding 24,000 separate experiments?

The chemical gradient of H^+ protons and OH^- hydroxyls embedded in the rocks is suspiciously similar to the gradient used by all cells to make ATP. Carbon chains naturally coat the walls of the chamber, reactions could use the stored energy to increase complexity, and, who knows? One fine day a self-sustaining carbon bubble could venture out from its rocky nursery and make a living in the wide ocean. Lane and Martin's story even works with suboptimal, leaky membranes that tighten up over time.

Rocks offer diverse chemistries and geometries in their craggy surfaces. When molecules stick to a rock surface, they are brought together in new ways that are improbable in water. The rocks reduce a three-dimensional problem of molecules combining to a two-dimensional problem that is easier to solve. The scientist Gunter Wächtershäuser showed the power of sulfide rocks in such reactions, such as how iron sulfide rocks at 100°C, along with CO (carbon monoxide) and H_2S (hydrogen sulfide), string amino acids together into protein-like chains.

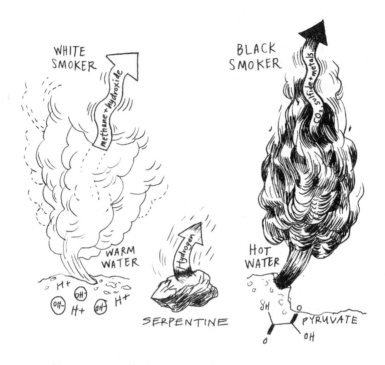

FIG. 5.2 How rocks might mold flows of energy and matter into materials for life. The white smoker underwater vent on the left contains pores with acid-base gradients; green serpentinite in the middle makes bubbles of hydrogen; and the black smoker underwater vent on the right may make the biochemical fuel pyruvate.

Ultimately, study of the origin of life takes chemists into some delightfully messy chemical possibilities. For the past century we've been studying pure solutions to keep experiments as standardized as possible. Pure solutions don't work for the origin of life. Life requires complex molecules in constant contact, and so large differences in temperature or energy run the risk of blowing life apart as soon as it comes together. This is one of the major flaws in Miller and Urey's spark discharge experiments: lightning rips apart as much as it puts together.

The problem with the spires near alkaline vents is that they may not be messy enough, and some biochemistry is missing from Lane and Martin's list of gradients. The missing biochemistry is called "redox" reactions, for "reduction/oxidation." These occur when electrons move from one atom to another. A redox reaction reduces/adds electrons to one atom and oxidizes/takes electrons away from another. This terminology comes from the fact that electrons are negatively charged, so adding them to something adds a "negative" to it and therefore "reduces" its charge.

Alkaline vents today don't have enough of the metals that are especially good at the redox chemistry of electron motion, and yet if there's one category of reaction that life inordinately loves, it's redox chemistry. The scant elements are the transition metals, especially iron and nickel, paired with sulfur.

You Are Wired with Iron Sulfide Rocks

Life loves redox chemistry because it is mild and controllable. If acid/base chemistry is smash-mouth football, redox chemistry is a well-placed goal in soccer. Redox chemistry takes electrons from carbon and moves them precisely onto oxygen, establishing a constant flow of energy with less explosive waste.

Redox proteins also hold a clue to the importance of metal chemistry to early life. Your oldest electron-moving proteins are not all CHON, but are bedecked with tiny rocks that look like they were ripped from the Earth. The pictures of these proteins in biochemistry textbooks show unusual colors for unusual atoms, orange for iron and yellow for sulfur, arranged in unusual cubes that look like sharp-edged rocks against the backdrop of the flexible CHON protein structure. In every cell, important redox reactions are catalyzed by tiny rocks.

A good example of this is a complex assembly of proteins called Complex III (Figure 5.3). This is one of four protein complexes that moves electrons to oxygen, all of which contain electron-paths made of iron, sulfur, and carbon rings (plus a few coppers at the end). Complex III has 12 structures that can hold electrons: two iron-sulfur chunks, six irons surrounded by a heme "belt" like in hemoglobin, and four carbon ring structures. (That's a lot of iron.) These are arranged in two paths of six structures each, so that the electrons hop from one to the next in a defined order like a bus on its route.

We know it's easy to build a new iron-sulfur-protein wire because we've built one in the lab. One group put two iron-sulfur centers at either end of a long, thin protein using a sticky sulfur-containing amino acid. This turned the protein into a small, conductive wire. If linked end-to-end, these proteins would pave a road for electrons. All it takes to capture electron-moving ability is a sulfur on a protein. Iron-sulfur clusters can be passed around and plugged in like video game cartridges.

Biological electron-paths and electronic wires can be joined together, because both are made from metals. One group put polarized metal electrodes into compost heaps and later found bacteria growing on them, pulling electrons from the electrodes. The researchers were fishing for electron-loving bacteria in the trash. These bacteria know how to bridge their biochemistry to our electricity, through metal-catalyzed redox reactions.

Some bacteria love fool's gold, known to geologists as pyrite. At the chemical level, fool's gold is iron and sulfur bound together in cubes, like in many electron-paths. These cubes are chemically "soft" enough that bacteria can extract the electrons for energy. (Is this mining or eating, or both?) These bacteria use wires like ours: stacks of carbon rings, iron, and sulfur. If bacteria can interface with metals now, then past protocells could have interfaced with metals then, which would have given them that metal's chemical power and a spark of life.

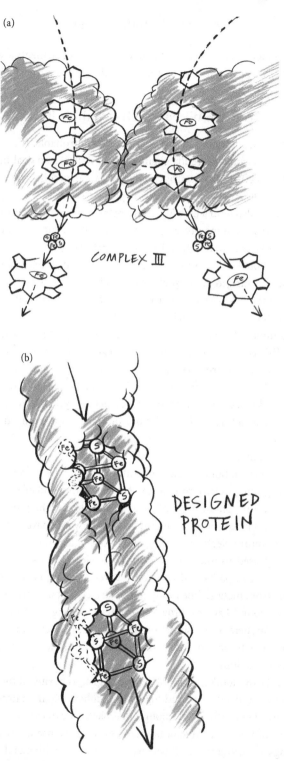

FIG. 5.3 Gallery of iron-sulfur (Fe-S) complexes and other electron wires. On the top is Complex III, a key part of life's machinery for transferring electrons to oxygen, showing two paths made by squarish heme molecules and FeS clusters. Below that is a designed protein with two FeS clusters that acts as an electron wire. On the next page is a closeup of pyrite, a mineral made from FeS cubes. All iron-sulfur arrangements are similar.

(c)

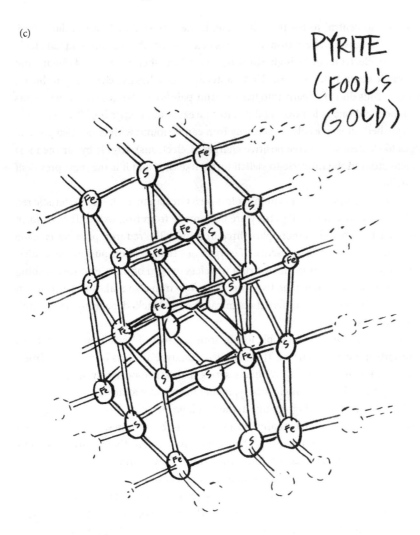

PYRITE
(FOOL'S
GOLD)

FIG. 5.3 (Continued)

Iron: Life's Switch-Hitting All-Star Element

Already we've seen iron carrying oxygen through the blood and moving electrons over distances small and large, but this only scratches the surface of its capabilities. All the things iron can do could fill another book. Iron is literally central to life: on a biochemical chart of metabolic pathways, even from a distance you can see a big central circle among the thicket of reactions, like the Arc de Triomphe roundabout in Paris. This biochemical circle is the citric acid cycle, which is central to life (and second-quarter biochemistry tests). Both this central biochemical roundabout and the pathway that moves electrons to oxygen are built around iron and iron-sulfur proteins.

Iron is also central to the periodic table. If the standard table was a dartboard, then its bullseye would be at iron. Iron is six boxes across the transition metal block, so it has six electrons in its highest-energy shell. This shell is the third one up and therefore we count 1, 3, 5, as we did in Chapter 3, to see it has five electron-pair boxes.

Six electrons don't fit neatly into five electron-pair boxes, because one electron fits in each box and one is left over and has to squeeze into another box. This means it can be pushed off iron easily. This is why iron can be found with either two positive charges (6 electrons) or three positive charges (5 electrons). Iron is by far the most prevalent atom in the universe to switch forms like this, and it is the most prevalent atom in life.

You can see iron's two forms in the colors on a Grecian urn. The characteristic red and black pigments in Greek pottery are both made from iron oxide. The red form is red like rust, which has more plus-three iron in it. The red iron pigment is made by firing the pottery in the presence of oxygen or other electron-pulling molecules. The black iron pigment is made by firing with less oxygen or more electron-pushing molecules, and it has more plus-two iron. (Plus-two iron also makes a cast-iron pan black.) Iron can act as two different chemicals, red or black, depending on its environment and electrons.

Because it can take two possible forms, iron is a sort of "switch-hitter" that can both absorb and eject electrons. This helps iron catalyze all sorts of reactions. Iron's central location in the Irving-Williams series makes it moderately sticky to oxygen, nitrogen, and sulfur. In chemistry, moderately sticky atoms are the best catalysts: they bind to something, change it, then let it go.

Iron is so reactive that it actually causes problems for protein storage. One group of scientists at Amgen noticed that their expensive protein therapeutics accumulated strange, small chemical additions over time. They found that the culprit was iron, which (in the presence of light) sliced up another molecule into small, reactive molecules. A little iron and light caused a big problem. Iron's target molecule was none other than citrate, the key molecule in the key citric acid cycle. Could such natural reactivity between iron and citrate have helped start the first life?

Another way to gauge iron's importance is to see how much organisms fight over it. The most valued toys are clutched most tightly. For cells, iron is clutched most tightly. Bacteria have whole arsenals of "siderophore" iron-binding proteins designed to hoard and steal iron. Even sugars floating in the ocean have inherent iron-binding activity (from their oxygens) that bacteria use to grab hold of iron so that they have a fully fortified breakfast. Iron-acquisition genes are located on DNA in places that evolve more slowly, meaning they are held back from changing and are likely more important to survival.

A bacterium with designs on colonizing your insides needs lots of iron, first and foremost, so bacteria have multiple intricate strategies to steal your iron. The bacterium *Staphylococcus aureus* has such a hunger for iron that it sidles up to your hemoglobin, yanks out the heme-iron square like a pickpocket, and rips it apart to steal the iron inside.

This thievery has a silver lining. If microbes need it so much, then scientists can block the process with drugs and cut off the supply lines to the bacterial armies. A group of scientists attached a protein that absorbs iron to a porous silica material. This soaked up all the iron and the bacteria nearby starved to death. Iron is so important that if you take it away, the whole bacterial colony collapses.

This idea can be taken too far. In 2012, a California businessman ran a large-scale experiment when he dumped 100 tons of iron dust into the Pacific Ocean northwest of Seattle. His hypothesis was that the iron would cause microbes to grow and pull carbon dioxide down from the atmosphere as their raw material for growth. (This is a little like the bacteria that pull electrons from the electrode.) Not much seems to have happened. The businessman wanted to save the world, but instead, the world swallowed up the iron and went on indifferently.

I think the experiment didn't work out because the natural system has different incentives and motivations than we do. In this case, microbes don't just take what they need and move on. Like the *Staph* bacterium that strips your hemoglobin of its heme, they greedily snap up as much iron as they can. Wild ocean microbes have never evolved self-control to say, "No thanks, I'm full." So algae probably lapped up the iron until they ran out of food, died, and sank to the ocean floor. (If they could feel regret, I'm sure they would.)

If that iron was dumped a few hundred miles south, off the Washington and Oregon coast, some of the dead, iron-bloated algae would have drifted down past another amazing formation of iron. Here is a hot, bubbling ocean vent with spires like those at the white smokers, but dark as cast-iron and caked with yellow sulfur.

This is our next clue, a second kind of vent and the second place where chemistry may have become biochemistry. The deep ocean seems like an unlikely place for life, but that intuition is clearly wrong. You can see with your own eyes that life is thriving here, in forms unlike anything seen above.

CLUE #4: METABOLISM ON THE BLACK ROCK

Right now, the best vehicle for finding new life forms isn't a starship, but a submarine. The Juan de Fuca ridge off the Pacific Northwest coast is a different kind of vent system from that of Clue #3. Here the sea floor spreads faster and the spires are more recent, with more sulfur, more acid, and much hotter temperatures. These temperatures both form and break bonds with alacrity.

The rocks here are darker, with more transition metals such as iron and nickel, so these are called "black smokers" in contrast to the cooler, alkaline "white smokers" of Clue #3. Black smokers are more chemically intense than white, in both good and bad ways. Rich lodes of copper, zinc, gold, and iron are deposited here, mostly paired with sulfur atoms as sulfides (and companies are looking to mine the results). Nanoparticles of pyrite, the same arrangement of iron sulfide found in protein electron wires, float around in abundance.

Despite the high pressures and temperatures, life survives and even thrives near the black smokers, big and weird, colored bright red and pale white. Underwater meadows of sinuous tube worms as tall as you and me coat the ocean floor. Near the plumes of black smoke, other bristly red and white worms bask on the edge of the thermal gradient like friends in a hot tub, with their tails almost boiling and their heads at room temperature, insulated against the heat by a sticky layer of goo that feeds a community of symbiotic bacteria. White fish, crabs, lobsters, and even octopi hunt among the black rocks. Compare this scene to the standard science fiction depiction of an alien as an actor with a forehead prosthetic: it's much weirder. And these species call our planet home.

As a final touch, white flakes float through the water, making this scene like a snow globe designed by Tim Burton. These flakes are clumps of fat and happy bacteria, which is somewhat less romantic but no less strange.

If life survives here, could it have originated here? If so much complex life can be supported now in the presence of sun, might a simple looping chemical reaction have developed here, where it is given energy and spun around by the heat and sulfur from the depths of the Earth? One study concluded that only about 125 reactions are "absolutely superessential" to all organisms in all environments, and that's for fully independent organisms. A pseudo-living complexity might be approached with a few dozen interlocking reactions.

The chemistry of the black smoker answers that question with a coy "maybe." Every element needed for proteins is here: CHON elements and dissolved iron sulfur chunks like those found in old, important proteins. The vent provides energy and flow as the rocks provide iron and nickel.

Although it's difficult to work with such high temperatures and pressures in the lab, some researchers have figured out technical ways to get closer and closer to black smoker conditions so they can be tested experimentally. A team led by Shelley Copley perfected a gas-inlet compressor valve system that approximates deep-sea vent conditions, and mixed simple chemicals in it to see what happens. Their initial results are encouraging. The more closely we approximate real places in the lab, the more the results look like life.

Copley started with a molecule located right above the central citric acid cycle roundabout of biochemistry: a three-carbon chain called pyruvate. Today, almost all life uses pyruvate as the entrance to the citric acid cycle roundabout. Several previous experiments showed that it's easy to make pyruvate under vent conditions when simple bubbling gases react with hot metal sulfides. So Copley started with pyruvate and subjected it to different temperatures, pressures, and metal-sulfide minerals.

She found that this cooked up some intriguing molecules. In the presence of metal sulfides, pyruvate gains complexity, picking up sulfurs, nitrogens, or even other pyruvates to form a full slate of complex molecules. Some minerals make the metabolic molecule lactate from pyruvate, and some even make the amino acid alanine, which is pyruvate plus nitrogen. Proteins make both of these molecules from pyruvate today.

Copley showed that rocks at vents make molecules familiar to any biochemist. More molecules have been made by mixing sugar into chemicals that simulate the ancient ocean, in particular with copious reduced iron. The scientists who did this were hoping that the iron would turn the sugar into a few new biomolecules. They got more than they were expecting when they counted 29 reactions, most of which involve the same molecules your body does.

In the game of metabolic connect-the-dots, the rocks could have provided dozens of dots and connections, but the complete connection of the dots is as yet incomplete. We need reactions that close the circle, products of one reaction that react in another to recreate the products again and mutually reinforce. The ancient alchemical symbol of the uroboros shows a snake eating its own tail—we need a reaction that does that.

Collective cycles have emerged with very different types of molecules and reactions. The ozone cycle uses simple molecules. Some reactive circles use complicated molecules, like nine peptides, or RNA. Something similar for small molecules in vent systems would be a true breakthrough.

Energy released randomly is dissipated as entropy, but energy released in a particular *direction* can push in that direction, which means it can do work. A self-propagating cycle of reactions that redirects and channels energy, if confined by a small compartment (whether rocky, oily, or both), might begin to live. Stuart Kauffman described how this type of situation is not well understood in chemical terms: "While we know all this is true, we seem not to have a language for it. The closest simple analogy I can think of is a river and river bed, where the river carves the bed, while the bed is the constraint on the flow of the river" (p. 319).

Eric Chaisson's energy rate density, by focusing on the watts of energy flowing through the mass of a system, could measure this flow. A system that is able to retain and increase its watts per kilogram could be favored. The question is how could it ever develop the chemical structures to maintain and improve itself against the random rush and jar of the heat around it.

CLUE #5: THE THERMAL SPRING CHEMISTRY
INSIDE EACH CELL

White and black smokers have a significant drawback—they are buried deep from the sun's energy. One special place at the Earth's surface hosts another possible biochemical incubator, where thermal energy from below meets solar energy from above. This place is in Russia, on the Kamchatka Peninsula, at the extreme northeast end of the Asian continent.

Follow the Aleutian Islands to the mainland and head south to the Mutnovsky volcano. Here geothermal heat bubbles up in hot springs and carves caves in the ice. The ice caves are captured in photographs by Denis Bud'ko as deep, scalloped passages that glow from sunlight above in mixed greens, blues, and browns, exactly like a Tiffany stained glass panel. The geothermal fields carved by the same heat are a bit uglier, being potholes of orange-gray mud boiling and bubbling like a stewpot. It's the ice cave photographs that are passed around online, but the geothermal fields have the hidden beauty of symmetry between geology and biology.

Each pothole is a separate bubbling compartment like a heated laboratory flask. The catch is that these potholes are too acidic to support life. However, in 2012 a team led by Eugene Koonin published a paper arguing that the acidity comes from oxygen in the atmosphere. On early Earth, with oxygen locked up in the crust, the pH would be more balanced, and metal-rich chambers would build up from the rock, like the black smokers, but above ground. Again I'm reminded of Mono Lake.

Steam would bubble through this network and condense where it's cooler, dripping down again, like a curlicue glass condenser in organic chemistry lab. This physical cycle could set up chemical cycles. In the lab we sometimes purify chemicals by dripping them through a porous column of sandy silica material, driven by the force of gravity. This separates chemicals because some chemicals stick to the silica while others flow right through. Koonin's hot springs turn that chemistry on its head: here, *evaporation* separates and purifies a solution by driving it *up* through a porous silica-rock network.

The extra benefit of doing this on the surface of the Earth is that the chemistry would be exposed to the sun. Ultraviolet (UV) light has enough energy to shuffle electrons and twist bonds. For example, every photon that your eyes detect hits a retinal molecule in your eye and twists it. Some complex chemical syntheses flow reactants through a chamber and flash them with light to flip bonds around. Both flow and flipping reactions were operating at the hot springs.

A more mundane example of the power of UV chemistry comes from a household trick my wife figured out for diapers. When our cloth diapers started smelling with what can only be described as a "funky barn" smell despite repeated washings, she tried hanging them out in the sun to dry, and it worked. Something about the solar energy destroyed the bad smell. But this shows that there's a double-edged sword here, because in the funky barn diapers, UV destroys molecules (in this case, stinky molecules). UV giveth bonds and UV taketh them away, breaking many molecules down to useless fragments.

So it's interesting that Koonin's springs include built-in sunscreen that protects against UV-mediated destruction. The rocky networks have thin shells of zinc sulfide, which is very good at absorbing UV light. Just a 5-millimeter layer of zinc sulfide is as protective against UV as 100 meters of water. With all the liquid movement from bubbles and steam, molecules could form in the UV light and drip or diffuse down under a zinc sulfide umbrella, available for chemistry.

Zinc sulfide can even transfer some of the light energy it absorbs to catalyze reactions. Another chemical, iron carbonate, can generate hydrogen like serpentine minerals do, but only in the presence of UV light, exposed to the sun. This could have been a fuel pump for Koonin's springs.

The best piece of evidence for Koonin's hypothesis is in every cell: these springs have a chemical profile that matches universal aspects of biochemistry. Every cell has five balances that it maintains between its inside cytoplasm and the outside environment. The biggest of these is the sodium-potassium balance. Because the ocean is so salty, maintaining this balance is costly. An array of pumps keeps sodium and potassium levels constant inside the cell despite fluctuations outside. If these levels get out of balance, the cell cannot stand the chemical pressure and it either bursts or implodes. This must be important because it is universal and expensive. (Another way to say this is that a lot of material and energy is *spread out* outside the cell in order to *gather* particular chemicals inside and outside the cell.)

The other four balances have already been mentioned, and all five are shown in Figure 5.4. Each balance has a chemical reason behind it that suggests that it is neither contingent nor arbitrary, but universal to water-based life. The five balances are as follows:

1. Electrons and hydrogens are brought *inside* for building big molecules; oxygens and oxidized materials are pumped *outside* to keep them from attracting electrons and breaking big molecules.
2. Phosphates are brought *inside* for DNA, energy, and signaling; sulfates are brought *outside* to balance phosphate's negative charge. (Electron-rich sul*fides* with hydrogens instead of oxygen are kept inside, because of rule #1.)
3. Plus-two magnesiums are pumped *inside* to fit with and balance the charge of phosphates; plus-two calciums are pumped *outside* to keep them from forming solid calcium phosphate with the same.
4. Trace metal elements (iron, zinc, nickel, etc.) are pumped *in and out* to keep their concentrations in line with their stickiness, as predicted by the Irving-Williams series.
5. Potassiums are pumped *inside*; sodium and chloride are pumped *outside*.

The reason for the fifth balance is that a new cell building big molecules with ATP and keeping records in DNA would have to hold all sorts of phosphates inside. This is so much negativity that it's too much to be balanced by just magnesium (the magnesium would start to stick and solidify, or even chop up the RNA). The cell needs to pump ions in and out to balance its charge, so it needs a small, charged chemical that won't mess up the works by sticking to other things.

Enter sodium and potassium. Sodium and potassium are abundant in rocks and oceans. (They are not in the Big Six of rock elements but they would be in the Big Eight.) Each ion has only one positive charge and is not very sticky as a result. Because sodium is much more abundant in the ocean, the simplest way for the cell to achieve the right balance is to reject the more abundant sodium and keep in the less abundant potassium, while also rejecting chloride. If a cell rejects table salt (sodium and chloride) like a patient on a low-salt diet, the cell can remain electrically neutral overall.

But it's a huge puzzle how life's distaste for sodium and preference for potassium could have gotten started. Seawater has about *40 times* more sodium than potassium, so a new cell has to pump out a tidal wave of sodium to achieve overall charge balance. That's too much to ask of a young cell. But what if the cell grew up in a low-sodium environment?

Lining up these five balances with the elemental profiles of different geothermal hot spots, the chemicals outside at hot springs match the chemicals inside of life better than seawater does. In particular, the hot springs are rich in potassium, phosphate, and zinc and (relatively) poor in sodium. On each of these four counts, hot springs are close to the inside of cells in ways that seawater is not.

Koonin and colleagues found that hot spring waters have equal amounts of potassium and sodium in the waters, and even more potassium in the waters condensed from steam. They speculate that potassium rises better in steam than sodium, so that the geological structure of hot springs pulls out the sodium but keeps the potassium as the solution percolates through the rocks. A cell could have learned to reject sodium more easily if there wasn't so much to reject.

The other chemicals that are specifically enriched at hot springs are phosphate and zinc. Zinc is found in many early enzymes, but it would have been scarce by itself in most water environments on the early Earth because it reacts so well with sulfur. A zinc-rich environment like the hot springs could have given the cell enough zinc to use in assembling proteins on the ribosome, for example. Potassium, phosphate, and zinc are three elements found at Koonin's springs that solve three chemical problems.

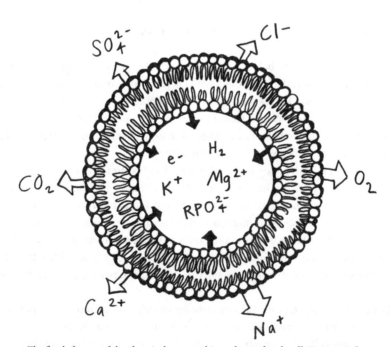

FIG. 5.4 The five balances of the chemicals pumped in and out of each cell. Trace metals are present inside at concentrations corresponding to the Irving-Williams series in Figure 2.3 in Chapter 2.

CLUE #6: PHOSPHATE SHAPES RNA

Clues #3–5 are grouped in a school of thought on the origin of life called "metabolism first." This school puts the "chicken" of protein-like cycles before the "egg" of DNA/RNA strand formation. Another school of thought, "replication first," puts the DNA/RNA egg before the protein chicken. Neglecting to mention "replication first" would be like teaching Aristotle without mentioning Plato. The "replication first" school has a strong chemical argument that the original catalysts may not have been protein at all, but phosphate nucleic acid chains.

RNA sits between DNA and proteins in life, as seen in Figure 2.2 in Chapter 2. DNA stores the information and copies itself for new cells, DNA makes RNA, RNA makes proteins, and proteins do the chemistry. Iron is central to the periodic table, and liquid is central between solid and gas phases, and each of those is important to the origin of life. RNA makes sense there as well.

Because RNA can form a double helix, it can replicate itself like DNA. Because RNA can form stable three-dimensional structures, it can catalyze reactions, like proteins. RNA is, chemically, between the two worlds of DNA and proteins, with a foot in each.

The biggest problem is making RNA in the first place. Clue #1 noted that nucleotides are complex three-part molecules. RNA chains are formed from nucleotides (and from nucleotides in a particular order, but we're getting ahead of ourselves here).

Nucleotides are complex, and any theory that uses them has to have a clue about how to assemble their three-part complexity. As of 2009, there is not just a clue; there's an experiment to show that the clue works. For decades the field was stuck, trying to make first one part (the sugar), then another part (the nucleobase), then stick them together the right way. In retrospect, they were doing too much. What works is not running three separate steps, but throwing the chemicals together in a big pot and letting *them* work it out.

A team led by John Sutherland has demonstrated that small-molecule precursors that look like cyanide and ethanol can self-assemble into three-part molecules that have sugar, phosphate, and nucelobase together. The sugar and the nucleobase form together at the same time and link up in the right way all by themselves. The third part, phosphate, is a versatile chemical in the mix that pulls triple duty. It is not just a reactant *but also* a catalyst that draws molecules together *and* a buffer that keeps the pH the same. Physical cycles of wetting and drying were used to concentrate and purify the reactions, a natural chemical cycle like the percolating chambers at Koonin's hot springs.

At the end of the synthesis, Sutherland's team didn't have too few nucleotides. They had too many. Two of the four natural nucleotides used in RNA were made with a lot of other nucleotides. They tried another natural twist of simulating sunlight by shining bright UV light on the nucleotides. This worked because light purified the mixture of nucleotides. The unnatural nucleotides were shattered by the light, but two natural nucleotides remained stable because they were more UV-resistant. Since then, the results were extended to some other nucleotides as well. The resulting chemicals are not perfect nucleotides but definitely chemically close.

Recently, Sutherland and colleagues published an update to this chemistry in which they found more intriguing chemical paths branching off the nucleotide-making path. These paths can turn just cyanide and hydrogen sulfide into several amino acids and carbon chains. UV light, meteorite impacts, and copper catalysts help push the molecules along these paths. This is incredibly promising, because all three of the types of molecules needed for life could have been made at once from common, simple materials processed by events that happened on the early Earth.

Because these syntheses use UV light, wet/dry cycles, and large amounts of phosphate, they match the geochemistry of the hot springs at Clue #5 better than the deep-sea vents of Clue #3 or #4. Koonin lists some other chemical clues that the conditions at the hot springs may be conducive to forming chains of nucleic acids as well.

(From my chemical perspective, Clue #5 is strong. Right now, I think this happened, but if so, it has a lonely side to it. If UV light and dry/wet cycles are needed to form life's molecules, then the other liquid places in the solar system, such as Titan and Europa, are barren without UV-illuminated waters. They would be just as desolate as the moon.)

Another strength of Sutherland's work is that it solves the chicken-and-the-egg paradox by saying that both the chicken and the egg came about at once, if chemical reactions fostered both replicating (RNA) and metabolizing (amino acid) molecules at the same time.

The "replication first" and the "metabolism first" camps have it out in the scientific literature. In particular, if RNA came first, then proteins that help RNA should be the oldest proteins, and independent proteins should be relatively young. But the labs that reconstruct ancient proteins consistently date independent proteins to be very old. The proteins that work with RNA show a pattern of evolution implying that proteins and RNA cooperated in writing the original genetic code, suggesting that RNA did not accomplish that fundamental task by itself. These findings favor the "metabolism first" camp, but the "replication first" camp are not yet conceding the fight. After all, Sutherland's pathways are best at making replicating nucleotides.

As amazing as Sutherland's nucleotide syntheses are, there's still no doubt that human fingerprints are all over those experiments. The mixing of the chemicals, the cycles of wetting and drying, and the application of UV light at a particular stage are all performed by human intervention. Some scientists point out that if this sequence is not strictly obeyed, then carbon-rich materials tend to react indiscriminately into a sticky tar-like mass (the "asphalt problem"). Sutherland found a path through the woods, but some are not convinced that the path is wide or automatic enough.

The road for the "replication first" camp gets much easier after this point. Once nucleotides are present in large amounts, phosphate chemistry makes the chains assemble in the right orientation by themselves. Even if they're not all perfectly oriented, the chains can still pair up to copy information. Several labs have made nucleotide chains that compete and even evolve, although the smallest of these is still pretty long to come together by itself (about 100 nucleotides long).

Sometimes complexity makes tar, but sometimes complexity makes nucleotides. Again, the experiments that work best in this field have multiple components that mimic the chemical complexity of the early Earth. These experiments fore-shadow the final clue, which is more of an overarching meta-clue that sums up the others.

CLUE #7: COMPLEXITY IS BORN OF COMPLEXITY

One of the frequent criticisms of origin-of-life chemistry is that biology is complex but chemistry is simple, so it's impossible to imagine how simple things became complex. This is a fair criticism, and is part of the "asphalt problem," but chemistry is not limited to simple reactions in big glass flasks. Chemistry is everywhere, in complex environments like newborn planets. Chemists themselves tend to start with purified, simple chemicals so that we can more easily narrow down what's going on, but as the field advances, simple experiments that are well understood can be built up and mixed into complex experiments, and new, surprising results emerge.

Take alchemy, for example. Complexity can make alchemy work. Don't get me wrong; most of the experiments described by the ancient alchemists are futile alle-gories of spiritual purification rather than chemical purification. We should be skeptical, but also skeptical of skepticism—that is, don't assume a priori that *every* alchemical experiment was *completely* baseless. Some of the experiments can be recreated today, and proto-chemists like Paracelsus conducted some undisciplined chemistry obscured by layers of charlatanry and fraud.

In his excellent book *The Secrets of Alchemy*, Lawrence Principe describes his own experiments following alchemical directions. He was able to make some of the old experiments work where others had failed by introducing complexity into the reactions in the form of impurities. One experiment failed because the instructions said to use an iron rod, but previous scientists stirred with stainless steel. When Principe stirred with an iron rod, the chemicals in the mix *reacted with the iron* and the alchemical reaction worked. Another reaction only worked when Principe used the form of metal mined precisely from the area in Europe that was the original alchemist's source, because that particular metal has an impurity in it that makes the reaction go. The previous scientists couldn't recre-ate the reactions because their materials were too *pure*. Natural messiness helped this chemistry.

Since the reaction of non-life to make life seems magical enough to be a branch of alchemy, perhaps we can take a page from Principe's notebooks and work with more complexity rather than less. Three different classes of molecules show how complex-ity can work together for good:

1. *Oily membrane lipids work with nucleotides*: Carbon chains tipped with a particular common "head group" chemical (phosphocholine) self-organize into multiple layers, and the nucleotide adenine will self-organize along with them, lining up between the layers like icing in a layer cake, looking for all the world ready to link into a strand of RNA.

2. *Metals work with RNA*: Chapter 1 described how magnesium sits between phosphates in an RNA strand, but recently scientists found out that the form of iron with two charges—exactly the form that would have been prevalent in the days before oxygen—will sit in the same places. Iron makes catalytic ribozymes work better than magnesium alone. An iron-RNA strand can use the iron's charge-switch ability to transfer single electrons, something RNA cannot do on its own but something that is crucial for metabolism. Iron's inherent reactivity can be corralled by an RNA shape, just like it is by a protein shape.

3. *Citrate works with metals that work with RNA*: Jack Szostak's lab, working with protocells made from simple membranes, found that magnesium in the protocells would actually start to cut apart RNA strands. This problem went away when citrate was added to the mix, because the citrate helped moderate the bad magnesium chemistry. Remember that citrate is central to the citric acid cycle and works well with iron, so it would have been very useful for early biochemical cycles. Here again, adding more early-Earth chemicals makes for more successful reactions.

Even more complex systems of multiple types of molecules can reinforce each other. Sarah Keller and a team mixed oily oxygen-tagged carbon chains with nucleobases, and sugars. This combination of three molecules was mutually stabilizing: the nucleobases and the sugars stuck to the lipids tightly, helping the lipids hold together in spherical vesicle membranes better, even in the presence of salt levels that usually blow the lipids apart. Not all nucleobases worked, but all of the *natural* nucleobases did. Also, the natural sugar ribose was the best of four similar sugars tested. RNA protects lipids, lipids protect RNA, and stability begets stability.

WHAT TO LOOK FOR NEXT

I expect that experiments will become more and more complex, with more encouraging results published. Because failed experiments are not published, we do not know how many graduate students' experiments in this area flamed out in a mess of sticky tar. But chemical cooperation and complexity can produce some very suggestive molecules that seem, at a low resolution, to approximate life. Both "metabolism first" and "replication first" camps continue to publish interesting results, and I don't feel qualified to declare this bout a victory in either direction yet (though my chemical heart is inclined toward "metabolism first").

As the years go by, these clues should converge. This may already be happening. Metal sulfides could produce metabolites at white smokers (#3) like they do at black smokers (#4). Nucleotide synthesis (#6) may succeed under the hot springs conditions (#5) that also have a chemical profile that matches life's biochemical profile. Rock compartments (#2) show up in clue after clue. If the clues continue to overlap, they may eventually converge. Even the "metabolism first" and "replication first" camps may converge if conditions are found that can make RNA strands at white smokers, black smokers, or hot springs.

This reminds me of the ending of the classic novel *Murder on the Orient Express.* (Warning: I'm about to spoil a mystery published in 1934.) There are 13 suspects to the murder, and as you read you're trying to figure out which one did it. Whichever one you pick, you're right—*all* of them did it, entering the room one by one to complete their part of the crime. Perhaps part of Clues #1–6 are true in some way, and there is a location where all these chemical processes can take place at once, in fulfillment of Clue #7.

But the theories have *not* converged yet, so we are left waiting for the conclusion to this story. It is immensely complex, and so as prominent a voice in this book as R. J. P. Williams writes that the chemical reactions that create life are unlikely enough that its existence here may be unique: " . . . we had to abandon the idea that life, in an initial reductive form, is necessarily a logical product in the sequence of evolution from the Big Bang" (p. 260). Williams thinks that all the other steps are chemically constrained and unavoidable, but this first one is a leap too far, a contingency even in an immense, chemically ordered universe.

On the other hand, scientists like the late Christian de Duve are convinced that because life makes chemical sense and similar reactions are catalyzed by such common rocks, life should be found across the universe. Others take a middle path, like Peter Ward and Donald Brownlee, authors of *Rare Earth* (2003). They conclude that simple life is widespread, but complex, intelligent life is rare or even unique.

You know what would help me make up my own mind? More experiments, whether collecting data from the stars, the rocks, or the labs. Feel free to do some of your own and send them my way. Try complex things mixed together in realistic conditions. If we're on the right track, the data will eventually converge, although it may take a century to try all the reasonable combinations.

This, more than anything else, is what convinces me that we have not reached "the end of science" in biochemistry. I suspect that, even with the recent spate in successful experiments, this topic will continue to be interesting for my lifetime and beyond.

If we do find conditions that could have built life on the early Earth, say, with Sutherland's light-driven reactions, then that would argue against Gould's "tape of life" thought experiment at the molecular level. Such a result would mean that, despite a vast distance of time, early-Earth chemistry could be deduced and repeated in a modern-day lab. This first song on the tape of life would be rewound, replayed, and recapitulated, even 4 billion years later. Life's most fundamental biochemistry would be explained by and predicted from the chemistry of the periodic table.

6

WHEELS WITHIN WHEELS

strange life from Antarctica to the Dead Sea // build your own microbe rainbow // the living stone in Shark Bay // the spark of life is the spark of fire // hydrogen cycles of chemical community // expanding energy wheels in predictable order // cobalt and nickel: useful tricksters // putting a ring on it makes a metal new // old genes agree with R. J. P. Williams // the great nickel famine

BLOOD FALLS, ANTARCTICA

Something strange and old lurks under the ice in Antarctica, at a place called Blood Falls. It is an echo of the early Earth.

Blood Falls is hard to reach and easy to find. Look through the seas of blue ice, white snow, and gray rocks for the bright-red frozen waterfall, spilling out of the ice around it in a gory cascade five stories tall. This is a red flag made from chemistry, telling that even the coldest environment on Earth is not completely dead. Liquid water can be found there, and in the water is life eking out an existence from the water around it and the dirt under it, just like it did a few billion years ago.

The "blood" at Blood Falls spills out of life, but it's not blood. Like blood, this substance is a form of iron bound to oxygen. In your blood, the protein hemoglobin hosts the iron, but Blood Falls is straight-up iron oxide, similar to rust. I saw some of this chemical last August near Mount Rainier. As we hiked up to Goat Lake, the frozen water looked dirty. The pure white ice was dusted with bright-red powder blown around from the iron-rich rocks surrounding it. The land was red as blood.

That was geological, but Blood Falls is biological. It shows that life in an extreme environment eats some pretty strange food—like John the Baptist eating locusts and honey in the wilderness—and outputs blood-red iron as waste.

A pocket of liquid water hides behind Blood Falls, sealed under the ice so tightly that air cannot penetrate. Even in solitude, away from the sun and oxygen, liquid water supports life. The microbes under the glacier get energy from adding oxygen to carbon to make stable CO_2, just like us.

The subglacial lake is sealed off from the air, so the oxygen must come from a solid or liquid source. These bacteria eat sulfate, pulling one of the four oxygens off it and producing the three-oxygen chemical sulfite. These bacteria "reduce" sulfate because this process takes oxygen away from it, the opposite of "oxidation." Sulfate could provide oxygen, but after a short time it would all be eaten up and turned into the electron-rich, reduced sulfite. Something else must be putting oxygen back on the sulfate, something abundant in oxygen-free waters.

That something is iron. Iron can accept an electron, switching from a plus-three form to a plus-two form. Because electrons are the "opposites" of oxygens, if iron accepts an electron it can give a sulfite an oxygen, turning it back to sulfate. The sulfate bank is replenished—let them (the sulfate-reducing microbes) eat sulfate!

It's a long chain of dominoes, and it only gives a little energy, but it works. Electrons pass through different forms of sulfur and iron, and oxygens move in the opposite direction, ending up on CO_2. The waste product is electron-rich reduced iron in a pocket of sealed water under a glacier. At Blood Falls, a little spout of this liquid worked up to the air, touched oxygen for the first time, and immediately reacted, rusting into bright-red iron oxide. A blood-red waterfall was formed (Figure 6.1).

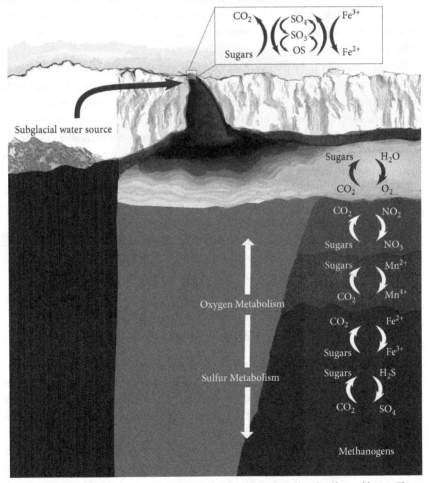

FIG. 6.1 Microbial chemistry in extreme environments produces ordered cycles and layers. The chemistry at Blood Falls (top) and the chemistry in ocean sediment layers (bottom).

Blood Falls shows that extreme environments contain extreme biochemistry. The early Earth was certainly an extreme environment. After life first awoke on a planet without oxygen, it probably looked more like the microbes sealed away under Blood Falls than anything up here on the surface. This shows the chemical minimum of life. An entire chemical chain can be supported by electrons moving through iron and sulfur, away from sunlight and oxygen.

The ocean floor is another oxygen-poor, dimly lit environment where iron and sulfur store and release energy together. These two elements are the reason you should be very careful when taking cast-iron cannonballs up from undersea shipwrecks. Some old cannonballs spontaneously explode when exposed to surface air. Microbial chemistry at the bottom of the sea turns iron cannonballs into bombs.

The reason is that cast iron is littered with holes, like the rocks around the deep-sea vents. In a cannonball on the ocean floor, the holes host sulfate-reducing bacteria like those at Blood Falls. These bacteria are like locusts that strip off all four oxygens from SO_4 sulfate, leaving H_2S sulfide gas behind. Other bacteria make H_2 hydrogen gas as well, and salt corrodes the iron, making yet more gases. The end result is that the cannonball is filled with a mixture of compressed gases that are waste products of bacterial biochemistry.

On the ocean floor, the gases are tamped down by the high pressure of the ocean. When removed to the surface, the ocean pressure is no longer holding the gases back, so they expand out of the holes and slam into the oxygen they've been missing for so long. These react with oxygen just like the iron at Blood Falls, but there's a lot more energy and the reaction is a lot faster. The cannonball explodes without a cannon.

These strange environments produce strange life from iron and sulfur instead of light and oxygen. In the absence of oxygen, iron and sulfur work together to provide enough energy to support a microbial colony. In the presence of oxygen, the whole scheme comes crashing down as the oxygen overwhelms the other sources of energy. In the days before oxygen, an iron-sulfur combo could have fueled simple life.

Blood Falls shows that biochemistry without oxygen or the sun can still squeeze energy out of the Earth. It also shows us there is a wide gulf between yesterday's and today's chemistry. When the two meet, chemical worlds collide and react.

THE LIVING ECONOMY IN THE SUPPOSEDLY DEAD SEA

Another extreme, oxygen-poor place that may give us a timeline of ancient life-chemistry is, ironically, the Dead Sea. As you may have guessed from all the water there, the Dead Sea is not quite dead. Its shore is the lowest point on dry land, but life can be found a short distance from that shore at the bottom of the lake, eking out a faint but definite, even encouraging, existence.

Like Mono Lake, the water flow to the Dead Sea is all inlet and no outlet, and it collects heavy, salty elements that cannot evaporate. The kind of life you'd take home as a pet is scarce, but the kind of life that keeps growing and clouding up your pet's aquarium is all around. In particular, the Dead Sea has a type of inlet that is not present at Mono Lake, which provides a foothold for a colony of microbes, including our recent friend, the sulfate-reducing bacteria.

This inlet is a few dozen craters at the bottom of the sea that let fresh water into the brine. It's not yet clear if these freshwater springs are providing nutrients or just pushing some salt away, but whatever they're doing, it works for life. Thick microbial mats of many species linked together grow in the freshwater currents.

These mats are hardy survivors in an unforgiving environment, but they are not rugged individualists. Multiple species cooperate to live in fluctuating oxygen and salt levels. One species' waste is another species' food, and vice versa. The microbes work best when joined together in tough biofilms, linked like an ancient Roman phalanx. Some of the hardest microbes to eradicate, the most resistant to antibiotics, are those that have joined in biofilms.

Often these mats will self-organize into complex layers, one species stacked on the next, collecting gases bubbling up from the species below and dropping carbon-rich food down in return. An entire economy will develop into a rainbow of chemical complexity.

Ordered stripes of microbes show up in surprising places. One common arrangement is the colors of lichens on a rocky seashore. Black lichens grow at the high-tide line, with orange lichens above and gray lichens on top. The patterns of tide and flowing water shape these lichens like great swaths of paint.

You can even make your own microbial community at home. If you collect a piece of an ecosystem and seal it up, it will develop automatically into a rainbow of layered bacteria. This experiment is called a Winogradsky column.

To make a Winogradsky column, take some mud and water from a local river or pond, and add a source of carbon food (newspaper is easy for them to "chew," but other sugary stuff will do as well, and a carbonate source like eggshells helps) plus a source of sulfur (rotten eggs, cheese, gypsum, or another mineral with sulfate). Put these ingredients together with a lot of mud in the bottom of a clear plastic bottle, pour a little water on top, and cover loosely with plastic wrap so it doesn't evaporate away. Most important, leave it out in the sunlight for a long time. As you wait, the ecosystem will develop like a photograph into a column of layered colors (Figure 6.2).

The top of the column forms a cycle of photosynthesis, some microbes making oxygen and sugar from light and carbon dioxide, and the others reversing the process (like us). In the mud at the bottom of the column, other cycles develop, depending on the microbes and minerals. If you get an orange band, then you've grown iron-oxidizing bacteria that are pulling electrons off iron and making rust, like at Blood Falls.

Below this, purple bands, white bands, or green bands may form. Lower bands have lower oxygen levels, and the dominant element here at the bottom is sulfur (that's why you needed to add it when you made the column). Sulfur-hydrogen cycles of moving electrons mimic the oxygen cycles at the top of the bottle, but with less energy. These microbes like to pull electrons off hydrogen sulfide and other sulfur compounds, putting out yellow elemental sulfur.

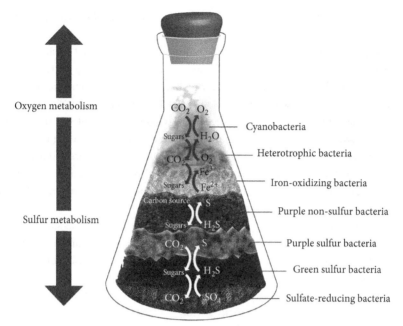

FIG. 6.2 A Winogradsky column made from placing mud and lakewater in a flask. Note the similarity between the layers here and in Figure 6.1.

After a poster from the 2012 Holiday Lectures on Science, Changing Planet: Past, Present, Future, www.hhmi. org/biointeractive.

Some microbes eat sugars and some eat carbon dioxide. Some use iron, like at Blood Falls, and some produce hydrogen gas, like in the underwater cannonballs. Some can even switch from adding electrons to iron to adding electrons to sulfur. All work together in a bottled ecosystem.

At the very bottom sits a dark layer of sulfate-reducing bacteria, tucked away farthest from oxygen. Like the sulfate reducers at Blood Falls, these hide from oxygen gas but pull oxygen atoms off sulfate ions, making the gas hydrogen sulfide, which carries electrons up to the layers above.

The biology is complex and the species are too many to count, but the chemistry is predictable: a gradient develops from oxygen at the top to sulfur at the bottom. Energy is extracted from sunlight at the top, but it is extracted from chemicals at the bottom. All of this happens automatically in sunlit mud.

The chemical loops that happen in a bottle also happen in the muck at the bottom of the ocean (see Figure 6.1). The top layers are today's world, depending on oxygen, but digging deeper reveals a world dominated by iron, sulfur, and methane—yesterday's world, with the chemistry of the early Earth.

The ubiquity of these chemical wheels has led some to argue that we should use a different fundamental symbol to represent life. Life is a loop, not a helix. The graceful curve of the DNA double helix is nice enough, but the turning arrows of a closed loop better represent the deepest essence of life. I'm intrigued. You'll certainly find more loops and circles than helices in this book's figures.

One thing about your Winogradsky column: don't bring your nose too close. It may have pretty colors, but its smells are not what humans prefer. The hydrogen sulfide from the sulfur metabolizers is light enough to float to your nose as a rotten-egg or fermented-kimchi smell. Oxygen-hating bacteria love hydrogen sulfide, but your nose does not, and it will raise the alarm at concentrations as low as 5 parts per billion. Either you will hate the oxygen and love the sulfide (e.g., bacteria), or you will love the oxygen and hate the sulfide (e.g., humans).

At high concentrations, this smell is the smell of death, and it is released by the microbes that decay dead flesh. It is also the smell of natural gas because of an odd connection. In the early twentieth century, the Union Oil Company noticed that they could find broken gas pipes by looking in the sky. Vultures would gather above where gas was leaking below because they smelled a sulfur molecule in the gas that is also produced by the bacteria growing in a dead animal. The vultures thought they smelled dinner below. The company added more of that sulfur molecule to their gas so that it would attract vultures more reliably.

In lower concentrations, sulfide is a bit easier to take. It's even part of the pleasant smell of breaking waves. The ocean holds deep reservoirs of sulfur that your nose can detect, churned by microbial cycles. When the air pressure drops suddenly, these gases bubble up from water and mud. If a sudden storm is on its way, you may be able to smell more sulfur in the air. This led to an old but accurate saying:

> If on your walks, you sniff a pong,
> Cover nose and head, rain won't be long.

These microbial rainbow layers form in sediments across the world, in the Dead Sea and beyond. Some chemical types will sit on top and use oxygen, while others will hide from oxygen and extract energy from the other chemicals around them. The natural tendencies to move toward or away from oxygen (and sulfur) are chemical forces that create the predictable, colorful layers.

HOW BACTERIA LEARN TO TRUST EACH OTHER

These colorful codependent communities develop because they are profitable for bacterial survival. Several studies have found that bacteria work together better than apart.

Cooperation works well if one species' trash is another species' treasure. One bacterium eats the three-carbon lactate molecule, breaking it down to hydrogen and small carbon fragments like CO_2. These waste fragments quickly build up and choke out the lactate-eater's chance for growth, like the waste accumulating in old London's streets.

So scientists added a waste-eating bacterium and sat back to watch. Chemically, this new species added H_2 (hydrogen) to the CO_2 to make CH_4 methane. This kind of microbe is a "methanogen" because it *generates* *methane*, but it could also be called a CO_2-eater. In six months, these bacteria that had never seen each other before forged a pattern of mutual assistance and rewrote their genes to formalize the arrangement.

The researchers repeated this bacterial introduction 24 times, rewinding this small bit of the tape of life and replaying it. What they saw was familiar. The new neighbors always encountered growing pains as the two species worked at growing and eating in sync. Most cultures had flickering moments where it seemed they wouldn't survive, and two cultures died off entirely. But the majority persisted and reached a state of greater stability. The most probable outcome of evolution was that the system's complexity increased as the pair of species became codependent.

Other studies evolved different types of cooperation. Even different kingdoms of life can cooperate, as shown in the association of a fungus with an alga, which the scientists described as "easy to establish" on the basis of carbon and nitrogen exchange. Cooperation can develop between bacteria with as few as three repeatable mutations.

Entropy, the tendency of the universe to spread out, appears to be a problem at first, but in fact, it may help cooperation come together. Some have worked out a thermodynamic theory for how making these complex cycles and *gluing them together* increases entropy elsewhere, so that complexity can increase within the bounds set by the Second Law.

Cooperations that move electrons need electron-moving metals available in the early Earth's oceans. The list of possibilities is short. Of the abundant metals, only iron, manganese, molybdenum, tungsten, and copper can exist in two or more states at energy levels useful to life, which is necessary for these "redox" reactions. Iron concentrations were thousands of times higher than the others on the early Earth, so life chose iron for its biochemistry. No other metal had the same combination of abundance and utility. To make a biochemical wheel, life would gather iron, sulfur, and small carbon molecules—and would *avoid* O_2 (oxygen), which would force all the iron into a solid, rusty mess like at Blood Falls.

You are cooperating with a community of oxygen-avoiding bacteria right now. Deep in your gut is a lively network of microbes that chew up the tough or complex molecules that your body can't absorb. Their waste products are small gassy molecules like CH_4 (methane), H_2S (hydrogen sulfide), and H_2 (hydrogen), like on the early Earth, because old habits die hard. The problem is your body can't absorb these, which means that foods with complex carbohydrates will indeed lead to gas. Did I mention that your nose is especially good at detecting hydrogen sulfide?

When this microbial community in your intestine gets unbalanced, odd things happen. One man got a yeast infection in his gut. The yeast did what they do in a brewer's vat—they fed on sugars and fermented them to ethanol. When this man ate bread, the yeast helpfully turned the carbs to ethanol and he ended up drunk, even though alcohol had not passed his lips. Like a beer commercial's King Midas, every carb he ate turned to alcohol, thanks to the yeast chemistry in his gut.

We are only now discovering how important—and alien—our gut bacteria are. Some chemists scoured the rainforests for new weirdly shaped chemicals, but their colleagues who stayed home found even weirder shapes made by the bacteria in the human gut. This interior microbial community can be shaped and even hurt by food chemicals like non-caloric sweeteners and emulsifiers. It is even possible that non-caloric sweeteners change your gut bacteria so much that they promote obesity!

The most striking thing about these communities is their self-organized striped layers. Your gut probably has a self-organizing layered microbial structure to it like a Winogradsky column. Even the lowly cockroach has a complex gut with multiple layers of structure, including compartments that make hydrogen next to compartments of methanogens that refashion that hydrogen into methane. One article suggests that we learn from such arrangements when designing synthetic biology reactors, so consider the cockroach, ye biochemist.

Bacteria may not be able to move quickly or sense the environment in detail, but they know how to eat and grow. Each bacterium is a specialized and evolvable compartment that can take in, transform, and excrete diverse chemicals. Specialization leads to organization as the waste of one is eaten by another. The self-organization that happens in the Dead Sea and in Winogradsky columns must have happened on the early Earth. More molecules processed means more energy produced means more complexity from cooperation.

Today these communities are shaped by predictable chemical trends, as oxygen-lovers move up toward oxygen and oxygen-haters/sulfur-lovers move down. Producers are next to consumers, with carbon builders next to carbon eaters and electron pushers next to electron pullers. A microbial city organizes itself into a model of efficiency, although to us, it just looks like striped slime.

STROMATOLITES: THE MEDUSA MATS

When a city takes hold, it builds walls and banks and temples to protect and reinforce the profitable relationships. Once a microbial community reaches a stable configuration, it also builds structures that glue it down. The Dead Sea microbe mats reinforce their layers by making sticky connections from carbon sugars bridged by oxygen. Fossils show similar mats from long ago, and the way it happened was a direct consequence of the five chemical balances common to all life (see Figure 5.4 in Chapter 5).

Fossilized stacks of mats, called stromatolites, are 2.5 to 3.5 billion years old. Some stromatolites look, from the side, like a stack of plates (I was going to say records, and then CDs, but both of those technologies are ebbing away). Sometimes geology deforms a stack into nested, curved swirls that bring to mind Celtic art or the labyrinth on a cathedral floor. Some are conical, and some are little rounded mounds like a dozen eggs preserved in rock. On Belcher Island, Canada, there's a series of them big enough to be seen from an airplane.

A pair on display in the American Museum of Natural History in New York has been split apart to show the structure inside, so that they look like a pair of lungs. Very old stromatolites have been found with sulfur isotope patterns caught in the layers that suggest that they lived off sulfur cycles like the lowest layers in a Winogradsky column. These stone lungs breathed sulfur.

Stromatolites can still be found in isolated spots around the world. The best known may be at the hypersaline Shark Bay, Australia, where dark footstools rise from the water like mushrooms. These are the same kind of mats as seen in fossils. Once they were successful enough to cover the globe, but now they hang on in distant, extreme places where other lifeforms can't compete. It's a very low-key Jurassic Park (Archaean Park, perhaps?).

Stromatolites once covered the planet because they result naturally from the chemicals that every cell keeps and ejects. The most important chemical balance is that cells keep hydrogen and electrons inside (making the inside "reducing"), while ejecting oxygenated material outside. The oxygen outside forms bonds and makes sticky networks adherent enough to hold the bacteria in a layered mat.

The sticky messes catch bits of rock, dirt, and other cells flowing by. The cell would have been ejecting silicon and calcium as well, and these stony elements would catch in the net of goo. Even more, external calcium binds the oxygen, forming a shell that would even serve as protection (so long as it doesn't get too thick and strangle the microbe). The entire assemblage would turn ever so slowly to a stony network, like one of Medusa's victims, tough enough to survive billions of years. The oldest are about 3 billion years old.

"THE WHOLE SHOW HAS BEEN ON FIRE FROM THE WORD GO"

These first microbes needed energy to keep themselves in a state of organized flow, against the onslaught of the environment and entropy. The Earth provided volcanic energy bubbling through deep-sea vents and hot springs. If these were the nurseries of life, then eventually life would leave the nursery and venture into the deep blue ocean.

A microbe that could make an energetic living in the cold open ocean would have a whole new field of exploration and thousands of descendants to help it explore. But first it needed to pull the plug from the easy Earth energy and develop something like batteries. Specifically, it needed reactions that would turn something unstable into something stable, releasing bursts of energy that the cell could use to form bonds and climb up the hill of organization. It needed food and fuel.

The problem is that all the best reactants had already reacted. Unstable things tend to stabilize on their own, losing their energy. This is why most of the reactive oxygen had stabilized and was locked up in the rocks, in water, or in stable carbon dioxide gas. The sources of energy that would have been around would be meager, providing at best 10% of the possible output of oxygen, but at this point, meager energy would be better than no energy.

One of the great things about being a chemist is that, as you examine how reactions give off energy, you are allowed to blow stuff up as part of your job. Even gummi bears hold energy and can be turned into a firework with the right chemicals. A surprising (or perhaps dismaying) number of YouTube videos can be found in which chemists set gummi bears on fire. The preferred technique is to dip a gummi bear in molten potassium chlorate in a test tube. A shower of white-hot sparks and smoke shoots out for several seconds, and afterward only a sticky black carbon residue remains.

The most active ingredient in fireworks (whether gummi or non-gummi) is not in the firework but in the air. The sugar in the gummi bear is a mixture of carbon-oxygen and carbon-hydrogen bonds. Each carbon-hydrogen bond is less stable than a carbon-oxygen bond. For one thing, carbon-hydrogen is only a single bond, and oxygen can form a more stable double bond.

When a spark of energy ignites a gummi bear, the oxygen in air replaces hydrogen in the gummi bear, until the carbon is full with oxygen bonds, which is invisible carbon dioxide gas. The substance is vaporized but the atoms are still there in the room (visions of vengeful gummi bear ghosts may occur at this point). Unless the energy is caught and channeled, it naturally spreads out as heat and light.

The carbon + oxygen reaction is the best way to release energy, gram for gram, so making carbon + hydrogen structures is the best way to *store* energy, gram for gram. Gasoline is made of carbon chains coated with hydrogens, waiting to be burned and replaced with oxygen. The carbon chains are holding electrons in your gas tank, which the spark plugs in your engine combine with oxygen. This releases the stored energy, which can either warm the environment or push things together.

Carbon-hydrogen chains make excellent energy storage molecules, which is why your body also uses carbon chains like gasoline to store its energy. Carbon-hydrogen bonds are stable enough that they don't break down spontaneously in storage—whether in a gas tank or a fat cell, it's pretty much the same chemically. Add oxygen and a spark, and you have fireworks, whether big or small.

Seeds, animals, fungi, and microbes all hold carbon-hydrogen bonds waiting to be lit, with ingenious schemes to capture every last drop of the energetic potential released when those electrons are fixed to oxygen. This means every sugar and fat holds the potential to explode like the gummi bear if the right spark is applied. As Annie Dillard wrote in *Pilgrim at Tinker Creek*, "The whole show has been on fire from the word go. I come down to the water to cool my eyes. But everywhere I look I see fire; that which isn't flint is tender, and the whole world sparks and flames" (p. 10).

The energy industry is a quest for such electrons. Gas companies dig into the Earth to find concentrated carbon-hydrogen bonds to drive the combustion reaction in engines. The mitochondria inside your cells perform the exact same reaction. For this reason, a universal measure like Eric Chaisson's energy rate density makes sense. Both a muscle car and a muscle cell are ultimately processing energy using the same overall reactions that release the same type of energy. This similarity may be caught by Chaisson's simple equation of watts per kilogram applied to plants, animals, and societies.

Flammable carbon-hydrogen bonds can be found in swamps in lesser quantities, as shown in another chemical YouTube mainstay, the Volta experiment. To run this experiment, invert a funnel over the shallows of a pond, stir up some sediment, and ignite the bubbles that come up. If you concentrate the gas, you can make a jet of flame from pond mud. (Was this ever in a TV show to help the hero escape from the bad guys, and if not, why not?)

Swamp gas is not geological but biological: the bacteria hiding in the depths from oxygen are methanogens, making CH_4 (methane). When the methane meets oxygen, and a spark jostles open a few of the bonds, the carbon trades its hydrogen for oxygen, makes stable CO_2, and energy makes a dramatic appearance.

Volta experiments work fine in our oxygen-saturated world, but they would not have worked 3 billion years ago, without free oxygen. Instead, life searched for single oxygen atoms or oxygen substitutes. The land was harder to till, yielding only enough energy for a microbe.

ENERGY BEFORE OXYGEN: THE FIRST HYDROGEN ECONOMY

If you want to know what low-oxygen environments are like, visit swamps and sewers. This shows that most people don't really want to know what low-oxygen environments are like. CH_4 (methane) itself doesn't smell like much, but it's usually accompanied by the other common hydrogen-rich molecules: nitrogen as NH_3 (ammonia) and sulfur as H_2S (hydrogen sulfide). These you can smell very well, and the smell pushes you away.

The natural human instinct is to flee from hydrogen-rich gases, showing that a hydrogen-rich ecosystem is not our home. Before oxygen, however, the situation would be reversed. Hydrogen-rich gases would be abundant, including hydrogen itself. The scientist Linda Kah describes this period before oxygen as "the smelliest time on Earth." (To be fair, imagine how we would smell to a methanogen.)

Free oxygen wasn't available, but these gases could combine with each other. A very rough measure of the energy available from a reaction is how far apart the elements are on the periodic table. Oxygen is up in its own corner. Nitrogen, sulfur, and carbon are closer on the periodic table, so the energy differences between them (and the corresponding energy rate density) would be smaller. On the bright side, smaller energy differences are easier to reverse, so making the molecules would be easier and it would be easier to set up a mutually beneficial wheel of reactions.

As these wheels turned, life pulled in the hydrogen and electrons and ejected oxygenated material like carbon dioxide or oxidized material like three-plus iron. Life's hunger for electrons led to reactions that oxidized the environment.

The problem is that the gas phase of matter is difficult to work with. I can tell you from experience that lab experiences with gases give students more trouble than those with liquids or solids—one little leak and your expensive reactant diffuses away. Chemically, working with a gas is like catching a cloud with a sieve.

Microbes have the same issues. To move electrons around with these small, smelly molecules, and to squeeze energy out of that electron motion, a microbe would need something heavy it could keep a handle on, a specific metal that would stick to that specific gas and help move its electrons around. Therefore, particular metals were very useful to life at this stage.

A lot of oxygen-eschewing microbes have an interesting metabolic pathway that would have been very useful in this situation. This process is called the Wood-Ljungdahl pathway, which is the metabolic equivalent of $1 + 1 = 2$. When using this pathway, the microbe takes two gases, one stable (like CO_2) and the other reactive (H_2 hydrogen), and bonds them together into the two-carbon molecule CH_3CO_2H, acetate. Acetate has hydrogen at one end to store electrons and oxygen at the other to help it react.

The Wood-Ljungdahl pathway may have been the first metabolism learned by life, 1 billion years before oxygen. It turns the reactivity of hydrogen into a reactive carbon molecule, although it's too weak to make large amounts of energy or ATP. Still, something is better than nothing, and the first independent economy was probably a hydrogen economy.

Today, microbes are still very good at making acetate. Acetic acid is the bitter taste of vinegar, which was ejected by fermenting acetate-making microbes. Fermentation also puts the sourness in sauerkraut and the alcohol in beer and wine. There is a story that the first monk who made bubbly champagne ran to his brothers to exclaim, "Look! I am drinking the stars!" His stars were the exhalations of tiny fermenters.

Fermentation reactions start with six-carbon sugar, which gives three times as much energy as making acetate from gases. Today, fermentations require shutting out oxygen, so that every person making beer or sauerkraut is mimicking the oxygen-poor conditions of the early Earth.

Your cells still break down sugars to lactate (and/or acetate) with a pathway called glycolysis. This pathway is remarkable because it, too, mimics the early Earth. The reactions only rearrange the sugar—no electrons move in or out, and no oxygen enters or leaves. It is a self-contained metabolic process, and it is also terribly weak. Glycolysis only squeezes 5% of the energy out of the molecule and produces two measly ATPs. To get the remaining 95% out, you need to move electrons onto oxygen, but that's for a later chapter in early history when oxygen is available. This central pathway to life uses suspiciously old and gentle chemistry.

In glycolysis, each of the 10 steps follows from the next with relentless chemical logic. Even though it's meager energy, it appears energetically optimized for a system that must keep all the electrons on the sugar overall (one pair is pulled off and then put back on). Glycolysis is simple enough to fit on a snippet of DNA and be passed around from microbe to microbe like a funny text message.

Glycolysis is surprisingly *evolvable*. The proteins that break down glucose sugar are nimble enough to break down other random sugars with different shapes. It takes only a few tweaks for a glucose-eating bacterium to eat another sugar. In a computational study, a glucose-eating network of proteins could shift a little to live on up to 44 different carbon sources. The authors of the study described this as "a latent capacity for evolutionary innovation." Another way to say it is that, when it comes to sugar, microbes are not picky eaters. They will find a way to get the energy out of that carbon even if they have to evolve a little to do so.

Beyond glycolysis, you have another thing in common with early-Earth microbes: every cell in your body builds molecules from, and breaks molecules down into, *acetate*. You will even mimic an acetate-producing bacterium at times. When your blood sugar drops, your body will make a "sugar substitute" from acetate that travels through your blood. These acetate-ish molecules are called ketone bodies and make your blood acidic, as if vinegar was mixed in. A few even turn to acetone and drift away through your breath. Readers with diabetes know to watch out for acetate-breath as a signal that energy balances are going wrong.

Other primitive pathways may have played a role in early life, especially those heavy in iron, sulfur, and hydrogen, and metals that react with hydrogen-rich gases. Iron-sulfur (FeS) can pull a sulfur off H_2S (hydrogen sulfide) to produce H_2 (hydrogen) and release energy. This makes iron with two sulfurs, chemically known as FeS_2 (pyrite), better known as fool's gold. A cell could use this reaction to release energy from the Earth, making small bits of fool's gold. In this way, chunks of iron-sulfur clusters would be portable hydrogen generators in sulfide-rich environments, pushing electrons around with the leftover energy. This could have been the battery that would allow a first cell to venture out into the ocean.

Other forms of sulfur might have provided energy, too. In the absence of oxygen, oxygen-rich sulfate would have been scarce, but small pockets of it might be found, and there, sulfate-reducing bacteria could thrive. Those bacteria would be eating Earth energy. Geological evidence suggests that certain rocks dissolved and made sulfate in patches around the globe, and isotope imbalances suggest that this was consumed in sulfate-reducing sulfur cycles. Sulfate stores as much energy as acetate, so it would have provided low-wattage energy for early life. It would have been a matter of feast and famine, depending on how much sulfate dissolved from the rocks.

Notice the recurrence of iron, sulfur, hydrogen, and carbon. Magnesium and phosphorus would also be needed to round out life's needs for DNA information storage. Metals other than iron that would be useful would be nickel (right next to iron, which also reacts with gases), and also cobalt and tungsten for the weird chemistry needed to process the molecules built from those gases (more details to come). Oxygen is not used in its free form, because it is locked up securely by tight chemical bonds.

These elements are the basement and foundation of biochemistry. They are the chemical basis of the oldest metabolisms, the most ancient proteins, and the lowest levels in both Winogradsky columns and ocean sediments.

BIG WHEELS ROLL

A young metabolism could eat unstable chemicals from the Earth, but eventually the Earth would be depleted, and they would have to move on or pass on. A more excellent way to process energy would be to set up a mutually reinforcing multi-species cycle, like in the Dead Sea mats.

Cycles can be simple or complex, so long as they loop around to produce what is consumed, forming a wheel of reactions. When a cycle of reactions repeats itself, low-entropy energy flows in and high-entropy energy flows out. This can be measured as the energy rate density of the cycle. Even a small energy difference can be exploited to create a structured, flowing cycle.

The most important wheel turns inside every cell. On any metabolic map, find pyruvate and acetate somewhere in the middle (probably in a large font) and then look down. Below these hub molecules, a large circle stands out as thick curved arrows. This metabolic wheel of eight reactions is as universal as DNA and proteins. The first reaction consumes two-carbon acetate and makes six-carbon citric acid, so it is called the *citric acid cycle.*

The citric acid cycle strips electrons and carbons off that six-carbon molecule, then loops around on itself to build citrate again. The electrons are sent to the mitochondrial furnaces to be combined with electron-hungry oxygen, and the CO_2 is exhaled from your lungs. Carbon-hydrogen bonds are replaced with carbon-oxygen bonds, and the energy from this is stored as ATP. When the ATP is used, some energy is lost as spread-out heat. Overall, concentrated chemical energy comes in and spread-out heat leaves, but the flowing structure of the eight reactions in the cycle continues turning.

Green sulfur bacteria, which avoid oxygen and form one of the bottom layers on a Winogradsky column, have the same wheel turning backward. If they like, they can *inhale* CO_2 and add it to four-carbon molecules, forming six-carbon citrate. They pump electrons and hydrogens into the carbon rather than pull them off, and use it to build rather than to break down. It would have been easy to run the cycle this way billions of years ago, when there was more CO_2 to push into it. This may be how the wheel got its start.

The ancient origin of this cycle is seen best in its second enzyme, aconitase. This enzyme uses an oxygen-sensitive iron-sulfur cluster that would have been right at home on the early Earth. This enzyme grew up when iron-sulfur clusters were abundant, and sees no reason to change now, despite the fact that the world has moved on. In some Italian towns, knife sharpeners still pedal their whetstones down the street, an old tradition that hangs on even though there are simpler ways to do it now. Aconitase is the knife sharpener of the cell, still carrying its iron-sulfur stone around, and the cell is happy to use its services.

These wheels are found not just inside, but also among organisms, in cooperative patterns like those evolved in the lab earlier in this chapter. Microbes use the chemicals put out by other microbes, shuffling bonds, moving electrons around, and keeping life going. A structured cycle spreads out energy and increases energy rate density. The ideal chemical arrangement is a wheel like this, where the material changes back and forth while energy spreads. If one species oxidizes iron, the other can reduce it.

Since each microbe has wheels of metabolism inside, when these wheels are connected in an ecosystem, this forms a big wheel made of small wheels. Imagine a huge Ferris wheel with each car being a smaller Ferris wheel (and some passengers throwing food from wheel to wheel). Wheels within wheels form an incredibly complex, flowing cycle that spreads energy out and concentrates material in complex patterns.

The methanogens used the Wood-Ljungdahl pathway to turn the air into sugars and carbon chains, making methane waste. Other microbes ate the methane, cracking its bonds open to get at the electrons inside, producing carbon dioxide and hydrogen. These microbes are methano*phages* (from the Greek word for eating—they digested a surprising amount of the oil spilled in the Gulf of Mexico from Deepwater Horizon). Fermenting microbes also join the party by breaking down the sugars made by methanogens.

This brings up the dark side of mutualism. At a certain point it becomes easier to break open another microbe to get the sugar inside than to make it yourself. Sometimes the resources one organism wants are being actively used by another one. One day an organism played Cain and destroyed its neighbor to get at the energy stored inside.

Even this part of life's circle has been fossilized, in a piece of chert from the north shore of Lake Superior called the Gunflint microfossils. A sample is on display in the Redpath Museum within sight of the glass cases of rocks. The fossilized microbes stand out ghostly white against the polished black chert, like stars in the sky. Recent imaging techniques allowed us to reconstruct their activities in three dimensions. Some were fossilized while eating the others, and they are now caught in rock with their last meal half-finished. "Predator and prey" goes way back.

To quote Annie Dillard again, "We the living are nibbled and nibbling—not held aloft on a cloud in the air but bumbling pitted and scarred and broken through a frayed and beautiful land" (p. 232). Even as these ancient microbes pulled methane and hydrogen from clouds in the air and turned them into life, they also wove them into cycles of eating and being eaten. All the while, energy was processed from low entropy to high entropy, energy rate density was maintained, and the big wheels rolled on.

And yet the cycles of eating and being eaten form part of a larger pattern. Three billion years ago the large pattern was only just opening up. All the cycles in this chapter, from methanogen cycles to iron-reducing cycles, can be arranged by the chemical reactivity of the cycling molecules. In chemistry terms, this is the *potential* of the electron-accepting molecule to accept electrons, measured in volts (Figure 6.3).

Molecules cycled in this chapter have negative potentials and sit on the left side of this graph. This means that the electron acceptors are not all that great at accepting electrons—they can do it, but they don't get much energy out of it. If we could show this graph to the microbes, they'd look longingly at the cycles with positive potentials on the right side of the graph: manganese, nitrogen, and oxygen accept electrons more readily and give off more energy. But these were not much available on the early Earth. Low-potential, low-calorie iron, sulfur, and carbon food was on the menu. The planet would have to change.

And change it did. The arrow at the top of this graph shows how the planet changed as it oxidized. When this happened, the potential of the planet moved to the right. As it did, the higher potential and greater energy of manganese, nitrogen, and oxygen became available with energetic help from the sun. The next few chapters tell how this happened.

But before we follow that arrow to the next chapter, there are a few more ways in which the old chemical cycles continue today, in the form of unusual chemicals (like the FeS rock in aconitase, but, if anything, weirder). The old ways still run through modern biochemistry in the strange trace metals that worked best with ancient gases.

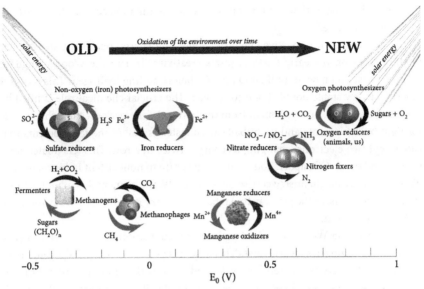

FIG. 6.3 Microbial chemical cycles arranged by "redox potential" (E). Older chemical cycles are on the left and newer cycles on the right.

After Figure 1 in P.G. Falkowski et al. "The microbial engines that drive Earth's biogeochemical cycles." 2008. *Science*. 320 (5879), p.1034. DOI: 10.1126/science.1153213.

We still use these metals in odd corners, like old family heirlooms that tell of shared history. These old ways maintain a deep, universal connection to the ancient biochemistry that worked without oxygen. Beneath the surface, this world was wasting away, and one day the bottom fell out of the whole scheme. Before this chemical revolution took place, an exotic world flourished, as different from our world as the days of the dinosaurs.

THE DEVILS IN THE DETAILS OF ARCANE MICROBES

Some microbes are just plain weird. Carl Woese was the first to put his finger on how deep this weirdness runs. Woese looked into the genes of simple microbes and read there a deep difference that separated them neatly into two groups, fundamental enough to be different domains of life. Before Woese, simple microbes were all called bacteria; after Woese, the weird group was called "Archaea" because they appeared to be the older group.

Archaea are like the strange Goth kid at the microbial high school, and as a former strange kid myself, I appreciate their distinct nature. I'd hang out with Archaea. The ways in which they differ from bacteria are manifold. Bacteria make membranes with simple phosphate-capped carbon chains, but Archaea have solid, harder-to-make (and harder-to-break) links, kind of like Doc Martens as opposed to regular shoes. Some Archaea even eschew straight chains and build carbon squares into crazy linked ladders called ladderanes. They twist carbon into unique, stable shapes that last for eons in rocks, leaving molecular graffiti reading, "Archaea were here."

Archaea are also innovative chemists. Squeezing energy out of the simple, stable molecules on the early Earth requires creativity. To run the Wood-Ljungdahl pathway, first you have to pull an O off CO_2 before adding hydrogen. This requires severing a very stable double bond to oxygen. The crucial chemical role is filled by tungsten, a strange, heavy element from the bottom of the periodic table that we use for light bulb filaments and yacht ballast. Tungsten is one of the few elements that can accept two electrons, changing its charges twice in a row. This lets it latch onto oxygen and break CO_2 apart. Today, tungsten is rare to nonexistent in biochemistry. Nowhere in my biochemistry course do we draw W, the elemental symbol for tungsten, but there it is, at the gateway to Archaea's primary metabolic pathway for using CO_2. W is for weird.

The rest of the Wood-Ljungdahl pathway requires another weird metal: nickel. Nickel is methanogen's favorite metal, because it is great at binding hydrogen, methane, and acetate. Nickel is useful in the lab today doing the same chemistry. Organic chemists made a small CHON framework that holds iron and nickel still, and this pulls electrons off hydrogen all by itself. Another hydrogen-electron reaction catalyst was built around a single nickel atom (and inspired by a nickel enzyme). Scientists trying to produce hydrogen are finding that the part of the periodic table containing iron, nickel, and cobalt can catalyze hydrogen production for less cost than traditional metal catalysts like platinum.

Primo Levi tells a story in his memoir *The Periodic Table* in which he had to purify nickel from a rocky sample, and his eureka moment was to blow hydrogen across the sample. Working alone late at night, he writes, "I felt part conspirator, part alchemist" (pp. 82–83). By using hydrogen to react with nickel, he was using a very old chemical pairing to accomplish some new chemistry. Levi was acting like a methanogen in reverse—instead of using nickel to grab hydrogen, he was using hydrogen to grab nickel. Chemically, the affinity is reciprocal.

The methanophages that eat methane use nickel to pull off the first hydrogen (the first hydrogen is indeed the hardest). Then they move the pulled-apart methane around with the next metal over on the periodic table, cobalt. Cobalt's chemical specialty is binding single-carbon units like this. It is so good at this, in fact, that you still use it for this purpose, in the form of vitamin B_{12}. Vitamin B_{12} is a special form of cobalt surrounded by a scaffold of CHON elements, where the cobalt does all the work.

Even if you have all the iron in the world, you can still become anemic if you don't get a few micrograms of cobalt in your diet. (It's not just humans: the salt licks that farmers and zookeepers put out for their animals include cobalt salts.) If you don't get enough of B_{12}, or the related molecule folate, you develop what is called "pernicious anemia," a disease that is as bad as it sounds. Without cobalt's carbon-moving abilities, you can't build enough red blood cells.

These metals are so important to methanogens that you can't grow this kind of microbe without the right kind of weird metal. Scientists couldn't grow human gut methanogens until they added nickel and manganese (manganese is one step over on the graph of chemical cycles, and it steps onto the stage in the next chapter). Another study shows that nickel alone can spur methanogen growth.

Although you use nickel in a few enzymes, it's more often a problem that you are exposed to too much. Nickel has high abundance along with its odd chemistry, since its place right next to iron guarantees its abundance in stars. This means that it's prevalent in the environment, and your body often has to say "thanks but no thanks."

You may know someone with "nickel itch." This person's immune system reacts to the nickel in stainless steel and attacks it—nickel binds to and activates immune proteins, a bit like beryllium does (in Chapter 2). This is a sign that, when it comes to nickel, your body has moved on and no longer craves it. We don't do hydrogen or methane-moving chemistry, and we don't need more than micrograms of the metals that help it along. Nickel's days as an important vitamin for life are over.

EXPANDING THE METAL PALETTE WITH HEME

Cobalt and nickel are chemical chameleons in the lab, which comes from their electron-moving abilities. Cobalt changes from pink to blue, and nickel changes from green to blue. For one demo, I use acid and heat to make a cobalt parfait in a tube, pink on top and bottom but blue in the middle.

Vitamin B_{12} contains cobalt, but a tube of it in the lab is not pink like naked cobalt—it is bright neon red, a touch more vivid than the blood red of iron in hemoglobin. The different color shows that its chemistry is also different. Vitamin B_{12} surrounds cobalt with a flat square belt of nitrogens. If this reminds you of the square heme in hemoglobin, it should, because the two nitrogen belts are almost exactly the same. The square of nitrogens is called a porphyrin and it is found attached to all sorts of metals, dating back to this early era.

Attaching a porphyrin to a metal will probably change its color and will definitely change its chemistry. Metals in porphyrin belts carry oxygen, move electrons, and capture light with chemistry that the metal by itself cannot match. R. J. P. Williams writes that when a porphyrin ring was added to a metal ion, "[i]t is as if a second kind of metal ion had been made" (p. 219).

For such a big molecule, porphyrins are surprisingly easy to make. Some labs made porphyrins in water from simple combinations like methane, ammonia, and a spark. Porphyrins have been found in meteorites and petroleum deposits, and there is even a porphyrin mineral. Abelsonite, a special kind of orange-purplish semi-transparent rock found only in Colorado and Utah, is composed entirely of nickel porphyrins.

Porphyrins are useful because the hole in the middle of the four nitrogens perfectly fits metals from the first row of the transition metal block of the periodic table. Iron, nickel, and cobalt fit into porphyrins, which we have talked about, and magnesium, vanadium, copper, and zinc do as well. Some of these are shown in Figure 6.4. In proteins, magnesium plus porphyrin is the green color of chlorophyll, nickel plus porphyrin is yellow (Coenzyme F430), and iron plus porphyrin can range from blood-red (hemoglobin) to a dark rocky green like the boulder near Camp Casey (Heme d1)

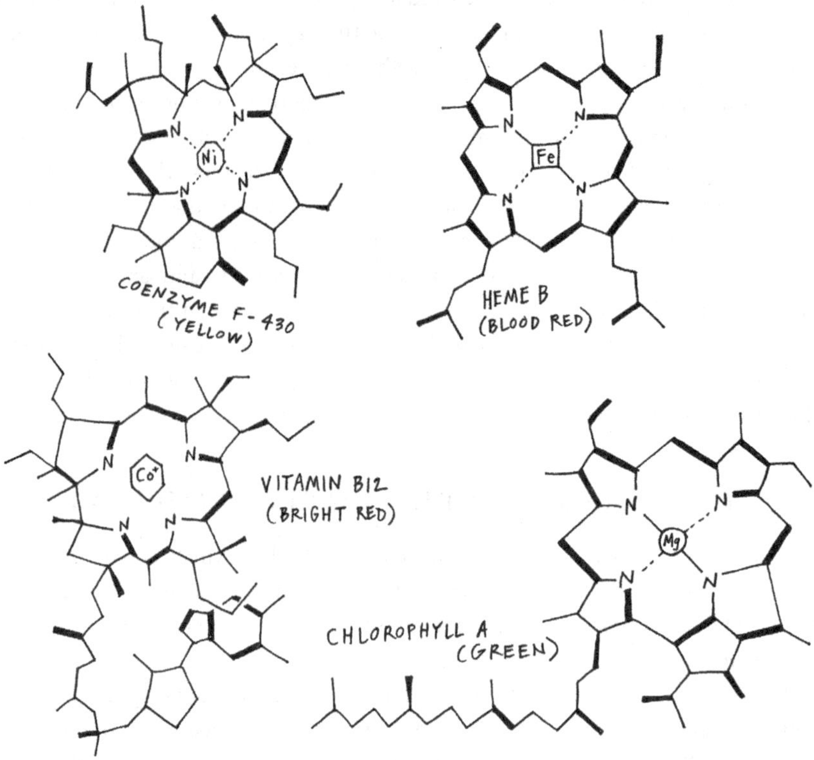

FIG. 6.4 Different porphyrin-metal combinations. Note how similar CHON networks with different metals inside make very different colors (and very different catalysts).
See Figure 1 in H. A. Dailey. "Illuminating the black box of B$_{12}$ biosynthesis." 2013. *Proc Natl Acad Sci.* 110 (37), p. 14823. DOI: 10.1073/pnas.1313998110.

to even a blue-gray (Siroheme), depending on which atoms are added to the edge of the porphyrin square. All of these different colors denote different chemistries.

We will see the magnesium porphyrin in the next chapter, but life does not use vanadium or zinc porphyrins. It's not because they don't fit—organic chemists who make useful chemicals with porphyrins have no problem inserting the metals. Vanadium is too far to the left on the periodic table, so it is not sticky enough for widespread use. But copper and zinc are on the sticky, useful right side of the table, yet are not found in porphyrins. In particular, life uses zinc in many places, but if it's found in porphyrins, something is terribly wrong. Zinc porphyrin in your blood is a red flag to doctors indicating lead poisoning. So why don't we use zinc porphyrin?

Zinc follows an odd pattern in life. In bacteria, it is found in old functions gluing things together (especially the ribosome), but not in the newer functions catalyzing useful reactions. We humans use zinc to help with many reactions that bacteria don't do, often in the oxygenated environment outside the cell, rather than in the reduced environment inside the cell. In this book, it can be seen in Chapter 5 and then misses most of Chapters 6 and 7 before coming back in Chapter 8.

R. J. P. Williams explains this strange pattern by looking at zinc's chemistry. Zinc binds sulfur so tightly that water cannot pull it apart. Basically, if there's a lot of sulfur around, then zinc is not available for life. About a decade ago, the old ocean was thought to have a lot of sulfur and no oxygen. If a protein needed zinc, it would need to hold it tight and never let it go (like the lyrics to countless pop songs). According to Williams, there is no such thing as a biological zinc porphyrin because when porphyrins were first invented, zinc was stuck to sulfur.

This point is contested by recent evidence found in rocks. The geologist Kurt Konhauser and colleagues found zinc in old black shale rocks formed from ancient oceans. They propose (from several other lines of evidence) that sulfur existed only in pockets in the early ocean, so zinc would have been widely available outside those pockets.

It's possible that some life developed in the sulfur-rich pockets because of its need for sulfur, when the rest of the world had zinc, and so the pattern of zinc missing out on porphyrins may refer to *local* unavailability of zinc—or that sulfur levels fluctuated, along with zinc levels, and these periodic "zinc famines" made zinc too much trouble for the struggling bacterium.

Copper's absence from porphyrins is more readily explained. Copper was simply unavailable on the early Earth. Its binding to sulfide is the tightest of all the first-row elements, and without oxygen around it would grab onto other elements and sit in rocks, never dissolving into the ocean. Fitting with this, no copper proteins date back to this early oxygen-free era, and they are found in later chapters.

Iron, on the other hand, would be much more available during this time because its sulfide dissolves into useful iron-sulfide chunks. Without oxygen, most iron would be in the plus-two form found in porphyrins. Plus-two nickel and cobalt fit snugly as well. We see all of these elements in the proteins dating to this era, in porphyrin and/or non-porphyrin-bound forms.

EARLY EARTH CHEMISTRY IS WRITTEN
IN OUR GENES

If all of this hangs together, then there should be a faint signal from the past inside each living thing. Billions of years have passed, but life begets life, and all life propagates with the same DNA-RNA-protein system. Some scientists have collected reams of gene sequences and put those together in a grand genealogy based on their commonalities. The oldest genes have been written and rewritten in DNA countless times like ancient manuscripts, and the signs of this can still be seen. These are indeed manuscripts, although written in DNA, not ink, and by polymerases, not hands.

Three scientists have looked at the genes in three different ways, and the results agree. It took until around 2005 for enough genetic data to be collected that it achieved a critical mass. Christopher Dupont and colleagues published two papers, one in 2006 and one in 2010, in which they searched for known metal-binding sequences in proteins and compared them to see which proteins had the most in common, and were oldest. The 2010 paper was done in collaboration with Gustavo Caetano-Anollés, whose molecular clock was seen in Chapter 5.

These studies fit the general patterns of this story. When proteins started to get picky about metals, the first protein folds bound iron, then manganese (soon), and molybdenum (also soon) appeared. Zinc, calcium, and copper came later, or in the case of porphyrins, never showed up at all. Zinc's line looks different from the others, appearing earlier but growing more slowly, showing a unique pattern over time.

Gustavo Caetano-Anollés assigned a specific age to all enzymes (whether metal-binding or not). For early events corresponding to this part of the story, Caetano-Anollés assigns second place to the porphyrin-using proteins, which fits with this story, too: porphyrins were very early, but not the very earliest. His data says that copper proteins don't come about until the first free-oxygen-using proteins, which don't appear until Chapter 8.

Then, the real tour de force was published by Lawrence David and Eric Alm in 2011. They designed a new method of analyzing genes that could account for the extreme over-sharing of genes among bacteria. This let them date many genes in unprecedented detail to about 3 billion years ago. David and Alm divided up their data by element for every small chemical that would be oxygen-sensitive. Specifically, they dated proteins that used free oxygen, seven transition metals, five sulfur compounds, and four nitrogen compounds.

Their results complete the story that R. J. P. Williams predicted. Proteins using oxygen were very few at first and grew over time, increasing by 300% since 3.5 billion years ago. The sulfur compounds and the nitrogen compounds follow the same basic pattern: the more oxygens they have, the later they were used.

The opposite's also true: proteins that preferred hydrogen-containing molecules were invented early and used less as time wore on. Sulfur with two hydrogens was used less over time, while sulfur with four oxygens was used more. Nitrogen with three hydrogens (ammonia) was used less, while nitrogen attached to oxygen (nitrate/nitrite) was used more.

The pattern is the same as the pattern of a Winogradsky column, where deeper means older. Oxygen controls the pattern of the Winogradsky column—so free oxygen could have controlled the pattern of chemical evolution over time as well.

David and Alm's analysis also provided data about the transition metals that mostly fit with the other two analyses. They say that proteins used iron, zinc, and manganese at first, followed by cobalt and nickel. Only later do molybdenum and copper proteins appear, so they appear in later chapters in this book. Oxygen chemistry correlates with these metals like it does with the nitrogen and sulfur compounds.

Zinc is the most inconsistent among these analyses. This may come from a patchy planet in which some areas had sulfur and no zinc, while other areas had less sulfur and more zinc. The broad outline of the overall story remains that hydrogen-rich food, iron, nickel, and cobalt were used early, while oxygen, molybdenum and copper were used later.

It all adds up to a world that was chemically very different at first. There was no oxygen and few oxygenated fuels. The oceans were full of iron (especially in its two-plus state) and manganese, with possibly enough zinc to use for building but not catalysis. Cobalt and nickel were in the oceans, but copper and molybdenum were missing due to their chemical reactivity with sulfur and other atoms. As a result, enzymes that needed certain chemistry moved down from molybdenum on the periodic table and used tungsten instead. This order of metals (except zinc) was predicted by R. J. P. Williams before any of these genetic studies were begun. These are not choices that could have gone any way, but they were predictable patterns recorded in stone.

HOW THE OLD WORLD STARTED DYING

Atom by atom, the world was changing. Hydrogen leaked into outer space. Cells hoarded electrons and hydrogen while ejecting oxygenated materials ever so slowly. Every ejected oxygen atom and escaped hydrogen molecule contributed to the world's oxidation. The old hydrogen economy of methanogens, methanophages, and fermenters was running quite well, all things considered, but these low-calorie gases couldn't support fast, complex life with a higher energy rate density. The wheels turned, but they creaked slower than they could. Little did they know that a hidden trend in the periodic table was about to undermine their entire chemical scheme.

This trend is a theory of Kurt Konhauser (of the zinc shale study), who proposed that there was a nickel famine 2.5 billion years ago, predating the first rise of oxygen 2.4 billion years ago. There is no element more important to methanogens than nickel, so if they couldn't find it in the ocean, they couldn't turn their wheels. They would be forced to retreat to pockets of the environment where they could scrape together enough nickel to survive.

Konhauser and colleagues compiled data from rocks across the world and inferred ocean nickel levels. Nickel went missing around 2.5 billion years ago. Most important is availability in the ocean, where nickel levels in the ocean dropped by more than two-thirds very quickly, enough to decimate nickel-hungry methanogens. But where did the nickel go?

The only place it could go—down. Remember that the mantle is constantly mixing with the ocean at subduction zones and injecting hot gases and metals through vents. The crust serves as a blanket to hold this heat in, but the blanket leaks, so the mantle is slowly cooling off. Konhauser calculates that, at this point in prehistory, the mantle cooled past the point where nickel would stay liquid. Instead, the nickel solidified and sunk down, away from the surface and out of the oceans (Figure 6.5).

The mantle is a complex mix of chemicals, but the melting points of the pure metals can give us a feel for what metals melt more easily. When looking at the melting point trends along that line of the periodic table, something interesting stands out in the bottom of Figure 6.5. Moving from iron through zinc, the melting points steadily drop, meaning that as the mantle's temperature dropped there was a point where nickel solidified but copper and zinc stayed liquid. These last two metals would remain available to the ocean—locked away as sulfides, to be sure, but able to play a part in future chapters when released from sulfur's grip.

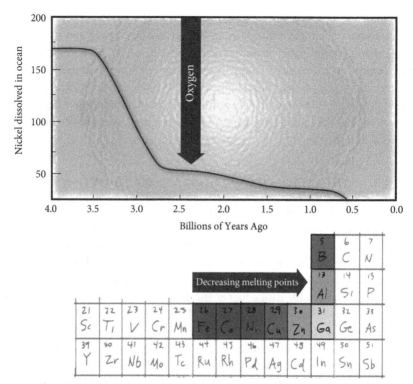

FIG. 6.5 The nickel famine and a possible chemical explanation. *Top*: The loss of nickel from ocean water predated the Great Oxidation Event 2.5 billion years ago. *Bottom*: Melting points decrease across the second half of the transition metals, which may explain why nickel solidified and entered the mantle, while copper and zinc remained at the surface.

Cobalt also could have dropped out around this time, and a "cobalt famine" may explain why Vitamin B_{12} is hardly used at all in modern, complex organisms. As for iron, it too would drop off, but its switch-hitting chemical capabilities were so useful (and its overall abundance high enough) that life was able to keep using it. For the series from cobalt to zinc, melting point could have determined what was available for later chapters.

Konhauser and the accompanying news article interpret this as "contingency." This implies that nickel went away with a random roll of the dice, and life's history was indelibly shaped by that randomness. In this interpretation, the methanogens stood in the way of oxygen-dependent life until the random event of the nickel famine knocked them out. Konhauser points out that oxygen-using chemistry was around for at least a few hundred million years before oxygen levels started to rise, and says that such a long lag indicates that nickel stood in the way by enabling and entrenching the methanogen world.

I disagree, not with the data, but with its interpretation. *Chemistry* determined the disappearance of nickel, and chemistry is not random like dice. Rather, the melting points follow a neat, repeatable, even predictable order within their series that may have caused this famine. If so, then run the tape of life again, and nickel would disappear at the same time.

Oxygen-based fuels and oxidized materials would become a better and better energetic deal as time wore on and any oxygen accumulated at all, even if it took a billion years. This is the trend followed by sulfate, nitrate, and oxidized metals in David and Alm. (This trend is backed up by geological data. For example, SO_4 sulfate metabolism before oxygen was a limited affair, but today, it is one of the major metabolic activities in the ocean.)

The trend of melting points dropping from left to right is built into the periodic table, which was built from math adding proton to proton and electron to electron. The place of nickel on the periodic table determined *both* its usefulness for hydrogen chemistry and its disappearance from the oceans after a few billion years. This is not a chance meteor falling from the sky, but a predictable transition ordered by the periodic table. If water-based life developed on other planets, it would have undergone its own nickel famine at around this time.

I interpret the nickel famine and the oxidation of the planet as inevitable and orderly, rather than random and contingent. I can agree that the nickel famine was a contingent event only if I may modify the usual use of that word to say that *it was primarily contingent, not on chance or choice, but on chemistry*, and chemistry is contingent on the order of the periodic table.

As time ticked on, life kept electrons and ejected oxygen, the Earth oxidized, some metals were locked up, and others were released. New metabolic wheels and new chemical fuels became available. But we have to think local before we act global. Oxidation of a corner of the world had to take place before the world could oxidize. With the help of the sun, a powerful new oxidative chemistry emerged, like a small, warm mammal in the cold shadow of the nickel-using dinosaurs.

7

THE RISK AND REWARD OF SUNLIGHT

pigments in pink chains and green squares // purple poets and green sculptors // orange aphids // manganese: the chameleon that makes oxygen // the artificial leaf // oxygen turns an old chemical into a new signal // with a little help from my (symbiotic) friends // green magnesium plastids and red iron mitochondria

THE ORANGE EARTH, THE PINK LAKES, AND THE RIVER OF FIVE COLORS

Three billion years ago, the Earth may have been a pale *orange* dot. The first atmosphere was rich in methane and related hydrocarbons, which covered the globe with an orange, smoggy haze like Titan's. That would change drastically over the next billion years. According to some models, this orange haze entered a period about 2.5 billion years ago in which the methane haze disappeared and then reappeared again, blinking on and off several times like a loose light bulb. Then the methane disappeared for good as another gas, oxygen, took over the atmosphere. Life changed from using methane to using oxygen, and the Earth was terraformed.

One thing about measuring the chemistry of an entire planet is that it's hard to fit billions of years and gigatons of material onto a two-dimensional graph. Yet with oxygen, we can come surprisingly close. Three billion years ago was point *A* with little oxygen in the air, a state of *anoxia*. Now we are at point *B*, an "oxic" state with 20% oxygen. The line between point *A* and point *B* had some peaks and valleys along the way, but its trend was uphill. (This graph will play a central role in the next two chapters.)

The geologist Donald Canfield and others have found evidence for wispy amounts of free oxygen as far back as 3 billion years ago. Oxygen increased in a trickle and then a flood. Canfield found that surprisingly high amounts of oxygen, possibly almost as high as today's amounts, existed 2.1 billion years ago. Once the chemical key was found that unlocked oxygen, the atmosphere filled with this reactive gas, allowing other organisms to respire and grow in turn. In *other* other words, if the Doctor shows up in his TARDIS and invites to you time travel back to 2.1 billion years ago, you might be able to step outside and breathe the air, even without a hastily inserted plot device.

Now, if this happens, make sure that you don't shave off that decimal and travel back to 2 billion years ago. At that point, oxygen concentrations hit a wall and dropped down precipitously, taking another billion years to recover. This chapter is about the meteoric rise of oxygen from 3 to 2 billion years ago, and the next is about the chemistry that placed its rise on pause for another billion years.

The chemical key that changed the color of the Earth's atmosphere involves a whole rainbow of colors with dual purposes: they both protect and produce. Despite their hues from red to green, most of these colors are chemical cousins, made from carbon chains with different shapes. Eventually, these colors caught enough light energy to change the planet, following a predictable chemical sequence. Complexity came from abundant solar energy—but use of that energy carried a dangerous price.

The sequence may have started with carbon chains acting as umbrellas. To see these colors in action, start at Lake Hiller, which sits southeast from the Shark Bay stromatolites in Australia. The lake is surrounded by green forest and sits near the blue ocean, which only serve to highlight the water's bright pink color. Not "slightly pink" or "squint at sunset and it looks pink," but Bazooka-bubble-gum princess-birthday-party pink.

This pink or red color can be found in other lakes across the world. You may have seen some in San Francisco Bay as you approach the airport, which are artificial ponds for making sea salt. The Dead Sea and Mono Lake are other salty reservoirs where salt-loving microbes bloom. The pink and red colors come from long carbon chains called bacterioruberin. The microbes weave this red pigment into their cell membranes, like the tiny red threads in a paper dollar.

The color and the function of bacteriorubin chains are related. The chains hold extra electrons lined up in double bonds. The electrons absorb high-energy light colors but let low-energy red light through. Bacteriorubin could act as a chemical umbrella, absorbing high-energy sunlight (reminiscent of the ozone layer). Or it could provide extra electrons that scavenge small reactive oxygen molecules produced by the high-energy sunlight. Or both.

The same pink color can be seen in other extreme places. At Yellowstone hot springs, scientists saw pink gel growing in the hot, flowing water, which was home to a variety of hardy microorganisms. The pink probably comes from a protective molecule like bacterioruberin. The bright pink color of Lake Hillier, therefore, is a safe, even protective, color. You can swim among these microbes. The most dangerous chemical in the lake may be salt.

On the other side of the world is a bright magenta river made by different carbon-chain pigments. Located in Columbia, this river is officially "Cano Cristales" and unofficially "The River of Five Colors." Since it was discovered by the Internet, it has graced countless computer monitors with pictures of green, pink, orange, yellow, and red mosses and plants growing under crystal-clear waters.

The magenta comes from the *Macarenia clavigera* plant, which changes colors like the autumn leaves. In the short span between the wet and dry seasons, the water level drops enough to let sun through to the riverbed, but still flows enough to support life and growth. This is when *Macarenia clavigera* turns as red as autumn leaves.

This intense color may be caused by the same pigment molecule that makes my Japanese maple turn the same bright red in the autumn, and turns other leaves deep purple. This pigment is named antho-*cyan*-in, and so the name itself contains a blue color. Anthocyanins have a lot of double bonds like the bacterioruberin, curled into three rings instead of strung out in a long chain. They also protect from light and stray electrons like bacterioruberin, and the plant produces more of them when it encounters environmental stress.

Yellows and oranges fill out the spectrum of the River of Five Colors, and of fall foliage. These colors come from chains not quite as long as the bacterioruberins. Such yellow-orange molecules are carotenes, such as orange "beta-carotene" in carrots and yellow "lutein" in marigolds. Similar but slightly longer chains are red "astaxanthins" that make shrimp (and the flamingos that eat the shrimp) pink, and make cooked lobsters red. The red, rusty pigments in the bottom of old lakes and drained pools are the old astaxanthin collections of microbes. Astaxanthin chains can even be twisted by proteins into configurations that are a bright blue color—and hence, it is perfectly natural to find the occasional lobster that is an unsettlingly bright blue.

The extra electrons along the length of each molecule can both absorb light (creating color) and can react with molecule fragments that have shattered from absorbing too much light. This is how different lengths of electron-rich carbon chains and rings can account for a whole rainbow of colors: pink, magenta, orange, yellow, red, purple, and even (with some effort) blue. On the surface, the colors are very different, but their underlying chemical structures are similar.

One obvious color is missing: How does chemistry make a green plant green? The most plant-like of all plant colors requires help from a non-carbon atom, a metal located far from carbon on the periodic table.

MAGNESIUM FILLS IN THE RAINBOW

A medicinal chemical may have determined my favorite color. I was five or six and living in Phoenix. I had some issues with asthma and allergies, so a specialist gave me a new drug to help me breathe. On the way home, my perceptions were somewhat enhanced by the drug. Looking out the window, I saw patches of green in the Arizona desert that were so vibrant to my fuzzy eyes that I had never seen anything so beautiful. Apparently all the way home I pointed out every patch of green and may even have composed a little song about it (which is lost to history). The drug wore off, but the impression of the green on my brain remained as my common answer to "What's your favorite color?" This is how chemistry shaped *my* biology.

Much later, I found out that the green in the plants and the gray and tan rocks of the desert backdrop included the same metal element, magnesium. The particular color green of plants is made by placing magnesium inside a square porphyrin ring. This metal, which is robotic gray as a solid and colorless when dissolved, turns vivid green when centered in a carbon ring—the same basic structure that, with iron, makes blood red. The chemistry of leaves is only slightly different from the chemistry of blood.

The green of a leaf is made from these magnesium + porphyrin = *chlorophyll* molecules. When a green plant doesn't get enough magnesium, its leaves will lose their green color, turning yellow in the spaces between the veins. In a healthy leaf, large proteins hold the green chlorophyll molecules in place, with other colors of carbon-chain pigments woven between. The chlorophyll squares are arranged in an apparent disarray of angles to catch photons from the sun, wherever it is in the sky. This protein-pigment structure is called the light-harvesting complex.

When light hits a leaf, some of its energy is captured by the electrons in the complex. This energy flows inward from metal to metal and pigment to pigment, like water swirling down a channel, until it reaches the very center of the complex, a disk of proteins called the reaction center.

The reaction center is abundant in iron (like so many proteins in these chapters). Iron ions, iron-sulfur clusters, and pigments form a circuit for electrons. The energy from the light-harvesting complex is plugged into the system and pushes electrons along the circuit. It is so much like a battery that its energy levels are measured in volts. The electrons are pushed by the light energy with enough power to break or make bonds. Light gives power to synthesize new chemical structures, and therefore this process is *photo-synthesis*. Once photosynthesis was developed, life could grow by capturing the free sunlight falling all around it.

Ultimately this power would even break the bonds keeping oxygen locked away. Paul Falkowski describes photosynthesis as "the last of the great inventions of microbial metabolism, and it changed the planetary environment forever" (quoted in Leslie, p. 1286). The planet would turn green, the same green I saw riding home from the doctor, but that would take a while yet. (First, actually, it may have turned purple.)

The full array of colorful molecules is used to catch all sorts of light. Each molecule by itself absorbs a relatively narrow band of sunlight. Light-harvesting complexes combine different colored pigments to catch multiple colors of light. A single species with a certain set of light-harvesting complexes will still only capture a fragment of the sun's available light. But when all pigments are added together, every wavelength of light is absorbed somewhere (Figure 7.1). The shadows of the light absorbed by all the pigments, when added together, match the shape of the sunlight falling upon the system.

Each species has its own collection of pigments, and chance may determine which species gets which pigment, but the overall energy available to the system is consistent and predictable. The species' ability to try different colors, in fact, is what allows the system to respond to the sun. Some species change to capture light that falls through the gaps from other species' pigments. Individual effects are difficult to predict, but the shape of the overall energy absorption matches the energy from the sun.

Some of pigments are colors that human eyes cannot see. You can see some of these invisible colors if you have an intense black light and a banana. Shine the black light on a ripe banana with brown spots. Around each spot will be a glowing halo of purple-white light. This halo is an invisible pigment—a broken, degraded chlorophyll that absorbs high-energy black light and releases it as visible light (a more spread-out, higher-entropy form of light).

Like the pink bacteriorubins, many pigments are protective as well as productive, because the same electron-rich chains that absorb light also absorb stray electrons. It

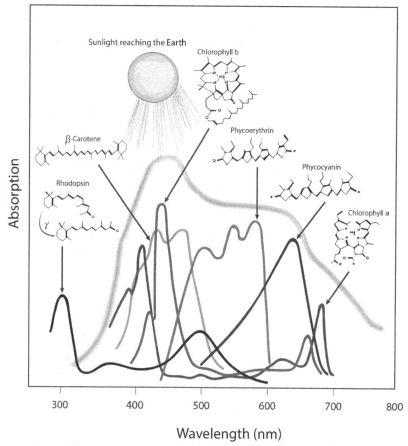

FIG. 7.1 Different carbon-chain pigments absorb different colors of light (dark lines). When the major pigments in an ecosystem are added up, they absorb most of the wavelengths of the sunlight that reaches the Earth's surface (gray line).

Data after Figure 19-41, p. 727 of Albert L. Lehninger and David L. Nelson, *Lehninger Principles of Biochemistry*, 4th ed. 2004, W. H. Freeman.

is easy to imagine a microbe randomly stitching double-bonded carbons together and then surviving better because of the protective role those electrons can play. Later, when this molecule absorbs energy, a stray electron may fall off onto a nearby iron. If that electron gets pushed to a productive place, forming a good bond or breaking a bad one, then, again, that microbe would survive better. Evolution may depend on molecules that can do two things like this, both protecting and pushing.

RHODOPSIN: THE SOLAR-POWERED PROTEIN YOU SHARE WITH BACTERIA

Perhaps the simplest energy-absorbing, electron-pushing protein is also the most universal. This protein is called rhodopsin, and aspects of it are found in every leaf and eye. Rhodopsin is found in the membrane of Lake Hillier's pinkish-purple salt-loving microbes, contributing to the pinkness of the pink lakes and the pink jelly microbes at hot springs. In microbes, rhodopsin has a pigment and a proton pump. The pigment is a short carbon chain, which makes a sample of rhodopsin as mauve as office furnishings from the late 1980s.

The color shows that it is absorbing light, and since the light energy has nowhere else to go, it flips a double chain in the pigment around (Figure 7.1, lower left). This movement presses on part of the protein, which presses on another part, which opens a trap door of sorts that is big enough for protons. After light hits rhodopsin, protons flow through, crossing the membrane.

Rhodopsin is a solar panel for the cell. If you shine a light on bacteria with rhodopsin, they will push protons out and acidify their surroundings. The only cost to the bacterium is the cost to make the protein (and replace it when it breaks). The energy from the sun is free.

Proton movement is a nice trick, but the bacterium has another protein in its membrane that makes the trick really pay off. This is a huge protein called ATP synthase that makes ATP from the energy stored in the protons. This makes the outside part of the cell a gas tank for protons, which store energy until they can flow back in. The road runs through ATP synthase, which the protons turn like a waterwheel, activating a machine-like motion that forms new phosphate bonds. Light energy is changed into the chemical energy of phosphate bonds.

The newly minted ATP can then be taken to any part of the cell and broken open. Its energy can send a signal, or build a new DNA strand, or pump out nasty toxins. If ATP is a $20 bill of energy, then this combination of rhodopsin with ATP synthase is a printing press powered by photons, all from two proteins, one pinkish pigment, and a handful of protons.

The design of rhodopsin is so useful that it has been duplicated in many places with minor twists. Inside you is a human version of rhodopsin that looks just like the bacterial one. Your rhodopsin also sits inside a membrane, is built from seven columns like the bacterial version, and binds a short carbon-chain pigment that flips when light hits it. The major difference is that your rhodopsin does not pump protons, but sends signals by changing its shape.

Your rhodopsin may look primitive, but it serves an advanced function. It detects light in the retina of your eye. As you read this, carbon chains inside rhodopsins are detecting photons by flipping back and forth and pushing a protein around, just like in a pink sample of light-sensitive bacteria.

Bacteria also use the energy captured by rhodopsin to send signals so they can tailor their chemistry to light and dark areas. Through proteins like this, day and night cycles are detected by the smallest organisms and passed throughout an ecosystem. In response to the rising and setting of the sun, ocean bacteria oscillate together, in tune with each other and with the hot star far away. These communications probably produce a dynamic structure of complexity, perhaps as structured as a fluid, unseen Winogradsky column.

Rhodopsin has enough power to pump protons and send signals, but this is all the energy that can be squeezed out of a single photon. There's only so much a single carbon chain can do, even if it is powered by the sun. At first, this couldn't give enough energy for the Holy Grail of our quest, breaking open water to make oxygen, but it was a start.

PURPLE-POWERED PROTONS (AND ELECTRONS)

Light-driven systems more complex than rhodopsin are built from similar pigments and proteins. Plants arrange their pigments in protein structures called photosystems. These photosystems combine light-harvesting complexes, which catch energy, with reaction centers that use that energy. These pigments are arranged so that energy harvesting is maximized while energy damage from loose electrons is minimized.

There are two types of photosystems with different colors, one the purple-pink of rhodopsin and the other the green of spring leaves (with some orange carbon-chain pigments mixed in). Most familiar plants have both photosystems, connected with a flow of electrons. Each system is built around chlorophyll, but it uses that chlorophyll for a different purpose. On the other hand, many photosynthetic microbes have only one photosystem.

Each of the two photosystems found in plants is found in a different "color" of bacteria. Green sulfur bacteria have photosystem I, but purple bacteria have photosystem II. The two photosystems are chemically distinct, even using different elements.

Purple bacteria photosystems look the most like the rhodopsin molecule described earlier. Both have similar light-catching pigments, and both ultimately move protons across a membrane. The photosystem adds an extra twist, using light energy to push electrons from one place to another. Chemically, this would not be hard to develop because the electrons are already there in the magnesium, waiting to be pushed around. Light gives the electrons enough energy to move down a road made of long carbon-chain pigments, irons, and sulfurs.

Recently, scientists found that these pigments flex in sync so that their quantum mechanical states become "coherent," allowing the electrons to move from one pigment to the next. I am reminded of a string quartet keeping tempo. The electron steps through the coherent pigments until it arrives at a pigment that is not attached to the rest of the protein. Like a ship sailing out to sea, this pigment dissociates with its excited electron cargo and sails through the membrane.

Another protein provides a safe harbor for the carbon chain and then sets the electrons on another road to travel. This electron path is paved with iron (an iron-sulfur center and hemes) instead of carbon chains, like the proteins from the previous chapter, suggesting that this protein developed around the same time. As the electrons move, the protons are pumped across the membrane, storing energy that an ATP synthase uses to make more ATP. Another iron protein, cytochrome c, closes the loop by moving the energy-depleted electrons back to the reaction center, where they can be excited by light again.

High-energy electrons are a useful commodity, and if the purple bacterium has pumped enough protons (and made enough ATP) for the time being, it can divert those electrons to a second use, toward the purpose of building things. It does this by making perhaps the second-most-important molecule after ATP, which is NADH.

The NADH molecule is a chemical box, made of CHON elements, that can hold a pair of electrons and can move around inside the cell to where it is needed, like ATP. After NADH moves, it presses its electrons into service, making new bonds and building pretty much any possible chemical configuration. (Technically, this function is taken by a chemical box named NADPH that is tagged with an extra phosphate, but I will not distinguish between the two because their chemistry is the same.)

The bonds provided by NADH-like boxes can tie down even flighty, stable gas molecules like carbon dioxide. Purple bacteria use the surplus of electrons and ATP from their solar-powered photosystems to tie together carbon dioxide gas into a three-carbon sugar chain, and that three-carbon sugar is just a few bonds removed from becoming six-carbon glucose or twelve-carbon sucrose (the sugar in your sugar bowl). The air has been captured and turned to sugar by color, light, and pigments. It's a trick worthy of Willy Wonka himself.

Remember that one of the five chemical balances is that every cell hoards electrons inside itself—NADH is how these electrons are held. Every organism is organized with electrons. Those electrons (and the positive hydrogens that balance their negative charges) can tie anything together and build intricate structures out of simple foodstuffs. ATP, NADH, and DNA all include sugar molecules, and any biochemistry student memorizing those structures needs to know sugar structures first.

The counterpart of having electrons inside the cell for building is that the chemical opposites of electrons (oxygen itself or chemicals with more oxygens) are pushed outside the cell. Therefore, all bacteria, and all life, oxidize the environment, by virtue of life's non-negotiable need for electrons.

GREEN PHOTOSYSTEMS POWER
MICROBIAL SCULPTORS

There is a second type of photosystem, so different it has a different color. Green sulfur bacteria have *green* photosystems that differ in chemistry from the purple photosystems just discussed. Green photosystems use more iron than their purple counterparts, and they use that iron to make more things.

Green sulfur bacteria use light and iron to make general-purpose NADH-like electron boxes. They, too, push electrons into bonds that tie down the gases CO_2 (carbon dioxide) and nitrogen. In green sulfur bacteria, CO_2 is tied down in a different way than in purple bacteria, making simpler, smaller carbon molecules that look more like proteins' amino acids than sugars, with the reverse citric acid cycle from the previous chapter. In this way, the green sulfur bacteria can make amino acids from air like purple bacteria make sugars.

Although sugars don't use nitrogen, amino acids do. Green sulfur bacteria get this important element from nitrogen gas. They push electrons and hydrogens onto the nitrogen, which turns the N_2 gas into electron-rich NH_3 ammonia, "fixing" it and nailing it down as liquid rather than gas. This is yet another way microbes can make life molecules from thin air. Electrons must be powerful to push onto stable nitrogen gas, so the fact that green sulfur bacteria can do this chemistry shows that they must make very powerful electrons. They can use these electrons to make other chemical structures, too.

If the purple systems tend to be poets and painters, using colorful pigments to turn ATP-making cycles, then the green systems are sculptors, laying down molecules of gas like bricks with electrons as the mortar. (As bricklayers themselves, the green photosystems may appreciate the blocky *ekko* sculpture from the beginning of Chapter 2 more than the purples.) This distinction is not absolute, but it's a helpful tendency.

Green photosystems begin by exciting chlorophyll electrons, and then pass the electrons through two pigments. Past that point, the carbon chains end and iron begins. The path for electrons goes through three iron-sulfur cubes before landing on an electron-moving protein called ferredoxin.

Ferredoxin uses an iron-sulfur center to carry electrons. It can move through watery passages to go many different places, passing its electrons along to form bonds where needed. Ferredoxin is a "mortar" supplier, giving electrons to diverse "bricklaying" proteins. The green photosystem is powerful enough that green sulfur bacteria use light to boost electrons up to ferredoxin's high energy level. Not bad for an energy source that is free for the taking.

Once the electrons have made the leap up to ferredoxin, they are powerful enough to be pushed almost anywhere. This is shown in the electron-pushing power of different molecules (Table 7.1). The molecules with the more negative voltage number (their "redox potential") are able to push electrons onto those with more positive voltage.

Ferredoxin is an electronic trump card. Sitting at the top of this list, it can easily push electrons onto carbon chains, other iron-sulfur clusters, CO_2, or nitrogen. The biological electron box NADH is also near the top of the list. This is the molecule that your body uses to supply electrons, so it has considerable electron-pushing power itself.

Ferredoxin was powerful enough to build new pigments, filling out the spectrum of light-catching molecules. One set of experiments shows that ferredoxin pushed electrons onto one version of chlorophyll to build a second version of chlorophyll with a different color. The microbe that figured this out captured new colors of light energy that allowed it to thrive.

Light is in plentiful supply falling from the sky, but the electrons must come from the Earth. Five molecules on Table 7.1 are earthbound electron sources. Hydrogen, at the top, is a great electron supplier, but its abundance was dwindling, even this early in the planet's lifetime, as it was consumed or floated away to outer space. Next on the list is hydrogen sulfide, and farther down, iron. Both of these can supply electrons, given power from a photosystem or two.

In line with this chemistry, green and purple photosystems can be found that pull electrons off hydrogen, sulfide, and iron. Many of these work better without oxygen, suggesting that they originated before that gas permeated the atmosphere. When someone says "photosynthesis," you naturally will think of plants and oxygen-generating chemistry, but these other kinds of photosynthesis came first and may

Table 7.1 Redox Potentials of Biological Molecules at pH 7 (with More Pushing Power on Top)

Molecule	Potential (V)
Ferredoxin	−0.430
Hydrogen	−0.420
NADH	−0.320
Sulfide	−0.240
Menaquinol*	−0.070
Ubiquinol*	0.1
Plastoquinol*	0.1
Rieske FeS protein	0.100–0.270
Iron	0.15
Peroxide	0.27
Water	0.815

*Electron-carrying carbon chain.

Data from Table 1 in M. F. Hohmann-Marriott and R. E. Blankenship, "Evolution of photosynthesis," 2011. *Annu Rev Plant Biol.* 62, p. 515. DOI: 10.1146/annurev-arplant-042110-103811.

still be important, although overlooked. About a tenth of the energy in the oceans is provided by these alternate photosynthesizers.

An amazing example of using light to make powerful electrons is found in aphids. I use the word "amazing" because the scientific paper that introduces these aphids also used it (technically, it said "amazingly"). The aphids in question stand out because they are orange, filled with a huge excess of orange beta-carotene, the carbon-chain pigment from carrots.

Beta-cartone sits in the aphid's cell membranes like the red molecule in the pink-lake bacteria. It collects light, which (through an unknown mechanism) excites electrons and pushes them onto an NADH-like electron box. Then the electrons are used to make ATP. Overall, sunlight hits the carrot pigment, and its energy is turned into ATP for the aphid to live on. Could this aphid be made an honorary plant?

This type of arrangement has been seen before with whole photosynthesizing *bacteria* living inside an animal as separate organisms. What sets these aphids apart is that the process is integrated into the aphid's own membranes. How did this happen? This may show that setting up photosynthesis is not as hard as it looks, and that once an excess of pigment is in place, new energy transfers and bond-making reactions can follow with just a little evolution.

Finally, notice how water is at the bottom of the list on Table 7.1. Even if there is "water, water everywhere," there is not an electron to drink. Water's electrons are held so tightly that special chemical power is needed to pull them off, from another part of the periodic table. Pulling electrons off water is so much trouble, in fact, that every life form that can do so will make the trouble pay off by recycling the electrons multiple times.

THE DOUBLE-EDGED POWER OF
PURPLE + GREEN = BLUE

If I'm ever called upon to be an advice columnist (an improbable hypothetical), I know one thing I could say. When letters ask, "Should I do X, or should I do Y?" I'll answer, "When in doubt, do both." I remember reading this in an advice column at a young and impressionable age (when I still read advice columns), and it actually worked.

In the case of how to get the most power out of a photosystem, the answer plants have chosen to the question, "Should I use purple, or should I use green?" is that they "do both": they use both photosystems at once. For the most powerful photosystems, purple and green combine forces and become blue. These double-photosystemmed, chemically powerful bacteria are even called *cyan*-obacteria for their blue-green color, which can be seen on rocks in some bluish lichens.

Cyanobacteria have fully realized the power of photosynthesis, channeling sunlight to pull electrons off water, running those electrons through a purple photosystem, pumping a few protons across a membrane, then running the same electrons through a green photosystem, and finally pushing them (via ferredoxin) onto an NADH-like electron box for building. That's a run-on sentence if there ever was one, but the cyanobacteria deserve it because they do all that at once. They are sculptor-poets that both pump protons and build new molecules.

We don't know yet exactly how the purple and green combined, but it's clear that photosystems can be found in unusual places doing unusual things (see: orange aphids), so the systems appear to be able to change and morph into different configurations. Purple and green photosystems next to each other in some lucky microbe could toss electrons back and forth, and that could be the beginning of chemical cooperation.

In a two-photosystem arrangement that splits water for electrons, the purple photosystem must come before the green. This is because if negative electrons are pulled off H_2O (water), positive hydrogens come along to balance the charge, and O_2 (oxygen) gas is made. On the way to making oxygen, a lot of electrons are moved around at high energies, and some of them fall off, reacting with nearby electron-moving iron and sulfur atoms. It makes sense to put some distance between the dangerous oxygen-generating site in the purple photosystem and the oxygen-sensitive irons and sulfurs in the iron-rich green photosystem.

Oxygen is a double-edged sword that reacts with and destroys the very structures that make it. We are accustomed to the benefits of oxygen and identify it with life, but it was not always this way. In a world where the oxygen was locked up behind chemical bonds in water and rocks, free oxygen would quickly react with and break down the more reactive elements, which were precisely the ones most useful for life.

Oxygen's reactivity was shown, or smelled, at the Apollo moon landings. When the astronauts returned to the lunar landing module and removed their spacesuits, they smelled gunpowder. The residual moon dust in the air reacted strongly with oxygen. As the oxygen pulled electrons to itself, each grain of dust underwent an atomic-level explosion, and their noses detected the products of this combustion.

The exploding gummi bear from the previous chapter and the other explosions that are most people's favorite part of chemistry class (and, I admit it, mine as well) demonstrate the power of oxygen. An oxygen-generating photosystem is a tiny munitions factory, and such factories are dangerous places to work.

Therefore, the photosystems that generate oxygen are scarred by the high-energy water-splitting chemistry they have to do. In fact, the protein complex that splits water is damaged so frequently that the cell doesn't bother to repair it. The cell throws out the protein and builds a new one every half hour. Oxygen and the chemical power that makes it are dangerous, even and especially to its parent photosystem.

Once oxygen is stored in the atmosphere, and once the ecosystem adapts to the appearance of this new chemical, it lends innumerable benefits. Sugars like in the gummi bear can be combined with oxygen in controlled steps rather than uncontrolled explosions, and the cell can catch the tiny "explosions" of energy that result from forming new bonds to oxygen. Whole kingdoms of life, like ours, can live off oxygen, which is a waste product thrown away by cyanobacteria-type double photosystems.

This process of burning sugars (or anything) with oxygen is called respiration and is the opposite of photosynthesis. Photosynthesis pushes electrons with light to build structures, leaving behind oxidized material and oxygen itself. Respiration, on the other hand, breaks down structures by recombining them with oxidized material, releasing energy put there by sunlight in the first place. Together, photosynthesis and respiration create a circle that takes in sunlight and puts out the movement and heat of life. The system takes in low-entropy sunlight and spreads the energy out as high-entropy heat, processing more energy and increasing the energy rate density.

To a biochemist, respiration is not just breathing. For one thing, respiration doesn't even require oxygen. Other organisms respire with anything oxidized that can accept electrons, stuff like plus-three iron, sulfur chains, even manganese. Respiration is ultimately about moving electrons so that they release energy. The best place to move electrons to is oxygen, but other materials will do if you need less energy.

Gustavo Caetano-Anollés analyzed the protein folds related to oxygen respiration, and dated the inception of the process to 2.9 billion years ago. This means that it took a billion years of life before enough oxygen was around to sustain oxygen/sugar-burning respiration. This fits with the detection of wisps of oxygen in 3-billion-year old rocks reported at the beginning of this chapter.

HOW DO YOU SOLVE A PROBLEM LIKE ANOXIA?

Over a billion years, wisps became torrents and oxygen filled the atmosphere. A specific chemical activity unlocked the oxygen from the oceans, solved the problem of anoxia, and repainted the planet—changing orange methane air and purple rhodopsin waters to today's globe of green and blue. But water was too tough a nut to crack at first, as its position at the bottom of Table 7.1 shows. The Earth first needed a more accessible, less stable source of oxygen's chemical power.

Fortunately, oxidation can come from any molecule with oxygen in it. The first key was not H_2O but H_2O_2, hydrogen peroxide. On Table 7.1, hydrogen peroxide is more like iron than water, meaning that its electrons can be removed like iron's. Peroxide, and its relative superoxide (O_2^-, with an extra electron) are part of the group of unstable oxygen-hydrogen molecules called reactive oxygen species (ROS). We saw these in the first few chapters, noting that these are dangerous inside a cell, and are ejected lest they cause permanent harm to the carefully ordered bonds. At this point in history, their reactivity helps rather than hurts.

Inside the cell, ROS cause oxidative stress, especially attacking redox-sensitive iron-sulfur centers. The stress would be especially acute for organisms that had never seen oxygen before, so they would scramble to get rid of the ROS. When ROS are ejected, they oxidize the environment and contribute to the inevitable march toward oxidation and oxygen itself.

ROS from life have been measured oxidizing the environment. Prior to 2013, we thought that sunlight created most of the superoxide in the ocean by splitting water, but a closer look revealed that living microbes produce at least as much superoxide as the sun. ROS change ocean chemistry, oxidizing iron and manganese directly, changing iron plus-two to iron plus-three and manganese all the way up to manganese plus-four. They also indirectly affect the concentrations of oxygen-sensitive metals like mercury.

Sunscreen molecules that absorb solar energy, like pigments, will move around nearby electrons, like pigments. Near popular beaches, sunscreens can generate enough ROS that the levels can produce oxygen stress in ocean algae. Scientists are working on sunscreen pigments that direct this excess energy into more harmless, less stressful outlets.

Metals react with ROS inside cells, too. Each element has a different style. Iron reacts best with peroxides to make oxygen radicals that, true to their name, perform some truly radical chemistry with their unbalanced electrons. Iron can also do the reverse reaction, turning radicals into peroxide. Iron's peroxide wastes with their loose electrons are a nagging problem for iron-using microbes.

Manganese, on the other hand, catalyzes a different reaction. It reshuffles the bonds in H_2O_2 (peroxide) to make H_2O and O_2. Chemistry professors impress their classes by mixing peroxide with manganese oxide in a flask, which creates an explosive plume of water. True to oxygen's double-edged nature, this is dangerous, so I am happy to show it on video.

In cells, iron and manganese work together in a chain reaction: iron turns radicals into peroxide, and manganese turns peroxide into the safer molecule O_2. If iron's dangerous waste product, peroxide, is directed to a manganese atom, O_2 gas is released in a controlled stream.

Gustavo Caetano-Anollés and colleagues have proposed that, with a sequence like this, manganese chemistry provided the first oxygen-using enzyme. Their molecular clock says this enzyme originated 2.9 billion years ago and used the oxygen, not for burning, but for *building*. It made a versatile chemical called PLP that comes from vitamin B$_6$.

Enzymes today use vitamin B$_6$ for at least three very different reactions, making it the Swiss army knife of enzyme cofactors. The active group on PLP is a reactive form of oxygen-carbon-hydrogen called an aldehyde (which has a memorable stink). PLP uses this aldehyde like a sharp blade to break and make bonds. Today, it is vital, as a vitamin should be.

PLP's most important atom is a reactive oxygen, so a cell making PLP would need to get this oxygen from somewhere. Exactly 3 billion years ago, about 100 million years before the PLP-making enzyme came on the scene, an enzyme called manganese catalase showed up that could have made that oxygen. This enzyme uses manganese to detoxify peroxide from inside the cell, and its waste product is oxygen gas. Some rocks from this time contain a form of manganese that fits with this story.

Here, a chemical element and a biological activity are closely intertwined. Manganese catalase served as a portable oxygen generator, recycling peroxide and producing oxygen gas, which in the blink of an eye (geologically speaking) was grabbed by a nearby enzyme used to make the Swiss army knife molecule PLP.

Once manganese was in place, taking in peroxide and spewing out O$_2$ (oxygen), the oxygen from manganese could have run into some carbon rings, making a reactive, potentially useful aldehyde. If this reactive molecule pushed some other atoms around in the right way, that chemistry would become a feature, not a bug, that would help that cell survive better.

Later a protein with the right-shaped pocket in its surface could form. This could make the aldehyde consistently, and its shape could be refined over time. It helps that protein pockets don't need to be precisely shaped to work, so a "good enough" pocket would probably exist somewhere in the cell. With these proteins, the cell would survive even better—and the rest of oxygen chemistry (really, the rest of this book) would follow. Once the oxygen-generating chemistry of manganese kicked this process off, in an ocean of billions upon billions of flowing and evolving microbes, this chain of reactions became a distinct possibility, predictable from the oxygen-generating capability of manganese.

MANGANESE: THE ELEMENT OF FIVE COLORS

The question becomes not how did this happen, but why did it take so long? Manganese is so innately powerful at this chemistry that it should have developed earlier. Manganese dissolves well in anoxic ocean water, so it wasn't a matter of its chemical availability. Rather, according to Caetano-Anollés, it was a matter of life working out the chemistry to snag it.

In the catalase, manganese is held to the protein with complicated nitrogen-ring amino acids that look a little like the nitrogens on the inside of porphyrin squares. This amino acid is the second-most expensive amino acid and would have been difficult to make. The Caetano-Anollés molecular clock agrees and says this particular amino acid was one of the last to be invented, some 3 billion years ago. This difficult amino acid was invented, then manganese was captured, then oxygen made vitamin B$_6$, and voilà, new chemistry was born.

Manganese is good at this special chemistry because of its position on the periodic table, to the left of iron. Manganese forms a plus-two ion just like all the other transition metals, but it can mimic its neighbor iron and lose different numbers of electrons. It can also lose one or two *more* electrons, forming a plus-three and even a plus-four ion. Since chemistry is the art of moving electrons around, and since electrons are little packets of negative charge, manganese's ability to shift its charge and reach a highly charged plus-four state (along with its ability to dissolve in ancient oceans) made it a valuable commodity on the microbial metal market.

Manganese's ability also makes it a chemical chameleon. Two-plus manganese is a soft pink in water and in rocks, three-plus a brown powder, four-plus a black powder, six-plus a lemon-green in water, and seven-plus a deep amethyst violet (amethyst gets its color from manganese ions embedded in glassy quartz).

Each color codes a chemical power. For example, the violet color is familiar to organic chemists as the permanganate ion, and it is used in countless fume hoods across the nation when organic chemistry students need to add oxygens to other chemicals. When this happens, manganese accepts electrons and slips down to one of its other colors.

Some go so far as to call potassium permanganate the "survival chemical." If you're stuck in the wilderness and need to start a fire, mixing permanganate and antifreeze may do the trick. It also is antiseptic, anti-fungal, anti-cholera, and purifies water in general (because this form is so reactive that it is anti-life). Permanganate is an aggressive oxidizer, a kind of chemical key that unlocks the power of oxygen. On the ancient Earth, manganese could have done a gentler oxidation.

Manganese and iron are charge-switching neighbors on the periodic table that work together well, for example, in glassmaking. Glass is melted silicate rock, so iron is frequently in the mix, and it gives glass a green tint. Two thousand years ago, Romans used a black powder called *sapo vitri* ("glass soap"), which when mixed into molten glass would remove the green. The black powder is the plus-four form of manganese, which reacts with the iron to remove its color.

Manganese itself can be an impurity in the glass, although most of its colors are too faint to be seen. When old glass is exposed to sunlight for centuries, the sun slowly and steadily removes electrons from the manganese until it reaches the plus-seven state. The intense purple of this form of manganese tints the glass purple. It is an artificial amethyst made from glass, light, and time. Light switches manganese's electrons.

Such electron-switching means that, like iron, manganese can donate electrons to life and can form the basis of metabolic cycles in underwater sediments. While iron can only change from plus-two to plus-three, manganese can go all the way up to plus-four. This plus-four form of manganese is very reactive and will pull oxygen atoms from its environment to make a blanket of black manganese-oxide crust around the microbe colony.

Mysterious fields of metal eggs on the ocean floor may come from this chemistry. These "eggs" were found around 1875 as the *Challenger* expedition sailed around the world and dredged the ocean below, pulling up dozens of nodules with intricate circular layers inside. These are essentially crystals made of manganese, with iron and other metals caught in the crystalline net, made by the cooperation of biological and geological processes. (You could mine manganese by dredging the ocean floor for these.) Such cycles mean that once a microbe caught some manganese, it could be used for a new biochemical wheel.

AN INTERNAL SHIELD OF MANGANESE

Manganese can also be used by life as yet another shield against chemical assaults. A manganese shield is the secret inside *D. radiodurans,* the microbe certified by the Guinness Book of World Records as "the world's toughest bacterium." Scientists discovered it in the 1950s growing in a can of meat that had been dosed with high levels of radiation. So the scientists upped the ante and zapped it with higher and higher amounts of radiation until they finally finished it off. It took a long time. When all was said and done, *D. radiodurans* endured a dose of radiation a thousand times that which kills humans, without even breaking a sweat.

D. radiodurans has two unusual sets of proteins inside: one set that pastes DNA fragments back together when they are broken apart, and another set that protects against ROS. This second set of proteins uses manganese to react with energetic superoxides and peroxides produced by radiation. It's the manganese, not the proteins, that's doing most of the protecting here. Small molecules stuck to manganese and free manganese help protect *D. radiodurans.* Manganese protects more complex animals under stress, too. One study injected diabetic rats with manganese ions and found fewer symptoms of oxygen stress.

As a result, many microbes eat large amounts of manganese as well as iron. The microbe that causes Lyme disease, for example, needs a huge amount of manganese for protection. Cutting off its manganese supply weakens its internal shields and makes it easier to destroy. Scientists are working on drugs that do just that.

One lab is working on a manganese-binding protein that is a sort of manganese sponge. This chemical would work by soaking up all the manganese so there's none left over for the bacteria. It shows promise as a single drug that could take out a broad spectrum of different bacteria. Chemistry meets immunity.

The balance of health inside a cell can be tipped by metals, and manganese tips it in a beneficial direction. Researchers showed that cells can tolerate high levels of sticky, toxic cadmium when they also added high levels of manganese, or the sulfur molecule glutathione, another molecule that protects from loose electrons and ROS. If the balance tips too much the wrong way, then the manganese might be out-competed. Toxic levels of zinc harm the cell by blocking manganese entry with their excessive stickiness (as the Irving-Williams series predicts) and raising the stressful ROS levels. High levels of cadmium and zinc poison cells by blocking the protective chemistry of manganese, loosening electrons and ROS, and breaking the manganese shield.

"MOONLIGHTING" PROTEINS MULTITASK
WITH METALS

The more energy life uses, whether by photosynthesis or respiration, the more toxic ROS it accidentally makes. Catalase enzymes help neutralize these loose electrons and are one of the foundations of life. The oldest catalases were probably built around manganese's special chemistry, but its neighbor on the periodic table, iron, can facilitate similar reactions (in the lab, one small change turns an iron peroxidase into a catalase). Heme-based catalases are very common today and would have been easy to make by early microbes, because the exact shape of the protein doesn't matter so long as the metal is present.

Since manganese and iron can do oxygen chemistry by themselves, it also makes sense that they can still do that chemistry when incorporated into a protein, even one originally built to use the metal for a very different purpose. Inside every iron and manganese protein is a catalase trying to get out. This can result in some surprising biochemical connections between electrons, light, iron, and manganese. It seems that many enzymes have more than one job:

1. The scientists purifying the iron enzyme that takes nitrogen groups off adenine nucleotides found that it had a double life. When they purified it in the lab, its activity was dead because its iron had reacted oxygen with nitrogen- and sulfur-containing amino acids. Basically, it acted as a catalase on itself.
2. The immune system proteins, antibodies, that hold onto heme rings also make peroxide, because the iron in the heme rings reacts with oxygen. Antibodies are only made to stick to things, but when they stick to iron, they become catalase enzymes.
3. Hemoglobin's primary job is to carry oxygen around with iron-heme groups, but its iron atoms have accumulated additional jobs. For example, the iron in plant hemoglobins can remove an oxygen from a nitrogen-oxygen molecule to make ammonia.
4. When a lot of respiration happens, ROS increase enough to kick a heme off one cytochrome protein and move it to an empty catalase protein, which then helps protect the area from stray ROS.
5. A fungal protein uses the same carbon pigment structure to sense both high-energy light and high-energy ROS.

This innovative and overlapping chemistry is how the cell responds to the risks and rewards of light energy. Light both makes ROS and triggers a protective response to ROS, even with a simple pigment molecule. Iron can make and protect from ROS, depending on context.

Dual purposes are found not only for pigment molecules but for metal ions, and those dual purposes could even influence the path of evolution. Bacteria with too much iron inside have more ROS and therefore more damaged DNA. But the bacteria that persist through this will evolve antibiotic resistance more quickly than those with normal levels of iron. DNA damaged by loose electrons can kill the cell, but it can also make a new protein that can help the cell survive.

The prevalence of moonlighting means that the line between enzyme functions can be a thin one. A metal enzyme's job may be "both/and" rather than "either/or." This can happen in the lab when scientists redesign metal enzymes. Just four changes next to the heme in hemoglobin make a protein that changes nitrogen oxide into dinitrogen oxide. These four changes changed the name of the protein from hemoglobin to reductase. Because metals are so powerful, it is easy to morph one enzyme into another, meaning that evolution can quickly change metalloenzymes from one category to another.

BABY STEPS TOWARD A METAL LEAF

The risk and reward of light is found in a leaf raised to the sun. It takes in our exhaled breath and free-falling sunlight, and it puts out oxygen and sugar (two of my favorite things). The leaf's chemistry rests on a foundation of manganese. Because it can tear a water molecule apart, manganese is sword as well as shield. Manganese seems eager to perform this reaction, even in an artificial laboratory context.

When protein designers add manganese to a protein, they can give that protein the special powers of manganese, possibly retracing steps that evolution took billions of years ago. One group took a purple photosystem and added a manganese-binding site next to the light-catching chlorophyll. The additional manganese ion switched its charge from plus-three to plus-two, which pulled an electron off superoxide and made O_2 oxygen—but only when it was *illuminated*. The researchers had made a light-driven oxygen generator from nothing more than a well-placed manganese.

Likewise, a typical manganese catalase has two manganese atoms that convert peroxide to oxygen. As simple as this catalase is, it is already halfway to being a useful water-splitting complex for the most difficult step of photosynthesis. In the catalase, two manganese atoms and two oxygen atoms are held in a half-cube by the protein's nitrogen "claw." A modern water-splitting complex has the exact same arrangement in its core, plus another pair of manganeses, two more oxygens, and a calcium (Figure 7.2). It looks like a second level was built on top of an old manganese foundation, and world-changing chemistry was invented.

The power of this arrangement comes from the chemistry of manganese. Manganese when attached to oxygen has a special electron orbital with a unique shape that fits together with oxygen's electrons like a puzzle piece. This is why manganese is so good for oxygen chemistry; it is the most abundant atom with this special oxygen-reacting orbital. Cobalt also has a similarly shaped orbital and can do some of the same oxygen chemistry, but iron, calcium, or nickel cannot.

Chemists have been inspired by this. If it really is all about the manganese (or cobalt), could we take manganese (or cobalt), build a chemical scaffold around it, turn on the lights—and have a device that splits water in half? Manganese works as a water-splitting axe in three promising ways: as nanoparticles, as a ship in a bottle, and as a metal leaf.

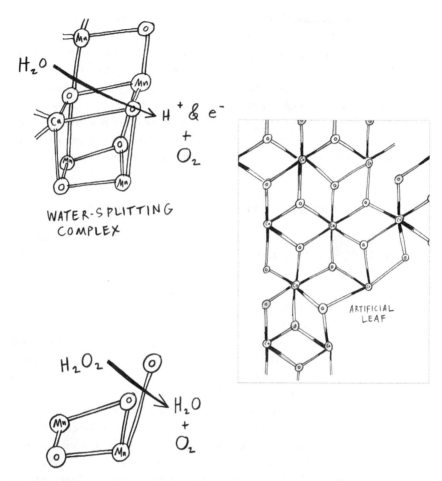

FIG. 7.2 Similar arrangements of manganese and cobalt are found in different molecules that react with oxygen. Shown are manganese catalase (*lower left*), the purple photosystem's manganese water-splitting complex (*upper left*), and cobalt and oxygen in an artificial leaf (*right*).

Manganese-oxygen nanoparticles (tiny metal-oxide chunks) by themselves will split water, but not very well or very quickly. Some researchers made the manganese work better by mixing calcium into the nanoparticles, like those seen in nature. These worked better because the calcium loosened the metals' grip on water, allowing the water to move through the reaction more quickly. This "impurity" of calcium helped the manganese split water faster.

Another group built a chunk of manganese the same way that a model-builder builds a ship in a bottle. One of the problems with manganese is that it's so reactive that adjacent manganese ions will react with each other, like schoolchildren crammed in a bus. This group kept them apart by giving each its own seat. They built a carbon-based lattice of small holes and put a manganese in each, then assembled a cage of carbons around the manganese. The resulting framework had holes big enough to let water in but too small for the completed carbon cage. Like a ship in a bottle (or a sofa from Ikea), it was assembled in the room and then was too big to leave through the door. This arrangement split water for seven *hours* under conditions in which the "naked" manganese complex didn't last seven *minutes*.

A third group of scientists both mimicked life's wisdom on how to make oxygen and also followed the truism that sometimes the perfect is the enemy of the good. Daniel Nocera heads a group of researchers that study metal catalysis of water splitting. Like other researchers, they struggled with the fact that water-splitting is risky business, producing so many ROS oxygen fragments that most great ideas are self-deactivated by half-reacted waters.

Nocera noticed that the purple photosystem has a simple strategy for managing this barrage of reactive oxygen fragments. Instead of deploying a complex shield, the purple photosystem treats its water-splitting manganese protein as disposable, retiring the old electron-scarred protein and replacing it with a brand new protein every half hour. The reward of splitting water is so great that it's worthwhile to throw away an entire protein if it keeps the oxygen factory running.

The scientists took the same approach as the leaf. They searched, not for chemicals that would prevent damage, but for chemicals that would *re-form* the correct bonds *after* they were damaged. They developed a cycle of reactions with cobalt that would heal broken bonds.

From this, they made an artificial metal leaf that split water into hydrogen and oxygen. The leaf is a shiny slab the size of a cell phone. One side has nickel-molybdenum-zinc to make hydrogen bubbles (note the use of the hydrogen-loving nickel), and the other has cobalt-phosphate to make oxygen bubbles. Put it in a beaker full of water, illuminate it, and hydrogen and oxygen bubbles spread out from its surfaces. This metal leaf may eventually turn sunlight and water into enough hydrogen energy to keep the lights on in a small house.

It's not clear if the best format for the metal leaf technology is the wafer, or another form like nanoparticles, or a ship-in-a-bottle arrangement. Even a hybrid microbe-electrode system is under development. The full fruits of capturing solar energy may still be a decade or more away (as they seem to have been for several decades now), but I expect that nature can continue to provide chemical inspiration as we move forward. The power comes from the peculiar chemistry of manganese, and the trick is figuring out how to channel it.

BUILDING WITH OXYGEN: PIGMENTS AND STEROLS

ROS are risky because they are reactive, forming and breaking bonds indiscriminately. What if that bond-forming power could be used for good by being channeled into forming useful bonds instead of channeled *away* from *breaking* useful bonds?

The surfaces of proteins are craggy and irregular enough to contain many places where molecules can fit. If an oxygen-reacting metal like iron or manganese is next to one of these accidental binding pockets, then the reactive oxygen can bond to the molecule in that pocket. Oxygen is powerful enough that even a single atom added can change the chemistry of the whole molecule. In this way, a well-placed metal and oxygen make a brand new molecule. Just like many iron enzymes can moonlight as catalases, many iron catalases can become oxygen-building enzymes.

Scientists recreated this pattern in the lab with a newly designed protein. They made a simple, four-column protein and put a few amino acids in that arranged two iron atoms side by side. The original protein had a binding pocket that would bind a carbon ring and pull electrons off it. By changing the shape of the binding pocket, they made it bind a carbon ring with a *nitrogen* on it and then attach an *oxygen* to that nitrogen. The protein was a malleable container that the scientists molded to redirect the natural reactivity of iron and oxygen.

This is how iron and oxygen together made possible the rainbow of colored pigments in light-harvesting complexes. In Figure 7.1, many pigments are similar to carbon chains inside but are decorated with different oxygens on the outside. This same chemical strategy was repeated eons later by chemists when they made new dyes. Eastman Chemical's dye library at North Carolina State University contains boxes upon boxes of test tubes in every shade of purple, red, and yellow, labeled with the chemical structure. Different colors come from different carbon shapes decorated with oxygens, just like plant pigments.

In plants, metal proteins act like dye chemists, making new colors by adding oxygens. The iron protein ferredoxin adds oxygen to one version of chlorophyll to change its color. Also, the process of making many different pigments is dependent on manganese. By providing these new chemical activities, iron, manganese, and oxygen allowed life to capture more colors.

This story is bigger than just pigments—iron and oxygen built entire new classes of molecules. The strongest oxidant in nature is an enzyme found in methanogens called methane monooxygense. This enzyme replaces a C-H bond in methane with C-O, which doesn't sound that bad until you realize this must break both two tough bonds at once (C-H and O-O). It does so with two iron ions arranged in a "diamond core" geometry that maximizes the power of the metals. Iron and oxygen, once again, go hand in hand.

The biggest step forward for making new molecules came when oxygen helped tie together a long carbon chain, making a stable, compact four-ring structure that formed the basis for advanced chemistry and signaling. Molecules built from this four-ring core, often by iron enzymes adding oxygen, are called "sterols." You are familiar with some already, as chole-*sterol* and *ster*-oids. Like the carbon-chain pigments, these four-ring sterol structures dissolve well in oil and tend to reside in the cell membrane. Wedged between fat molecules, their linked rings provide rigidity, making the membrane both stronger and more fluid.

Cholesterol in particular strikes a balance between stability and fluidity. Cholesterol is only bad in excess. The right amount of cholesterol makes membranes strong enough that cells can grow larger and contain more structures. For example, when yeast evolve to withstand higher temperatures and harsh chemical conditions, the first thing they do is to make their membrane stronger with more and different sterols.

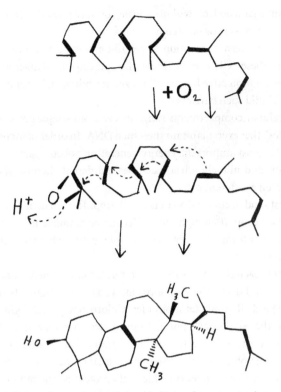

FIG. 7.3 A carbon chain and an oxygen can be molded into a four-ring sterol structure. The later steps in cholesterol synthesis require higher oxygen concentrations.
Konrad Bloch, *Blondes in Venetian Paintings, the Nine-Banded Armadillo, and Other Essays in Biochemistry.* 1994, Yale University Press, p. 20.

Since cholesterol is a sterol ring with just one added oxygen, it is easy to form when oxygen levels are low (Figure 7.3). Caetano-Anollés found that molecules like this were made soon after oxygen became available.

Cholesterol allowed cells to grow bigger and to flex in new ways, making appendages like the squiggly pseudopods on amoebae. These in turn allowed microbes to surround and even absorb other microbes at an increasing rate. Cholesterol is a special kind of glue that holds membranes together in all sorts of organisms, and cholesterol was only made possible by oxygen.

SUBDIVIDING THE CELL WITH OILY DROPS

Iron's ability to help make these molecules comes with a hidden chemical cost. Whereas unbound manganese ions tend to be protective, unbound iron is the opposite. Left to its own devices, iron will unbalance the electrons on oxygen and generate ROS. In fact, some cells have a simple self-destruct program: release a wave of iron ions and die from the ROS that the iron automatically makes. The sword these cells fall on is made of iron. Every time iron is used, a little ROS will be made accidentally.

Every organism that uses iron will die a little younger from this ROS onslaught. Yet iron is so useful that its pros outweigh its cons.

Some cells don't have the option to rebuild a new protein every half hour and get on with it. These cells compartmentalize, putting the dangerous oxygen-using enzymes behind a protective barrier. The size, flexibility, and strength given by cholesterol helped build such barriers.

These subcellular compartments range in size from simple protein cages to complex "organelles" that even maintain their own DNA. In order of increasing chemical complexity, the most important energy-generating subcompartments are chlorosomes, plastids, and mitochondria. Each has a different chemistry, different metal profiles, and even a different color.

The simplest kind of compartment is an oil drop. The cell membrane itself is a hollow, spherical oil drop. If enough oily molecules accumulate in a corner of the cell, they spontaneously form a light-brown oily sphere, like what you can see in shaken Italian dressing.

Green sulfur bacteria have subcellular microcompartments only a step more complex than an oil drop. These "chlorosomes" have oily pigments that are stacked together inside a shell of oily fat molecules. Chlorosomes catch light and give green photosystems (the "sculptors") power to push electrons onto iron. It seems disorganized, but it works well enough for those simple bacteria, because it assembles naturally as oil dissolves oil. The chlorosome membrane holds both pigments and light-generated reactive fragments inside itself, protecting the rest of the cell.

However, this oily sack of pigments is destroyed by oxygen. Subcompartments need to get stronger than this. But chlorosomes show that even messy, simple subcompartments can help microbes survive in low-oxygen conditions.

BUILDING PLASTIDS FROM THE OUTSIDE IN

Most compartments inside cells are more complex. These have membrane walls that look like bacterial walls, enclosing enzymes that look bacterial, and circular DNA, bacteria's favorite DNA shape. These compartments look like miniature cells embedded within the larger cells like Russian nesting dolls. Some even divide independently within the cell, budding out like bacteria. All of this lends support to the idea that these once were independent bacteria that now live symbiotically inside other cells, an idea called endosymbiosis (roughly, "inside-together-life").

These complex compartments come in two types: plastids and mitochondria. Plastids capture light, are found in plants and algae, and can be green or red. Mitochondria break down chemicals for ATP, are found in animals and plants, and are a brick red color, like blood. Each of these specializes in a type of chemistry that the larger cell cannot do, yet cannot do without. Both compartments also involve oxygen and appeared at this stage of the story, about 2 to 1.5 billion years ago. Because they require strong, flexible membranes, cholesterol helped form them, and their enzymes require copious iron. Therefore, oxygen and iron are chemical requirements for these compartments.

Plastids look like cyanobacteria and act like better-organized chlorosomes. They harvest light energy with intricately organized pigments, then run it through green and purple photosystems. They split water and put out ATP and electrons. They are especially rich in iron and manganese, the stars of this chapter. In leaf cells, plastids contain 80% of the iron.

Because they are individually wrapped by cholesterol-rich membranes, plastids can be shuffled around from cell to cell. Some plastid-containing plants and algae have been found in various states of capturing, engulfing, and digesting other plastid-containing organisms, making their family trees more complicated than those from the War of the Roses. (Would this be the War of the Plastids?) They can even be picked up by animals—one sea slug ingests algae but keeps the algae's plastids intact and functional. Is this slug solar powered?

The endosymbiosis event that brought about plastids was originally thought to be a unique event. But then a different kind of plastid in a microbe called *Paulinella* upended this assumption. Its inception is dated to .5 billion years ago, rather than 2 billion. Whatever microbial rendezvous intermingled plastids with host organisms, it happened at least twice.

In the lab, we are close to recreating plastid evolution. Researchers have evolved pea aphids and *Buchnera* bacteria into an intimate symbiosis in which both organisms require the other. Something like this with a cyanobacterium would create a plasmid. Another group evolved a cyanobacterium so that it is optimized for the rich chemical environment inside a different species' cell. This brings the cyanobacterium right up to the threshold of becoming a plastid, not in another organism but in a test tube.

So why haven't plastid-like compartments evolved more often? For example, sulfur-metabolizing bacteria make excellent endosymbionts at vents because they can make food out of the sulfurous waters. In the same way, it seems that green sulfur bacteria could make sugar from CO_2 living inside another cell, and slowly evolve into a new organelle. (The catch is that the air can't contain oxygen because it will kill them, but there are plenty of ways to hide from oxygen.) Yet no one has found green-sulfur-like organelles providing cells with sugars from CO_2. It might be that only *oxygen* gives an energetic payoff high enough to drive deep endosymbiosis, or it might be that we haven't looked hard enough in the right anoxic places.

The answer to this question would be to search anoxic vent environments for green sulfur bacterium organelles. If they continue to elude us, we can be forgiven for believing they don't exist. For now, the most successful metabolic organelles either make or consume oxygen.

THE SECOND ORIGIN OF LIFE

The second type of organelle is the mitochondria. These complex compartments are packed into the muscles and brains of complex animals like us, combining carbon and electrons with oxygen, making energy in the form of ATP. The most important organs are filled with the most important organelles.

Chemically, mitochondria are not too different from Chapter 6's hydrogen-eating bacteria. Mitochondria even have iron-filled hydrogenase enzymes (working on NADH electron boxes) that look like Chapter 6's nickel-iron hydrogenases (working on H_2 electron boxes). Like Chapter 6's bacteria, mitochondria pass these electrons through a series of iron- and sulfur-rich membrane proteins while pumping protons outside, and have an ATP synthase protein fueled by those proteins. Someone who can't make iron-sulfur complexes right has broken mitochondria and Frataxin's disease. If it looks like a bacterium, makes energy like a bacterium, and uses iron and sulfur like a bacterium—it might be a mitochondrion.

Like plastids, mitochondria have circular DNA that is eroded down to the bare essentials, and like plastids, mitochondria are biased toward iron-sulfur chemistry. Mitochondria take up a quarter of a cell's volume but more than half of its iron. Remember that all this iron comes with the cost of iron's inherent tendency to spew out ROS, so mitochondria are chemically assaulted from within. They have a tight double membrane to hold it in, but despite this protective measure, some ROS invariably leak out. Brain cells in particular are at constant high risk of ROS damage from their abundance of mitochondria—they are industrious but inherently polluted. Likewise, brain cells have a huge energy rate density and require floods of ATP and oxygen.

Manganese in mitochondria forms the basis of a fleet of superoxide dismutase and other metal-based protective enzymes. What's interesting is that mitochondrial "shield proteins" protect using manganese and iron (from this chapter), while outside the mitochondria, copper and zinc enzymes do the same job (from the *next* chapter). Also, inside the mitochondria an electron box (the alternate version of the NAD electron box called NADPH) is used to neutralize the ROS peroxide, so that this important and versatile biomolecule is consigned to what seems like clean-up duty. In both these cases, mitochondria use outdated technology as shields, which suggests that they come from old bacteria that predate the oxidized world. Mitochondria use the chemical equivalent of an old pickup truck to clean up their extra oxygen, but it seems to work for them.

Nick Lane and Bill Martin, who proposed theories about white smokers in Chapter 5, have another interesting chemical theory explaining why mitochondria are so useful. Cells with mitochondria are much bigger and more complex than cells without, including bigger and more complex genomes. Lane and Martin think that mitochondria provide energy for bigger genomes. According to Lane and Martin's calculations, mitochondrial energy allows genomes to grow 200,000 times bigger and provides the capacity for the intricate complexity of life as we know it. Lane and Martin put it this way: "If evolution works like a tinkerer, evolution with mitochondria works like a corps of engineers" (p. 929). This expansion is built on the power of the oxygen that mitochondria burn.

The simplest organisms with mitochondria actually have a slightly lower energy rate density than those without, but this masks a different kind of advantage. The cells with mitochrondria are physically bigger (thanks to cholesterol) and hold more *information* in much bigger, more complex genomes. This prepared those organisms to take later steps forward in energy rate density—but those steps would wait a billion years, until after the chemistry of the environment gained more oxygen fuel.

Scientists have found an alternate plastid, but no one has yet found an alternate mitochondrion, so the endosymbiosis of mitochondria might have been a unique event. Given how much potential for energetic gain and complexity resulted from this kind of endosymbiosis, it seems like it should have happened multiple times, but so far it seems it didn't.

If it was unique, then the creation of mitochondria may be as rare as the origin of life. It was the origin of *complex* life, or the *second* origin of life. Once it happened, the chemical motor for increasing complexity became the chemistry of oxygen.

Mitochondria and plastids worked together in a circle of carbon reactions turned by the light of the sun and the chemical power of iron and manganese. If Canfield's geological data hold up, then the manganese oxygen-releasing chemistry worked so well that oxygen levels rose to almost 20% of the atmosphere 2.3 to 2.1 billion years ago. This was the "Great Oxidation Event" that changed the planet.

But seemingly as soon as it had begun (or within a few hundred million years), the chemical party was over. Canfield's same data show that about 2 billion years ago, oxygen levels crashed back down. There was not yet enough oxygen to bring the planet into a new chemical age, as oxygen's double-edged nature showed its cutting edge. Oxygen's reactivity worked against itself as Earth resisted the chemical change. Change would still come, and it would span the whole planet, but change would take another billion years and another chapter to complete.

8

ONE STEP BACK, TWO STEPS FORWARD

when the planet rusted // a breath of toxic air // molybdenum unlocks nitrogen energy // an arrow through the ocean // the creation of gems // mercury and copper unleashed // surviving by losing genes // zinc guides DNA // then things get really interesting

CLOUDY WITH A CHANCE OF RUST

In seventh grade, I picked up *The Eye of the World*, the first in a series of Tolkien-esque novels. The author, Robert Jordan, had built a world with several creative innovations, but at heart it was the familiar story of the farm boy who grew up to be king that drew me in.

It was clear to all readers that this boy would be the hero of the prophesied Last Battle, but most characters refused to see it. I was frustrated by this by book four and graduated to acceptance by book eight. (Did I mention this was a long series?) It took 14 books and a second author to reach the Last Battle. Even though the direction was clear, the plot was anything but an upward march.

The chemical history of the Earth was also anything but an upward march. Once photosynthesis made oxygen and mitochondria used it, a cycle of oxygen-making and oxygen-breaking generated cheap energy for exploring and processing the planet, spreading out energy in sun-driven cycles. The direction of this story was set, toward oxygen and oxidation. But like any long story, the chemical story of Earth's development took twists and turns, encountering obstacles and opportunities, as the Earth oxidized and grew in complexity.

Of these twists, the biggest may be geological rather than biological. In this story, precipitation changed the early Earth. Not "precipitation" meaning rain, but "precipitation" in the chemical sense, as when a solid precipitates inside a test tube. Chemists work and think in the liquid phase, where the "rain" that falls is solid particles, which happen when two atoms discover that they are more stable together as solid than apart as dissolved ions.

For chemists, precipitation is usually a disappointment. Some of the most elegant experiments have been ruined by the chemicals "precipitating out" into a soggy mess. In a sense, it was a disappointment when it happened on the early Earth as well, although this disappointment threatened the continued existence of life.

Chemical precipitation is one of the easiest things to see in past history, because it forms solid puddles of rock. At this point in our story, 2–3 billion years ago, an unusual set of precipitates rained down as red jasper and other iron-oxide rocks, laid down as neat red layers alternating with gray silica. These banded iron formations (BIFs) are evidence of a planet-wide geochemical shift. Complementary methods like uranium isotope patterns back up this story.

These BIFs are found close to the pink Lake Hillier and the stromatolites of Shark Bay in the Pilbara region in northwest Australia, and also in the Iron Range mountains of northeast Minnesota, in northwest Russia, southern Africa, and southern Brazil. All have mining operations built on top of them that harvest the iron-oxygen rocks. These ancient puddles of rust tell us that long ago something pushed together a lot of iron with a lot of oxygen, and the Earth rusted. The drastic geological change that made all these BIFs was also a drastic chemical change.

BIFs formed at two distinct times: most formed 2–3 billion years ago, and some formed 600–800 million years ago (Figure 8.1). The first BIF peak fits with the biological invention of photosynthesis and oxygen-eating metabolism described in the previous chapter. On the chemistry graph on the wall of the Redpath Museum, an

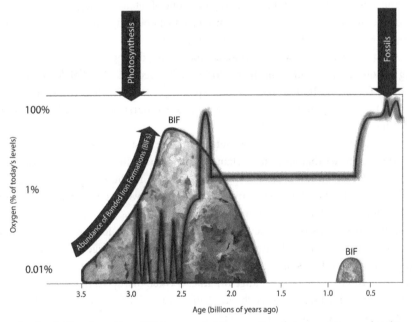

FIG. 8.1 Banded iron formations (BIFs) appeared at the same time that oxygen increased in the atmosphere. The two stages of BIF formation imply two stages of significant oxygen increase, the first following the invention of photosynthesis and the second preceding animal fossils.

BIF data from C. Klein, "Some Precambrian banded iron-formations (BIFs) from around the world: Their age, geologic setting, mineralogy, metamorphism, geochemistry, and origins." 2005. *Amer Mineral*. 90 (10), p. 1473. DOI: 10.2138/am.2005.1871; oxygen data from T. W. Lyons et al., "The rise of oxygen in Earth's early ocean and atmosphere" 2014. *Nature*. 506 (7488), p. 307. DOI: 10.1038/nature13068.

orange line expands at this point to show that BIFs were formed at this time in a huge, multi-million-year burst.

One story unites biology, geology, and chemistry to explain BIFs: manganese-using cyanobacteria produced oxygen, which stuck to iron and formed a rusty solid that precipitated out from the ocean waters. As this happened, other processes put other elements in. Deep-sea vents injected reduced iron into the oceans from below, and microbes turned sunlight from above into sugars and water into oxygen. Geological and biological cycles worked in the same direction as the iron and the oxygen met, forming a useless orange residue, and the cycles continued.

I have friends that purify their well water with this reaction. When they haul up the water, it is rich with iron and low in oxygen. To purify it, all they have to do is leave it out, exposed to the air. Oxygen reacts with the iron and a quarter inch of rusty sediment forms at the bottom of the bucket. The water is scrubbed clear by oxygen. This happened across the Earth when oxygen filled the atmosphere.

BIFs get their stripes of alternating orange iron and gray silica from a cycle that could be seasonal and must be biological. In the lab, a mixture of typical microbes from this era laid down orange iron stripes in the summer and gray silica stripes in the winter, as regularly as if they were knitting a sweater.

The period from 2 to 3 billion years ago was one of intense geological activity, centered around oxygen, iron, and sulfur mixing and combining, laying down brick-like bands around the world. The world was in turmoil.

And then everything went quiet for a billion years, during what is called the "boring billion." It's a period eventful only for its *lack* of geological and biological events, roughly from 2 to 1 billion years ago (to be precise, 1.8 to 0.7 billion years ago). At the end of the "boring billion," we find a second, small burst of orange BIF formation, and the boredom stops with a bang.

Right after the first burst of BIF formation, the first few fossils of complex life appear. It appears that something changed that allowed life to become a little more complex. However, the biggest "explosion" of fossils in the rocks does not appear until later, right after the *second* burst of BIFs. Then the world developed true biological complexity and diversity, in the next chapter.

Those two bands of rust, separated by a billion years, form orange bookends to this chapter's events. The two bursts of BIF formation fit together with our best graphs of oxygen levels in the atmosphere. At the Redpath Museum, a blue oxygen line increases along with the orange BIF line. Oxygen disrupted the world, reacting, rusting, and precipitating. This chapter is the story of that disruption, and how, at first, the oxygen did more harm than good.

HOW OXYGEN ALMOST KILLED US ALL

The time when oxygen increased is such an important event that it has been named several times over. Most scientists call it the Great Oxidation Event or the Great Oxygenation Event, but some call it the "Oxygen Catastrophe."

How can the element we breathe invoke a catastrophe? We breathe oxygen to perform "controlled burns" on the food we eat, breaking it down and atomizing it into carbon dioxide. Uncontrolled, oxygen burns *everything*, ruining the organization of an organism. Introducing oxygen into an unsuspecting world would indeed be a catastrophe.

Even for us organisms that have had 3 billion years to get used to it, a little oxygen in the wrong place can be dangerous, especially as reactive oxygen species (ROS). When ROS attack a protein, they don't have to hit the active site, but can break and loosen parts distant to the active site and still deactivate the protein. When blood cells burst, the leaked heme iron makes ROS, which eat away at the rest of your body.

Parasites have picked up the power of ROS, too. The tropical parasite *Leishmania* can only cause Leishmaniasis after it takes in iron and uses it to generate ROS. This means that too much iron from iron supplements might give microbial invaders ROS-generating chemistry to use against *us*.

Several diseases are caused by oxygen damage from the patient's own mitochondria, which are hot spots for ROS generation. In one of these diseases, Leber's hereditary optic neuropathy (LHON), the mitochondrial proteins that carry electrons and pump protons are broken. The first cells to die are the energy-burning optic nerve cells. The cells die because their broken mitochondria leak too many ROS.

Oxygen can even turn your own molecules against you. This happens in age-related macular degeneration, when ROS (possibly from active nerves near the eyes) attack the cell membranes. These oxygens add onto the carbon chains indiscriminately and make dangerous, sticky aldehydes that gum up nearby proteins and cause macular degeneration.

Heart diseases can be caused by similar oxidation of carbon chains. The heart uses so much energy and has so many mitochondria that it needs state-of-the-art defenses against the very oxygen that it pumps through the body. Proteins called perilipins act like sheepdogs and round up rogue oxygenated chains into a little droplet of fat, keeping them away from everything else.

Oxygen-related damage must be important, because every living thing spends expensive amounts of energy to defend against it. Every bacterium has stockpiled techniques that neutralize ROS. They seem to be very nervous about oxygen. From reading their genomes, we can see how microbes have been sharing notes for oxygen defenses through gene flow and evolution.

Some microbes build a sticky sugar shield called alginate to slow oxygen's penetration. Others decorate sensitive proteins with sacrificial atoms, often oxygen-sensitive sulfur or iron-sulfur chunks, which intercept ROS before they hit the rest of the enzyme. Still others burn oxygen in bursts, turning its chemical energy into flashes of light energy. (One prominent theory suggests that such bioluminescence developed to deliberately waste oxygen by setting it alight before it could damage the microbe. Later, animals and fungi set up a partnership with those bacteria to signal other animals, as if they were picking up a flashlight.)

Three billion years ago, microbes with thick sugar shields or sacrificial sulfurs would thrive in the presence of oxygen-generating cyanobacteria. As oxygen increased, those that didn't develop such defenses would have to retreat to anoxic environments or die. Today, many of the bacteria in oxygen-poor environments have simple oxygen sensors based on iron and/or sulfur. Oxygen in the air activates the sensor and tells the bacteria it's time to move out.

As oxygen was produced, it filled the air quickly, but because it does not dissolve well in water, it only slowly filled the oceans. The oceans have hundreds of times the mass of the atmosphere and so there is a lot more stuff to saturate with O_2. Organisms without defenses moved deeper and deeper in the oceans as oxygen pushed them out. In their versions of history, the invention of oxygen was not a good thing (but it's worth pointing out that they never evolved the complexity to record versions of history).

As the oxygen filled the planet from the top down, the oceans became patchy. Three major areas developed, each dominated by its own element, which was mutually exclusive with the other two:

1. The sunlit surface was oxygen-rich (but iron-poor).
2. The dark depths of the ocean were iron-rich (but oxygen-poor).
3. In the middle, some sulfur-rich areas persisted, with neither oxygen nor iron.

It's not entirely clear yet whether the sub-surface ocean was mostly sulfur (as was long believed) or mostly iron (as evidence now suggests), but there was some of each. The ocean was striped and stratified like a parfait, or like a BIF. Mixing would take more than a billion years. Even now, after long mixing, the cold deep waters cannot hold much oxygen, and are largely devoid of complex life.

As oxygen moved through the gas and liquid parts of the Earth, it encountered the solid part of the Earth and caused geological change there. Oxygen broke up exposed rock so that it dissolved in the rain and its elements flowed through the sea.

The step up in oxygen was followed by a swift step down in temperature, an event called "Snowball Earth" in which ice covered most of the Earth. The mechanism is not entirely worked out, but each Great Oxidation Event was accompanied by a Great Freezing, and each time, the planet came disturbingly close to becoming a lifeless chunk of icy rock. Oxygen turned Eden into Hoth.

MISSING BRANCHES FROM THE TREE OF LIFE

It's bad enough that oxygen attacked life's structures and almost froze the planet, but it gets worse. Oxygen indirectly harmed life by stealing away crucial metals and molecules. The most crucial of these was iron. Oxygen strips iron of electrons, changing it from the plus-two form to the stickier plus-three form through a process called oxidation. Together, oxygen and plus-three iron form tight, solid bonds that are good for little else beyond sinking to the bottom of the ocean. Oxygen removed the most important metal from the liquid oceans where life lived. (Plus-two manganese was also diminished, but demand was lower and the new forms of manganese more useful.)

We can see where proteins shifted away from using old forms of metal toward other forms. Some old DNA-copying proteins use manganese, while newer DNA-copying proteins look just the same, except that their manganese-binding site has been papered over and converted to a new use, like parents remodeling their teenager's room after she leaves for college. Also, the pathways you use to make oxygen-sensitive iron-sulfur cluster proteins incorporated extra sulfur-moving proteins to mitigate oxygen's threat.

In tough economic times, sometimes a set of proteins could not be streamlined and would have to be let go. One study of iron-poor parts of the Indian Ocean found that many microbes have jettisoned their iron-dependent genes because they weren't worth the trouble. Likewise, at this time in prehistory, genes were removed from the tree of life.

If the environment changes too quickly, there may not be time to innovate. One study saw this by watching bacteria respond to a deadly antibiotic. Too little antibiotic and they didn't change at all, but too much antibiotic and they all died. A moderate, slow threat allowed them to alter their genes effectively and find the best possible biochemical solution.

We are the result of a similar moderate path. Too much oxygen may have destroyed life before evolution had the chance to defend against oxygen's toxicity and find ways to use its chemical potential. We may even see such a pattern in the rocks in Canfield's oxygen spike 2.1 billion years ago that crashed back down into the low oxygen levels of the "boring billion." This moderation shows up today when scientists try to make propane using bacteria: a moderate amount of oxygen gives the bacteria energy to make more propane, but too much oxygen and the bacteria start to die.

Such moderate margins would persist in a stratified ocean with iron, oxygen, and sulfur layers. A microbe could live on the border between the oxygen layer and the iron layer and could have energy from oxygen on one side, and iron to channel that energy on the other, gaining the best of both chemical worlds.

THE PROS AND CONS OF WORKING WITH NITROGEN

Oxygen is a transformative element, and the first thing it transformed was the air. First oxygen reacted with the hydrogen-rich fuels, the favorite food of ancient organisms. CH_4 (methane), H_2S (hydrogen sulfide), and NH_3 (ammonia) were stripped of their hydrogens and replaced with more stable bonds to oxygen (or atoms like oxygen). Hs became Os: CH_4 (methane) become CO_2, H_2S (sulfide) became SO_4^{-2} (sulfate), and NH_3 (ammonia) become N_2 (nitrogen gas) or NO_2^-/NO_3^{-2} (nitrite/nitrate). David and Alm's genomic analysis in Chapter 6 found these exact patterns, in that older enzymes use hydrogenated foods and newer ones use oxidized food.

As it reacted with the planet, oxygen took methane from the mouths of methanogens. Oxygen destroyed the food source of countless other microbes, leaving them to adapt or starve. Some retreated to oxygen-poor nooks and crannies. Others adapted to eat the new, oxygenated food, but this wasn't easy either, because the new oxidized foods were more stable and had bonds that were harder to break.

To this day, oxygen-using microbes have to use more energy to build the same structures as their oxygen-avoiding cousins. Despite the fact that the oxygen gives them more energy, it doesn't give them the materials to build with. They have to use up that extra energy to reduce the materials to their old, reduced forms. For many microbes, this can result in a net loss of energy, as they must work harder to maintain the same microbial lifestyle.

Of the three main sources of food (carbon, sulfur, or nitrogen), nitrogen has an advantage at this point. Carbon is abundant, but CO_2 is so stable it's hard to squeeze energy out of it. Sulfate is the opposite, being useful for sulfate-reducing microbes, but not abundant enough. Nitrogen is the happy medium where nitrogen-oxygen foods would be both abundant and energetic enough for a microbe. So a community of organisms grew up cycling nitrogen, changing it back and forth between its oxygen-rich and hydrogen-rich forms for energy.

Nitrogen also has an advantage here because nitrogen and oxygen, being neighbors on the periodic table, can combine in multiple ways. Some forms dissolve in water, others float in the air, and all have different stabilities and can absorb or release energy when transformed or recombined in a sort of "nitrogen math." Some microbes make N_2 from N_2O, some make NO_3^- from NH_4^+, and some can even make O_2 and N_2 from NO.

When these complex systems fall out of alignment, disorder or disease is the result. Deep in the human gut, where oxygen is scarce, some microbes "breathe" with nitrogen-oxygen compounds—that is, they get energy by passing electrons to NO_3^{-2} (nitrate) instead of O_2 (oxygen). This has similar drawbacks to breathing oxygen, because it comes with the side effect of electron-imbalanced reactive *nitrogen* species, or RNS, instead of ROS.

Excessive RNS are tangled up in the microbial imbalances and overactive immune signaling that leads to colitis, and heart attacks generate RNS that damage proteins. RNS toxicity has forced the evolution of chemical shields against RNS. It's like the ROS story, but with nitrogen—yet even the nitrogen system pales next to the energy that oxygen itself can provide, and it is only used in the absence of oxygen.

Energy is also stored in manganese and can be unlocked by switching manganese electrons around. As the environment oxidized, it became easier to switch manganese to its oxidized plus-four form, and enterprising microbes could make energy from cycling manganese between plus-two and plus-four, possibly with sunlight. The manganese wheel on Figure 6.3 in Chapter 6 became more active in the presence of oxygen.

The energy that manganese provides is about on the scale of nitrogen metabolism: good enough for a microbe, but not much competition to oxygen's chemical power. Both nitrogen and manganese have more power than the reduced foods that came before, but neither nitrogen nor manganese has enough chemical power to drive complex life, as shown in Figure 8.2.

The biggest problem with nitrogen metabolism is not RNS, but the N_2 nitrogen gas in the middle of all the reactions, mixed up with all those N-Os and N-Hs. Some reactions lead to this molecule, and the ones that do are practically dead ends. N_2 is

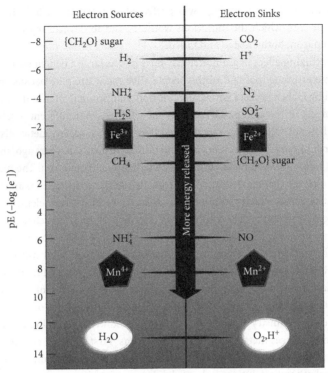

FIG. 8.2 Nitrogen cycles and manganese cycles release less energy than oxygen reactions. The energy in a transition from the electron source on the left to the electron sink on the right is depicted as a version of redox potential called pE.
Data from UC Davis ChemWiki, "Electrochemistry 5: Applications of the Nernst Equation." chemwiki.ucdavis.edu.

joined with an extra-strong triple bond. Double bonds are bad enough: CO_2's double bonds make it hard enough to use as an energy source, and even those are easier to break than nitrogen's.

To use N_2 in an energetic cycle, its triple bond must be cracked open like a tough nut. Eventually, nitrogen-using microbes evolved a nitrogen nutcracker. This difficult chemistry wasn't worth it when hydrogen-containing gases were abundant, but as those disappeared, and before there was enough oxygen in the air, N_2-breaking chemistry became worthwhile.

This wouldn't be the last time the problem of N_2's triple bond would be solved. It would recur in human history, and reshape the twentieth century, although not necessarily for the better. A scientist named Fritz Haber heated and pressurized nitrogen and hydrogen under high-flow conditions with an iron catalyst. Through this display of brute chemical force, Haber forced the triple bond of N_2 apart and made NH_3 (ammonia). Using Haber's reaction, German factories made tons of ammonia from nitrogen.

Since ammonia is a fertilizer, the Haber process cracks open nitrogen for plants, supplying them with an easy-to-use form of this important CHON element. This iron-based chemistry provided the equivalent of plowshares. Then it provided the equivalent of swords. Ammonia is easily reacted into energy-laden nitrates for explosives and munitions. This technology was perfected in Germany a half-decade before World War I, in which the munitions would be put to devastating use. Ammonia, made from iron, both created and killed.

The fertilizing effect of ammonia may have had as large a benefit as its destructive side. Some estimate that half of the people alive today are here because of the food provided by Haber-process ammonia. It is not clear how calculus and chemistry add up the effects of war and peace (and anyone who tells you different is selling something). History's web is more complicated than the most arcane metabolism.

In Earth history 2 billion years ago and human history 100 years ago, the center of nitrogen catalysis was a metal, even possibly the *same* metal, iron. The multibillion-year-old enzyme that makes ammonia from nitrogen and hydrogen is an iron protein called nitrogenase, which contains iron, sulfur, and a brand-new element.

Nitrogenase is expensive, requiring the sacrifice of 16 ATPs, channeling all that energy into breaking the one nitrogen triple bond so that hydrogen can be added. The trouble is worthwhile—a study in Panama found that trees with nitrogenases grew much better, accumulating carbon nine times faster than trees without nitrogenases. Nitrogenase involves a complex assemblage of genes, but the genes for it are arranged in a well-ordered genetic cartridge that could be passed quickly from organism to organism. Once this chemistry developed, it was passed around the planet like a viral hit video.

Nitrogenase uses metals to accomplish its aims. First it moves electrons to nitrogen through a chain of iron-sulfur centers pointed at a larger iron-sulfur complex. This complex has a catalytic metal—molybdenum, vanadium, or iron—next to where the nitrogen bond is cracked. In many of these high-powered chemical complexes, molybdenum (Mo) is grafted into an iron (Fe)-sulfur network, so the whole complex (Co) is called FeMoCo (Figure 8.3).

FeMoCo has attracted a lot of attention in labs because making ammonia is so important to national economies and world histories. One key point is familiar: the metal is the center of the action. If FeMoCo is removed from its enzyme, it keeps splitting nitrogen by itself, although oxygen soon destroys it. Chemists have made several metal complexes that mimic the shape of FeMoCo, some with molybdenum, some with iron, some with other central metals, all working to various degrees. The most impressive at press time may be a special arrangement of molybdenum, iron, and sulfur in a "chalcogel" that can turn nitrogen to ammonia at room temperature.

The biggest problem is not breaking nitrogen's triple bond, but keeping pushy, reactive O_2 oxygen away. Even a small amount of O_2 will gum up the works. On the early Earth when there was less oxygen to avoid, nitrogenase was easier to make.

MOLYBDENUM'S UNIQUE PLACE IN HISTORY

Nitrogenase was probably the first major use of molybdenum. Molybdenum is oxygen-sensitive and is more available in the presence of oxygen, so it probably wasn't widely available until after the Great Oxidation Event. Nitrogenases follow this logic and are dated to 2–2.5 billion years ago by both Caetano-Anollés's group and by David and Alm. (Some recent rock evidence shows molybdenum-based nitrogen processing as early as 3.2 billion years ago, so some may have been available from periodic wisps of oxygen.)

Iron may have been used more at first, but nitrogenases would have switched to the new element whenever possible because molybdenum is so good at nitrogen-cracking chemistry. This special ability of molybdenum comes from its unusual place on the periodic table, one level below the other elements. Because it is so much bigger, it is less abundant and harder to find, but ultimately worth it. Its large radius contains

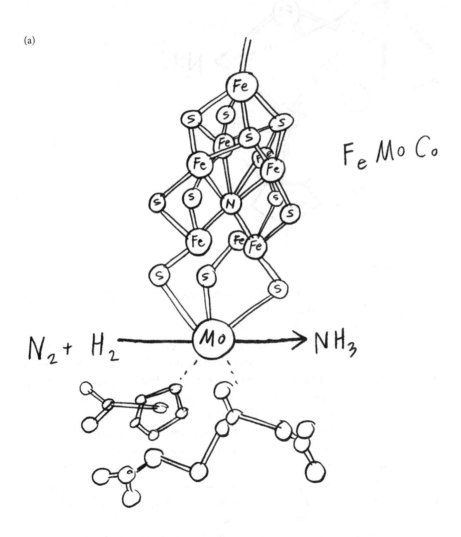

FIG. 8.3 Molybdenum structures and nitrogen chemistry. *Above*: the Fe-Mo-Co complex that makes ammonia from nitrogen. *Next page*: an iron-sulfur complex that also makes ammonia. O_2 (oxygen) inactivates these complexes by substituting for the less reactive N_2 (nitrogen).
Y. Li et al., "Ammonia formation by a thiolate-bridged diiron amide complex as a nitrogenase mimic." 2013. *Nature Chem.* 5, p. 320. DOI: 10.1038/nchem.1594).

(b)

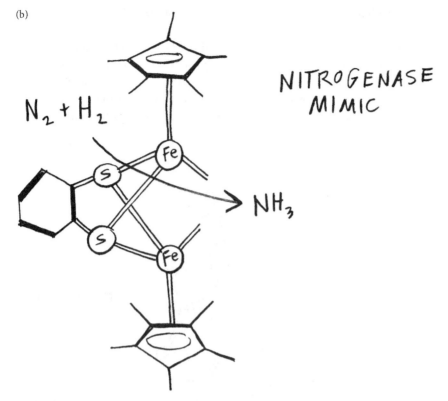

$N_2 + H_2$

NITROGENASE MIMIC

NH$_3$

FIG. 8.3 (Continued)

layers of electrons that are easy to peel off, so it can reach a *plus-six* state. No other element available to life today can do that.

Molybdenum is not only powerful but also versatile. A small chemical change makes it perform a completely different reaction. In 2011, a group of scientists made a hole in the nitrogenase protein near the molybdenum, the right size for fitting a CO (carbon monoxide). This changed the entire enzyme. Instead of adding hydrogen to N$_2$, the protein added it to CO repeatedly, which made the carbons stick together in short chains. A small change turned nitrogenase into a carbon-knitting enzyme.

Other enzymes hold molybdenum in place by a chemical framework called pterin that directs its chemical power in ways that other elements cannot match. Twisting the pterin framework shifts the electron-pushing power of molybdenum, so its chemistry can be tuned by the shape of the protein around it. Molybdenum-pterin centers are found breaking nitrogen-oxygen bonds, removing oxygens from sulfate or nitrate for energy (so it is often found in sulfate-reducing bacteria), and even transferring single oxygen atoms. Oxygen and molybdenum have a special chemical relationship.

AN ARROW THROUGH THE OCEAN

All of these observations can be tied together by an arrow of increasing oxidation over time. This is a consequence of the fact that oxygen is the chemical opposite of electrons, and all life hoards electrons to build complicated structures, so oxygen is inevitably ejected into the environment.

As oxygen entered the atmosphere and ocean, it did what oxygen does, thanks to its position near a corner of the periodic table: it reacted. It added oxygens to some things and pulled electrons away from other things. Non-metals were converted from the hydrogen form to the oxygen form, as the foods available on the planet changed from hydrogen-rich to oxygen-rich. Some metals, on the other hand, can hold different numbers of electrons, so oxygen changed how many electrons those metals held. As oxygen increased, metals were stripped of their electrons ("oxidized") and their positive charge increased as the negative electrons slipped away.

Some elements react more readily with oxygen than others. The remarkable observation is that the general pattern of reactivity we see in the lab matches the general pattern of element use that we see in genomes, back through time. Oxygen is the chemical key to this scheme. As oxygen was released over time, the rest of the planet was pulled along after it.

Generations of chemists and even alchemists have noticed how different chemicals react in water. Now we have a single number for the tendency of each chemical to react, called its reduction-oxidation potential. This is a mouthful even for chemists, so we call it "redox potential."

R. J. P. Williams then noticed that that these numbers told a story, in that the electron-pushing power of each element matches the historical pattern of when they were used. The history of genes by Alm, Caetano-Anollés, Dupont, and colleagues tells the same story in finer detail. Primitive microbes use the hydrogen-rich elements and the reduced metals with lower redox potentials. Our modern world of complex animals uses oxygen-rich elements, especially oxygen itself, and oxidized metals with higher redox potentials. This gives the chemistry of biological evolution a particular direction, as life followed a chemical arrow (Figure 8.4).

As the planet oxygenated, it followed a chemical sequence moving from left to right along this arrow. The Earth's redox potential increased over time. The lines on the arrow show the chemical redox potential of each element, which approximates the time when each element switched from a hydrogenated/reduced form to an oxygenated/oxidized form. As the planet crossed each line, the element changed its form from the reduced form on the left to the oxidized form on the right—and that element's chemical abilities changed, too.

Figure 8.4 is read from left to right, like a book. For example, NH_3 (ammonia) sits far to the left. When oxygen increased, ammonia became nitrogen. Molybdenum is nearby, so its plus-six form was unlocked at the same time, perhaps a little later. This would predict that most early nitrogenases should be based on iron or heme.

Vanadium sits even farther to the right, so its highly charged plus-four form would not be available until *after* molybdenum was oxidized. This predicts that vanadium nitrogenases would be more recent. Research by the lab of John Peters agrees with that sequence.

This chemical timeline also lets us see what happened next, after molybdenum gained its plus-six power. In the middle of this sequence, iron switched from plus-two

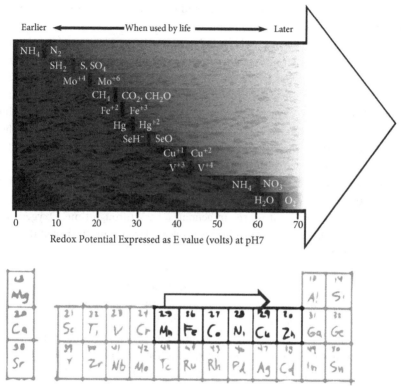

FIG. 8.4 The arrow through the ocean. *Top:* The redox potential of different reduction-oxidation pairs corresponds to the order in which reduced and oxidized forms were encountered and used by life. Elements with lower redox potentials were oxidized earlier than elements with higher redox potentials. For example, ammonia/nitrogen, sulfide, and molybdenum (IV) are more easily oxidized than copper (I) and vanadium (III). *Bottom:* Because iron (III) readily precipitates out of solution as iron oxide, over time iron concentrations decreased. Because copper (II) compounds are generally more soluble than copper (I), oxidation increased overall copper availability. This trend can be summarized by an arrow pointing away from iron and toward copper.

Data adapted from Figure 1.14, p. 28, R. J. P. Williams and J. J. R. Frausto da Silva, *The Chemistry of Evolution: The Development of Our Ecosystem,* 2006, Elsevier.

to plus-three, combined with oxygen, and rusted out of the oceans. Both iron and sulfur underwent transitions before the halfway point of the arrow and would have been easier to use in the pre-oxygen era. They are more characteristic of the previous chapter than the next.

The sheer quantity of iron and sulfur (as members of the Big Six of geological elements) means that most of the oxygen produced by early life would have reacted with the iron and sulfur in the Earth itself. Accordingly, oxygen levels may have been knocked back down after each increase. Canfield's early oxygen spike may have been stopped when the new released oxygen reacted with abundant iron and sulfur. It could have taken a billion years to react with such a huge quantity of iron and sulfur—and this is a ready chemical explanation for why oxygen stayed low during the "boring billion" era.

One chemical experiment that constantly tries students' patience is a chemical titration. You titrate a solution by slowly and steadily dripping acid into base, seeing no change, maybe dozing off a little, until suddenly the color changes at the point where acid = base. (Then you blinked and overshot the mark so you had to repeat the experiment. I've been there.) On the early Earth, photosynthesizing microbes titrated the planet with oxygen, but, as with student titrations, it took a while. Because of all the iron and sulfur absorbing the oxidation and resisting the change, it took a billion years for the tiny microbes to add enough oxygen to change the world.

Manganese concentrations edged down, but most important, iron shifted and rusted wherever it contacted oxygen. Nickel and cobalt react less with by oxygen, but their favorite molecules, methane and hydrogen, were oxidized away, making them chemically obsolete even in areas that avoided the nickel famine at the end of Chapter 6.

Iron's oxidation and removal was a major chemical "step back." Oxygen's reactions after this point began to "step forward" and produce new elements. Molybdenum was the early first-fruit. Farther to the right, selenium and copper dissolve better and can do more chemistry in the oxidized forms. These were unlocked later, when oxygen started to take their electrons away. Once iron and sulfur were converted, extra free oxygen collected in the atmosphere, available for use by life. Given the power of mitochondria, oxygen itself might be the most important newly available chemical of all.

Oxygen turned copper into a more useful plus-two form that released it from the rocks. Its neighbor in the periodic table, zinc, was used more and more as well. (Note, however, that zinc does not fit neatly into this story, because its concentration was dependent on the patchy, local sulfur conditions.) The two most important elements are the iron and copper, at either ends of the arrow on the periodic table. Iron dropped out of the ocean and copper dropped in.

Overall, oxygen and the reactions that it caused gave the planet a chemical shift. This made for an arrow through the ocean—a rightward move in the first row of the periodic table in the ocean's metal concentrations, from less iron to more copper—a shift that mirrored the arrow through the sky from hydrogen to oxygen in Chapter 4.

NEW ROCKS AND THE END OF THE STROMATOLITES

The Earth changed with life, in a process geologist Robert Hazen describes as co-evolution. Earlier in this story, Hazen described how *water* formed a thousand different minerals on the Earth's surface. At this point *oxygen* extended the process of mineral evolution, multiplying the different minerals even more: "We suggest that fully two-thirds of the approximately forty-five hundred known mineral species could not have formed prior to the GOE, and that most of Earth's rich mineral diversity probably could not occur on a non-living world" (p. 177).

Hazen goes on to list some of the new colors made possible in the Earth by oxygen: "semiprecious turquoise, deep blue azurite, and brilliant green malachite are unambiguous signs of life." Like a prism refracting light, the Earth transformed oxygen into new chemicals and bright colors.

As oxygen from biology began to affect the geology of the world through chemistry, oxygen-bound rocks were created. About 90% of the world's known iron reserves are laid down in oxygen-rich BIFs. Manganese, copper, nickel, and uranium ores are oxides that appear along with or right after the Great Oxidation Event (GOE). The mining industry makes money by finding metal ores, so it has learned to retrace the path of Earth's oxidation and to look for the places where oxygen and water made metal oxides.

Oxygen's reactivity also increased weathering and other geological cycles, mixing more chemicals into the ocean and forming new sediments on the ocean floor. Sulfate in particular dissolved from rocks and entered the environment for hungry sulfate-reducing bacteria. In fact, some sulfate-reducing bacteria appear not to have evolved at all since the middle of the "boring billion," 1.8 billion years ago. There, a very stable geological environment may have made very stable biological forms. The same general pattern probably occurred during the "boring billion," when geological stability led to biological stability.

Copper was also brought out into the ocean by water and oxygen working together. Hazen notes in "Mineral Evolution" that 256 of 321 copper minerals were formed by the action of water, which means they were laid down by rain, erosion, and sedimentation after copper was released by oxygen.

The chemical changes also brought previous forms of life to an end, such as Chapter 6's stromatolites. Two billion years ago was the heyday of the stromatolites, and then they were destroyed by other more diverse species. One protozoan called forminifera has long, flexible pseudopods that claw into the stromatolite mats, taking the mat's resources for itself. Those pseudopods were probably made possible by the invention of cholesterol, which was made possible by an oxidized environment, which was produced by stromatolites. Stromatolites carry the seeds of their own preemption, a Greek tragedy if I ever saw one.

The weathering and oxidation at the surface of the Earth could have changed the color of the planet, as iron oxide coated the continent(s) in an echo of the BIF formation in the oceans. The Earth turned red as blood, as if irritated and inflamed by the changes taking place. It wasn't just the oxygen that was a problem, because many of the new metals were toxic as well. Before things would get better, they would get even worse.

WHEN VENUS AND MERCURY ATTACKED

The blue oceans changed as dramatically as the red Earth, as two new metals dissolved in the waters: mercury and copper. In Greco-Roman myth, copper was associated with the planet and god Venus (because both the metal and the goddess came from Cyprus), and mercury was associated with the planet and god Mercury. Both metals lie to the right of iron on the arrow of oxidation, indicating that they were released by oxygen at this point in the story, and they wreaked havoc on the biosphere by sticking tightly to life's favorite elements.

Mercury makes ROS that can damage DNA, and it sticks to sulfur and selenium, blocking their use. Bacteria can unstick it by reversing what oxygen did to release it—they add electrons to it, reducing its charge from plus-two to zero. (One of the reasons that life collects electrons is to help cleanse itself from toxic chemicals.) The enzyme that adds electrons to mercury is part of a larger cassette of genes that work together and are passed around among bacteria like a note in the back of a high school class.

Other toxic metals are solved in a similar fashion: bind, add electrons, release, repeat. One chromium-cleansing enzyme is proactive, intercepting the chromium *outside* the cell. Sometimes just phosphate will do the same job. If yeast cells detect the heavy metal cerium in the environment, they eject phosphate and make a cerium phosphate crust around the cell, while the insides stay fluid.

Copper plus-two is about as sticky as toxic plus-three ions like aluminum. Earlier in Earth's history, cells ejected aluminum because it was too sticky. Copper also sticks to everything and turns the fluid intricacy of the cell's interior into a randomly stuck-together mess, unless the cell packages it with care.

One of the first pesticides works because copper is sticky. This is the Bordeaux mixture, made by mixing bright blue copper sulfate with lime (the chemical, not the fruit). Centuries ago it was sprayed on grapes near roads to color them blue so passers-by wouldn't eat them. Then French scientists noticed that the spray killed fungus without harming the grapes. The active ingredient is the sticky copper, which tangles up fungus proteins and keeps the fungus from spreading. A billion and a half years ago, this "pesticide" was released into the environment as oxygen reacted with the planet, and much life reacted like the fungi and died from it.

The good news is that copper's chemical power—its extreme stickiness—is also its Achilles' heel. Something sticky will keep copper out of the way. Today, plant roots in urban areas have surfaces like flypaper for copper, which stick to all the excess copper from the polluted surroundings. Something like that must have evolved long ago. A microbe could handle an influx of sticky copper by ejecting copper-fouled proteins. Over time, it could develop a class of sacrificial proteins whose whole purpose would be to stick to copper, and these proteins would be moving throughout the cell carrying coppers. This would form the basis for new steps forward, when the potential of copper was unlocked.

FROM CATASTROPHE TO EUCATASTROPHE

What would happen if one of these copper-proteins had a sugar-binding site right next to the copper? The copper would react with the sugar, and with a few tweaks, what was a copper-carrying protein would become a copper enzyme. The cell would have a new reaction based on copper to help it survive, built on copper's chemical power. The cell eventually decided that this copper protein was worth keeping, and it gained a new enzyme. An event that began as "one step back" (toxic copper) was turned around to become "two steps forward" (a new copper enzyme).

J. R. R. Tolkien worked with stories of the literary kind rather than the scientific kind, but he has a term that fits for both. Tolkien called this kind of happy turnaround a "eucatastrophe," adding the Greek prefix for "good" onto the "overturning" that is a catastrophe. The hobbits seeing the eagles coming to the rescue is a eucatastrophe; so is the sleeping princess awakened with a kiss. In this story, the catastrophe of losing iron and releasing toxic copper was followed by the eucatastrophe of oxygen-based energy and copper-based catalysis.

The new metals provided new chemistries that allowed for new possibilities. Just encountering a new metal can evolve new protein shapes. Also, old protein shapes can find new uses by picking up a new active metal center, even accidentally, and even if they are a little broken at the time.

Now that scientists can routinely read all the genes in an organism, they've found many broken, dead genes sitting between the live ones. In these "pseudogenes," the DNA looks erased or overwritten (similar to what Tolkien studied in old manuscripts of *Beowulf*). Pseudogenes are missing a crucial piece so that the enzyme cannot complete its own job.

But, to quote an article title about this, "'Dead' Enzymes Show Signs of Life." Pseudoenzymes no longer carry out their originally intended reaction, but they can do other things, from binding DNA to blocking other enzymes. In the end, the second function of the "dead" enzyme may become as important as its original function. The enzyme has experienced a eucatastrophe.

Life responds to periods of extreme stress by repurposing dead or broken enzymes. For example, stressed fruit flies change their genes in response to stress, but they only develop completely new enzymes about 1% of the time. More often, they *repurpose* old enzymes to serve new functions.

Sea urchins do the same thing. Acid stresses sea urchins, and they double their rate of evolution in response, first changing their outer membranes to compensate for the changing environment. Stress provokes evolution.

Above the genetic level and at the protein level, cells have an entire array of proteins that emerge under stressful conditions. These proteins are called "heat shock proteins" because the cell makes them when it gets too hot and its proteins start to unravel, revealing their oily insides. Heat shock proteins fill the cell and hold the big proteins apart from each other, like when two angry people lunge at each other but peacemakers step between them. Heat shock proteins have a calming effect on the rest of the cell and allow life to continue, even when its proteins are falling apart.

Susan Lindquist and colleagues found that heat shock proteins help evolution. When proteins first evolve, they are often less stable and a little broken, with their oily insides showing. A cell that turns on its heat shock proteins at this point gives the broken proteins room to be broken, and to change more. Lindquist calls heat shock proteins "evolutionary capacitors" because they allow for more rapid change in times of stress. Heat shock proteins gave fish swimming underground for generations the ability to evolve a drastic change and lose their eyes entirely. (Come to think of it, Tolkien had something like this happen to Gollum.)

Over and over again, evolution follows the same pattern: respond to a toxin or stress by binding the problem *deliberately*, before it binds something else *randomly*. Sugars are especially good for this purpose because there are a lot of them inside the cell. Worms, bacteria, and fungi detox toxins by bonding them to sugar molecules.

Sometimes a fungus will bond a toxin to sugar, you will eat it and be OK, but then the hungry bacteria in your gut eat the sugar and release the toxin. The bacteria don't know they're poisoning you—they're just hungry.

A binding protein like a heat shock protein can readily evolve into an enzyme. In one case, a long protein that binds fatty-acid carbon chains was the starting point, with a long, oily stripe for binding the carbon chain. The protein added reactive chemical groups next to the stripe and massaged its shape. After this makeover, it stuck to a different (but still oily) molecule decorated with a few extra oxygens, and pressed it together into a ring. The binding protein became a ring-making enzyme.

This new ring molecule reacted with ROS, so it probably was an ROS scavenger at first. But eventually, its unique shape was sent to other cells as a signal. Now plants make a whole class of molecules based on this ring structure for development and defense. This new ring molecule first protected and later signaled.

This happened with non-metal elements as well. For example, sulfur carrier proteins were transformed to become proteins that built sulfurous sugars. Like the accidental enzyme from Chapter 2, proteins that stick to something are very close to becoming enzymes that stick to something and then change it.

COPPER LOVES ELECTRONS, OXYGEN, AND MAKING BONDS

Oxygen introduced the chemistry of copper into this evolutionary fluidity. A home experiment shows how copper has a special relationship with oxygen. For this experiment, you need a penny minted before 1982 (before that, pennies were real copper; now, they're mostly zinc). Heat the penny up, and dangle it on a wire over a few tablespoons of nail polish remover, which contains acetone. In a few seconds, the penny will glow as red as molten iron. But there's little danger so long as you don't catch the acetone on fire or touch the penny. Once the penny cools off it will be as good as new, even cleaned up by the ordeal.

What happens with the penny is the same combustion reaction that happens when logs burn or when mitochondria break down sugar: carbon combines with oxygen and releases energy as red light and heat. The penny's copper surface sticks to oxygen, bringing acetone and oxygen vapors together more effectively.

A single atom held in place by a protein catalyzes in the same way, by bring reactants together through copper's natural stickiness, especially to oxygen. This kind of catalysis does not work as well with zinc or other common metals, and became available when oxygen released copper.

Copper's special abilities were soon incorporated into enzymes and microbes. For example, a group of scientists looking at human skin microbes found a surprising number that break down ammonia by using copper to combine it with oxygen. No one's sure why your skin needs so many—maybe they protect your skin from ammonia fumes or participate in nitrogen cycles.

Copper's good at moving electrons and at binding oxygen, so it's great doing both at once. Copper reacts directly with the electron-oxygen combinations that are reactive oxygen species (ROS), grabbing onto them without losing hold of stray electrons, so that it can avoid iron's risky ROS-producing chemistry. Copper is safer because it is stickier. Older organisms clean up ROS with iron and manganese, but animals do the same chemistry more safely with copper and zinc.

The champion enzymes of oxygen chemistry may be laccases, which use *three* coppers to make and break oxygen bonds. The laccase coppers push electrons so effectively that they don't need to hold finely tuned shapes to work. They react with anything in the general area. As a result, it's difficult to classify laccases like other enzymes because they adopt so many shapes. The general reactivity of laccases makes them highly designable. Even if the protein is crudely shaped, it still works, because in the end, it's all about the copper ions. "Close enough" works for horseshoes, hand grenades, and copper enzymes.

TO BREAK, BLOW, BURN, AND MAKE NEW

Chemically, anything that is good at making bonds is also good at breaking them. Copper is no exception, especially with its companion molecule, oxygen. Breaking and making are tied together in John Donne's sonnets, one of which gave its title to Camille Paglia's enjoyable book about poetry, *Break Blow Burn*: ". . . to break, blow, burn, and make me new." Breaking, blowing, and burning bonds are themes of this historical period (meaning 1.5 billion years ago, not 400). Copper was versatile enough to accomplish all four of Donne's verbs.

Laccases are part of a larger group called multicopper oxidases. Many of these are outside the cell in the oxidized environment of the blood or ocean and have a wide range of jobs based on the power of copper and oxygen. For one thing, copper is good at breaking the carbon nets that cells weave to protect themselves. Plants protect themselves with one such net called lignin. Multiple coppers in a laccase can slice up the lignin net and break down the plant's protection. A copper enzyme can also break down chitin, the long, tough sugar in lobster and beetle carapices. Biofuels researchers are following nature's lead by investigating the power of copper catalysts that "break" plants' tough carbon structures for fuel.

The same multicopper oxidase framework is found in human blood, in a protein that pulls electrons off iron using the electron-pulling power of copper and oxygen. Perhaps we can think of it as "blowing" the electrons away with oxygen.

In growing plants, multicopper oxidases "make new" things by adding oxygen onto vitamin C as part of a cycle that expands the cell walls. Copper can add oxygen onto CHON frameworks with ease, turning old molecules into new signals with oxygen, much like iron but with fewer ROS-related side effects. In this way, copper is iron 2.0.

Growth is not always good. One study found that copper makes tumors grow because it helps mitochondria burn sugars—so if copper can break, blow, and make, it can "burn" as well. Also, copper's "breaking" power is active in cancer metastasis, when copper proteins cut away the ties outside a cancer cell that hold it in place, allowing it to colonize a distant site in the body. But then again, this can be used for good: therapies that remove copper may starve the cancer of the copper it craves.

Most of the mitochondrion is built from iron, sulfur, CHON structures, all from previous chapters, but one final, crucial component where the electron road meets oxygen is copper. Copper lets mitochondria live up to their potential, making them "burn" all the brighter, like a hot penny over acetone.

Yet another copper enzyme, called polyphenol oxidase, blows, burns, breaks, and makes, as it catalyzes a familiar reaction and yet hides a mystery. Polyphenol oxidase is what makes cut apples and roasted tea leaves turn dark brown, and it uses copper to do so. The copper brings oxygen together with "phenol" carbon ring structures, adding oxygens on multiple times and linking rings together, so that the phenols quickly tangle into a brown, sticky mess (or in scientific terms, they *poly*merize into *poly*phenols).

This explains the "what" but not the "why." Plants and fungi have many polyphenol oxidases for different purposes, but they haven't told us what those purposes are. Why do plants keep multiple genes for this enzyme, like different outfits in a closet, and bring them out at different times? We know for a few examples but not all:

1. Some very picky polyphenol oxidases specifically help build new molecules, "making new."
2. Others are less picky and make networks quickly when a plant is wounded, making a polyphenol web that traps insects (which would be painfully ironic for any caught spiders).
3. Some fungal polyphenol oxidases make a brown melanin network around the cell that strengthens the cell wall and helps the fungus withstand higher pressures. This superfungus then drills into other organisms, infecting them by "building" its own membrane and "breaking" others.
4. Finally, polyphenol oxidases may absorb stray electrons from ROS oxygen fragments, reducing the "burning" irritation from the oxygen "blown" in the air.

Copper's ability to do all this means that it blurs categories and breaks down walls. It is a most poetic element.

COPPER IN IMMUNITY AND DISEASE

At first, copper was used for simple purposes, but its application grew. Today, copper is used in many advanced functions outside the cell, which fits with its advent late in the planet's history. Its ability to both break and make means that it was employed as both sword and shield against other organisms. The immune system, as a consequence, uses copper.

Both infective and infected organisms scavenge and hoard copper and iron. From bacteria to humans, molecules called siderophores traverse the environment and snap up iron—and similar molecules called chalkophores do the same for copper. These elements must be valuable because every species is locked in a race to hoard them as much as possible. The chalkophores have an additional benefit in that inside your cells they help push the copper toward its new, useful plus-two form and away from the old, toxic plus-one form.

Once the copper has been gained, it is put to multiple uses. Wherever ROS are used as weapons to help or hinder infection, a copper or iron ion is somewhere behind the scenes. Primitive immune systems in insects and crabs have a version of polyphenol oxidase called *pro*phenoloxidase that defends against infection. Humans have immune cells called macrophages that use copper to build up a lot of ROS inside a special box. When you are infected, macrophages open up that box and lob the ROS at offensive bacteria, sometimes with a few dangerous plus-one copper ions in the mix. Your immune artillery is copper-clad.

The flip side of this is that, if copper is dangerous to bacteria, then it can be dangerous to you, too. Since copper is such a sticky metal, it may be at fault in diseases where things stick together when they aren't supposed to. The prime example of this kind of disease is when proteins stick together in large, globby amyloid plaques in brain cells. Amyloid plaques are a part of Alzheimer's, ALS (Lou Gehrig's disease), and Parkinson's diseases, among others, and copper is implicated in many cases.

Copper plus-two can bridge two nitrogen-rich rings in proteins and staple them together, forming a nucleus for an amyloid glob. One instrument literally pulls proteins apart and has been used to test this. It's hard to pull amyloid globs apart, and when copper is added, it becomes nearly impossible—copper links the protein together with a strong, visceral interaction.

COPPER AND ZINC WRITTEN IN GENETIC HISTORY

Remember that in the Irving-Williams series, copper and zinc are the stickiest of the metals, and are the farthest to the right in the transition metal block of the periodic table. Zinc is almost as sticky as copper. For example, zinc helps blood clot by sticking together protein backbones. Zinc can increase amyloid stickiness, like copper.

Copper and zinc's partnership can be seen in history as well. Cyanobacteria, the first oxygen producers, show an appetite for different metals than more advanced organisms. Cyanobacteria crave iron more than any other metal, followed by manganese, nickel, and cobalt (all "old" elements in this story). At the bottom of their menu, with least priority, are copper, zinc, and cadmium.

Algae are more advanced organisms, which come in red and green versions, each with its own metal preference. The green version prefers the "old" elements iron, cobalt, and manganese, while the red uses more copper and zinc. This implies that chemistry of the green algae branch was established earlier, and they would have been dominant a billion years ago, while the red branch became more dominant after more copper was available. Along with copper came an increased reliance on zinc.

Studies that search microbial genomes for copper proteins find stark differences. Organisms that live in oxygen have copper genes, and organisms that live without oxygen do not. Land plants are more exposed to oxygen, and their genomes show a huge increase in copper proteins, many of which are copies of minor variations on the same laccases or copper-binding proteins.

All those copper-gene copies could be a sign of accelerated evolution, prompted by oxygen stress. The evolution itself may be a random exploration of the possibilities, but the stress that caused it was ordered by the chemical story told in these pages. The oxygen stress carried an increased cost in terms of defending against the stress and making all these new genes, but also carried an increased advantage in that the genes can adapt to novel situations, breaking, burning, and making new molecules—increased risk hand in hand with increased reward.

A SONATA BY ZINC FINGERS

The scent of a flower is built from a series of interactions tied together with copper and zinc. Plants, with their duplicated copper and zinc genes, signal animals through the sense of smell, which also depends on copper and zinc. Copper is outside the cells in your nose, where it sticks to molecules that smell (odorants) and changes their shape. Some odorants have two sulfurs in them, which the copper binds, winding the molecule into a circle. This circle shape is what triggers the nerves that signal a particular smell.

Zinc participates in smell in a different way. Zinc is the central active ingredient in cold remedies like Zicam, and some clinical evidence says it may actually work. It's thought that the zinc sticks to the immune receptor that cold viruses use to enter the cell. This could lock the virus out of the cell and also calm down the immune system's overreaction.

But zinc can kill nose cells, possibly by sticking too much. Zicam nasal sprays are no longer sold in the United States because of reports that they made people lose their sense of smell. Zinc's stickiness may destroy the flow required to sense the environment. (Remember how one of my colleagues killed snail cells with too much zinc in Chapter 2?)

Zinc affects other advanced systems, such as development and immunity. Zinc deficiency in humans can delay puberty, which involves an advanced, complex web of signaling throughout the whole organism. This complex problem can be treated with a simple solution: spoonfuls of zinc sulfate.

Most advanced functions have a connection to zinc ions. Plants watered with a zinc sulfate solution better survive drought. Mice without enough zinc don't make enough antibodies. Finally, when a mammal's egg is fertilized, it sends out a series of "zinc sparks" as a signal.

Unlike copper, zinc is used by organisms both simple and complex, and may have played a role in the origin of life (as described in Chapter 5). The difference appears to be in *how* zinc is used, because there seem to be two levels of zinc biochemistry, simple and advanced. Simpler organisms without subcompartments use tightly bound zinc in protein-based enzymes outside the cell. More complex organisms use zinc held more loosely in protein structures called zinc fingers, which help proteins interact with DNA.

Zinc fingers control when and where DNA is active, and they are duplicated with abandon in higher organisms. Zinc fingers hold almost endless possibilities, considering how much DNA can encode. It's like my wife playing a sonata on the piano—there are only 88 possible notes, but her fingers control when and where they sound, creating a complicated and multilayered sound. (Sometimes I turn off the music in my office just to hear her practicing in the next room.) Zinc fingers are associated with complex, even musical, movements of DNA, such as when your immune system shuffles DNA to make new antibodies.

Once again, we know that zinc fingers could have worked well long ago, because they work now, in the lab. Zinc and DNA go together naturally—even a set of random proteins with added zinc contained several proteins that bound DNA. Protein designers can attach specific zinc fingers to DNA-cutting enzymes, and the result is a precise and efficient set of scissors for DNA. A colleague who has used them in the lab told me the biggest problem with zinc fingers is that they are too sticky, another fact that makes sense in light of the Irving-Williams series.

THE END OF THE BORING BILLION

In this chapter we saw many changes—molybdenum up, iron down, copper up— which undoubtedly happened with some pulses back and forth in the short term, but an inexorable direction in the long term. Oxygen forged a path of chemical development and pulled the rest of biology after it, through spikes and valleys of local variation and patchiness. Oxygen plays a key role in what is called "long-term climate forcing" as its pressure changes the density of the atmosphere, so that oxygen levels affect even temperature and rain.

The "boring billion" was boring in a good way: there were no ice ages to smother the Earth in sheets of solid ice. Glaciers coated the Earth 2.2–2.4 billion years ago, around the time of the Great Oxidation Event, but then they vanished. According to Hazen, "For the next 1.4 billion years—almost a third of Earth history, including the boring billion—no trace of an ice age has been found. Earth's climate seems to have remained in remarkable balance, not too hot and not too cold" (p. 224).

Then, a little less than 1 billion years ago, both oxygen and ice increased together. The planet convulsed with ice ages again, and oxygen jumped up from less than 1% of the atmosphere to 10% or more.

The immediate cause of this jump was geological. Most prominently, the supercontinent Rodinia split in half, making many miles of new shoreline in its middle. The shallow, oxygen-rich waters and increased erosion from this move released a wave of phosphorus that built cells and carried energy for life, and a wave of molybdenum that split and moved nitrogen. The chemicals caused a short-lived boom of life, but as will happen with building booms, this boom went bust.

This life-building bust was caused by a simple and unavoidable fact: the two-by-fours for building life are made of CO_2 (carbon dioxide). When a tree or alga grows, most of its carbon comes from the air. Oxygen is pulled off so that the carbon can be used for a cell wall or sugar molecule. The glut of extra phosphorus and molybdenum meant that carbon dioxide was pulled from the air in huge quantities and no longer served as a greenhouse gas. The atmosphere lost its billion-year balance.

Some more CO_2 was pulled out of the air by geology. Rodinia's split made shallow seas, which caused more rain clouds, and the rain eroded the exposed Earth. This released rocky elements like calcium, which bound CO_2 and pulled it from the air even more, as described in Chapter 4. Plummeting CO_2 caused a global freezing (the logical opposite of global warming). The chemistry of oxygen in the rocks suggests that as much as 5% of the liquid water on Earth solidified.

Thankfully, there were feedbacks stopping the feedbacks. When the planet froze over, a lot of microbes died, and they stopped holding down all the carbon. More CO_2 bubbled out through volcanoes and rifts. The Earth warmed, and melted, until the atmosphere lost its balance and the cycle repeated itself. Rodinia split 800 million years ago, and the Earth froze 720, 650, and 580 million years ago. But as with the rest of this story, riches followed storms, and life followed ice.

Other geological processes could have contributed to oxygen's release. The fact that different rocks break down differently in the rain could have caused oxygen to rise when combined with other factors (increasing sunlight and a cooling planet decreasing seafloor spreading).

Behind these cycles and feedbacks, three inexorable trends change the planet, each from a different scientific discipline:

1. The physics of heat flow cooled the Earth's inside (causing the nickel famine in Chapter 6 and bringing less reduced material out at this point).
2. The geology of plate tectonics and rain mixed the planet (dissolving more rock and bringing more calcium, molybdenum, and oxidized phosphorous into the oceans).
3. The biology of photosynthesis made oxygen itself from water and sunlight.

All three of these combined into a single chemical trend, best summarized by a single chemical word: oxidation.

Oxidation brought order behind the fluctuation. As the planet shuddered through multiple "snowball" or "slushball" episodes, the back-and-forth variation contributed to a long-term increase of oxygen, and a second, small rise in the striped orange banded iron formations. This rise in oxygen was significant enough that it could be called the Second Great Oxidation Event. The higher oxygen levels suddenly meant that new chemical possibilities were available on a global scale. The multiple ice ages were birth pangs that preceded the biggest and weirdest explosion of life in history.

9

CRACKED OPEN AND KNIT TOGETHER
BY OXYGEN

the Cambrian explosion and the Great Unconformity // ancient gene network complexity // oxygen and sea sponges // making with oxygen and breaking with copper and zinc // calcium in rocks, shells, and bones // algae's dopamine rush // plants evolving quickly // the Carboniferous explosion // the line between cause and effect

AN EXPLOSION OF LIFE, SET IN STONE

The happy insight that biology and geology meet through chemistry has been seen throughout this book when life and rocks interact. A chemical called water transformed this planet's rocks and opened them to give life its elemental building blocks. The energy in the Earth became the energy in simple cells through chemical wheels. Sunlight split the water with the help of dissolved rocks, and the oxygen from that reaction brought yet more elements out of the rocks and into life.

That insight addresses a long-standing mystery here. Long ago, the biggest biological change in the history of the planet created plant and animal life. What caused the seas to teem with weird new life?

I think the periodic table connects that biological event to a previous global *geological* change. If so, then once again, chemical reactions opened up geology to provide new possibilities for biological complexity. Chemistry shaped the flow of geology and biology at once.

The evidence for this connection is like something that happened with the *ekko* sculpture in northwest Scotland from Chapter 2 (Figure 2.1). After the sculpture had been built, an archaeologist dropped by and found incisions in *ekko*'s rocks. The archaeologist read the shape and depth of the incisions and concluded that the stones were older than everyone thought, and must have been used for a structure now lost.

Like in *ekko*, there are "incisions" on the Earth made by massive geological processes. Geologists have read these and have concluded that a worldwide event altered the planet's surface. This geological event was also a chemical event. Soon after, a profusion of fossils filled the rocks. This biological event was also a chemical event. The common denominator of chemistry connects the geology to the biology.

The geological event provided chemicals that life used in new ways: especially oxygen, phosphorous, and calcium, resulting in new energy, shells, and signals for life.

This hypothesis is that chemical availability drove the evolution of life, and that the periodic table shaped the timing of life's greatest expansion.

The expansion was first glimpsed and dated during the fossil-hunting heyday of Victorian England. Fossils were dug up in the 600-million-year-old rocks found in Wales. Because Latin sounds more impressive than English, these were called the "Cambrian" rocks, after the Latin word for Wales, "Cambria." But, curiously, they found no fossils in older rocks.

The English fossil hunters quickly came across this mysterious pattern in all their collections. Complexity does not gradually ramp up, but all sorts of life forms suddenly appear. The first of these are unlike anything we've seen, bulbous and elongated, spindly and plated, spoked and spiky. The horns, stalks, and armor on these creatures outshine even the imagination of Guillermo del Toro (director of *Pan's Labyrinth*).

This remarkable pattern in the history of life is called the Cambrian explosion, which was a jump in new biological families about 500 million years ago. To find it in Figure 9.1, look for the steepest part in the families line (the darker line), where the line rises to more than 500 families. Note how this was a long, sustained explosion, lasting at least 50 million years. During this time, evolution produced new species at a clip 5–10 times faster than usual. This production was broad-based, happening in all lineages at once, rather than just a few lucky ones. All branches of life won this lottery—which suggests to me that the lottery may have been rigged by an external factor.

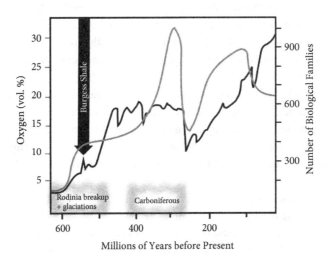

FIG. 9.1 Oxygen increased as families of life increased. Biological diversity is represented as biological families of crustaceans, shown by the black line. Chemical estimates of oxygen levels in the atmosphere are shown by the gray line. Geological events and eras are listed at the bottom.
Families data from J. J. Sepkoski. "Crustacean biodiversity through the marine fossil record." 2000. *Contrib Zoology.* 69 (4), Fig. 6; oxygen data from G. Rayner-Canham and J. Grandy. "Did molybdenum control evolution on Earth?" September 2011. *Educ Chem.* p. 144, Figure 1.

My answer to what external factor may have prompted this explosion, as you may have guessed, is chemical in nature (did I mention I'm a chemist?). Other possible answers are out there, as varied as the forms of life that radiated out from that point in time.

LAYERS OF ROCK, JOINED, GAPPED, AND EATEN AWAY

Geological patterns usually follow "slow and steady" patterns, but occasionally they explode into remarkable jumps, or unconformities. Standing out from the rest is Powell's "Great Unconformity." This unconformity can be found across North America. If you fly over the Grand Canyon, you may be able to glimpse it. Look for flat tan layers stacked on top of greyish walls tilted down toward the Colorado River below. The tan layers on top are 525-million-year-old Cambrian Tapeats sandstone sitting on top of the grayish 1,740-million-year-old Vishnu schist. Between those layers, a billion years of rock are missing.

The Great Unconformity is a geological counterpart to biology's Cambrian Explosion—and, oddly enough, immediately precedes it. The biggest discrepancy in geological history immediately preceded the biggest event in biological history. It is an incision in rock made across the globe, and like the incisions in *ekko*, we can use this incision as a window into the past.

At the time of the Great Unconformity, some combination of geological events caused a continent-wide erosion that swept the surface of the Earth into the ocean. Dry land was scraped down and cracked in two by assaults from above and below. From above, glaciers during snowball Earth events scraped rock into the ocean. There were so many glaciers that the period two steps before the Cambrian is called the "Cryogenian," as if the whole Earth was cryogenically frozen. From below, jets of upwelling magma pried the supercontinent Rodinia apart, exposing fresh new rock to hungry waves.

Dissolved rocks brought more elements into the ocean, and raised oxygen levels changed the elements to their more oxidized form. Four crucial elements stepped up: calcium, phosphate, oxygen, and molybdenum. In particular, a class of rocks called skeletal minerals—calcite, aragonite, apatite, and opal—abruptly rose in the Cambrian period, along with the calcium-based fossils observed by Victorian rockhounds.

In the seas, banded iron formations once again painted the bottom with iron and oxygen, like in the original Great Oxidation Event, creating the second BIF peak in Figure 8.1 in Chapter 8. At least three or four cold snaps passed over the planet like an inverse fever. The chemistry of the oceans changed dramatically, and soon after, so did life.

Life took the extra phosphorus as a gift, making it into DNA, RNA, ATP, and membrane lipids, which birthed new photosynthetic organisms that took in CO_2 and put out O_2. Life took the extra molybdenum as a similar gift, growing from its nitrogen and oxygen chemistry, and the deep oceans finally filled with life-giving oxygen.

On the other hand, life took the extra calcium as a *threat*, ejecting it, but in the process, new calcium chemistry happened outside each cell that was key to the whole Cambrian explosion.

As earthquakes, tides, and rains mixed the dirt into the ocean, the Earth was also cracked open in another significant way, this time biologically. Beginning a little before the Cambrian explosion, fossilized mats show tiny cracks and fissures that look like little chew marks. Some animals figured out how to burrow and crack apart the tough bonds holding the mats together.

These cracks and crevices promoted diversification and specialization. On your skin, each physical niche holds a different community of bacteria. Earwax bacteria are different from armpit bacteria (true, though gross). This is true on the planet as well, as a more broken environment creates more space for life. Saying that an organism created a "niche" for evolution is often a metaphor, but here, it is literal. Burrowing created physical niches that protected and sustained life.

We can find burrow-holes in the fossil mats, but we can't find evidence of scars or notches on the thick shells and hides of the first Cambrian animals. It seems that for a time, everyone was a vegetarian and fed off the mats, and it was a world without scars. That would end soon enough.

THREE INTERPRETATIONS OF FOUR EVENTS

Four major changes happened during this period, and ever since the Victorian age, scientists have wondered what ultimately caused the final event of the four:

1. Rocks show us that oxygen rose as the planet froze and thawed (900–600 million years ago).
2. The Great Unconformity shows us that Rodinia cracked and continents eroded, sending calcium, molybdenum, and phosphate into the oceans (800–550 million years ago).
3. Fossils show us that animals burrowed through mats but did not prey on each other (580–550 million years ago).
4. Fossils show us the Cambrian explosion (550–500 million years ago).

The September 20, 2013, edition of *Science* magazine contains two essays that define a range of interpretations for the Cambrian explosion. I propose a third way (which is a distilled version of the second response).

A response to the first, inadequate interpretation is found in a review of Stephen Meyer's book *Darwin's Doubt: The Explosive Origin of Animal Life and the Case for Intelligent Design*. This review's author, Charles Marshall, does something rare for a scientific journal—he tells you how the book made him feel:

> I like to read the arguments of those who hold fundamentally different views from my own in the hope of discovering weaknesses in my thinking. . . . However, my hope soon dissipated into disappointment . . . while I was

flattered to find him quote one of my own review papers . . . he fails to even mention the review's (and many other papers') central point: that new genes did not drive the Cambrian explosion. (p. 1344)

I could have written that myself (except for the part about "my own review papers"). Meyer's argument focuses on the sheer scale of the biological change, constantly telling us what mechanisms *don't* work, but never providing mechanisms that *do* work. Rather than mechanism, Meyer suggests that a personality caused the changes, which he terms an Intelligent Designer. Meyer's agnosticism about any mechanism is one kind of response to the Cambrian Explosion, but it is simply inadequate as science.

A second, more acceptable interpretation is found a few pages to the right in another *Science* article. This article, by Paul Smith and David Harper, is straightforwardly titled "Causes of the Cambrian Explosion." The plural must be deliberate. Smith and Harper lay out a dazzling array of mechanisms and show how *many* causes there *could* be to this explosive event. The article's figure contains 15 boxes of causes and effects connected with 25 arrows. Understanding this figure is like trying to follow one of those children's puzzles in which you're supposed to follow one long line through a whole tangle of lines—it's easy to lose track.

The basic point of Smith and Harper's review is sound. A lot of scientists say a lot of different things about what may have caused the Cambrian explosion. This review describes more actual experiments in two pages than do two typical chapters of Meyer. Yet I can't help but feel that it cuts off the limb that it sits on. Smith and Harper's article offers so many explanations that they seem to add up to no explanation after all.

As a chemistry teacher, my point is not to bewilder, but to explain. I think Smith and Harper's morass of lines can be reduced (or, more specifically, can be *oxidized*). Each of those 15 boxes of causes and effects relates to the four events listed earlier, so that's a start. And looking through them further from a long perspective, I reduce each of those further, to a single element, even a single letter: O for oxygen.

No doubt you saw this coming, Dear Reader. And in making these simplifications, I'm open to charges of oversimplification. My interpretation is as qualitatively different from Smith and Harper's as it is from Meyer's. I consider it a third way to approach this problem—not with Meyer's inscrutable simplicity, nor with Smith and Harper's inscrutable complexity, but with, well, *scrutable* chemistry.

Oxygen is related to every one of the four events listed above. Geology suggests that, were you to have a time machine but not a breathing apparatus, that you could go back and explore the Earth as far back as 650 million years ago, but no further. (The oxygen spike 2.1 billion years ago before oxygen levels crashed back down into the boring billion did not persist long enough to drive evolution.)

After the Second Great Oxidation Event had done most of its work, the planet reached a stable threshold of oxygen levels, near today's, with oxygen present from the air to the deep sea. While surface ocean oxygen increased 10 times, deep ocean oxygen increased a *million times*. A hundred million years later, after the events of the Great Unconformity, the Cambrian Explosion exploded. If we scale the history of life on this planet to a calendar year, then the Second Great Oxidation Event was centered on November 1, the Great Unconformity dates to November 3, and the Cambrian Explosion was November 10.

Precedence is necessary but not sufficient. Here's where chemistry helps connect the dots. Oxygen is logically connected to each of the four factors listed at the start of this section. Oxygen is involved with each new chemical function that life obtained at this point. This is not a *chain* of evidence—it is a *network*. The chemical elegance of this network points to oxygen's increase as being the single biggest cause for the developmental quantum leap that was the Cambrian explosion. After oxygen took a leap upward, the chemical possibilities changed and the world followed suit. Here's how I think it happened.

INTERNAL NETWORKS OF COMMUNICATION

As Marshall wrote in his review, "new genes did not drive the Cambrian explosion," but the new genes were already there. Genetic networks expanded well before even the glimmers of the Cambrian explosion 540 million years ago. The proliferation of fossils seen in the rocks did not invent new genes, but repurposed old genes.

These old gene networks allowed cells to live together in communities. Stromatolites were a first step toward this, but they are really cells that live as individuals while stuck together haphazardly. It took another billion years for algae mats to evolve, near the end of the "boring billion" and the beginning of the Second Great Oxidation Event. Underneath the "boring billion's surface, genes were connecting and developing networks that would be put to good use in the Cambrian explosion.

These algae mats are held together with oxygen, too. The sliminess of algae contains strings of "alginate" sugars, which is sugar with extra oxygens on it for extra negative charge. Alginate's extra oxygens attract and absorb water as the sugar chains tangle together into a coherent mat. Alginate is a useful additive for thickening ice cream, and it is all natural, albeit from a somewhat slimy part of nature.

At some point, the algae changed in essence, from being individual cells stuck together with oxidized sugar to being a true many-celled organism. Comparing the genome of a multicelled alga named *Volvox* to that of a single-celled green alga named *Chlamydomonas* shows surprisingly little difference between the two: most of the about 15,000 genes are similar enough that we can't tell a difference. The multicellular *Volvox* has different genes for making the alginate-like sugars outside the cells, and for governing reproduction (so that many cells can grow from a single germ). Such genomic similarity was also found among a single-celled animal ancestor called a choanoflagellate and simple multicelled animals. Only a few genes separate the single-celled from the multicelled.

Cyanobacteria were in on this transition, becoming pioneers in multicellularity, as they were in oxygen production. Cyanobacteria learned to stick together in strings as far back as the first Great Oxidation Event, then they refined their technique over time. At first they formed disorganized clumps or strings of cells (like stromatolites made of sugar rather than sand—and like stromatolites, the stickiness of cyanobacteria and algae ultimately came from oxygen-linked sugars). Later they formed more complex structures.

Underneath the turmoil of the snowball Earth episodes, a new kind of creature was learning how to coordinate its stuck-together cells, providing a controlling and cohering personality, a Theseus to its Athens-like collection of cells. This new colony of cells may have come about from a cellular failure to "cut the cord" holding parent cell and child cell together. When cells divide, they disperse and find another niche in the environment, but a dividing cell that was too sticky to its parent and had sticky children would form a nearly homogeneous collection of cells that could then learn to work together.

For both gene networks and inside-out stickiness, oxygen's involvement would be indirect. Microbes with mitochondria would get the energy to run large genomes from oxygen, with extra "disk space" on their DNA "hard drive" for encoding complex and ever-evolving networks. Oxygen also would protect these networks from UV radiation through the ozone layer, which increased with the Second Great Oxidation Event.

Oxygen concentrations inside the cell, however, would not change that much, because cells keep their insides low in oxygen. Oxygen levels changed dramatically *outside* each cell, and it's there that new oxidized structures were built.

EXTERNAL NETWORKS WITH OXYGEN LINKS

The simplest multicellular organism, a sea sponge, is a passive collection of cells that does not move but lets its food move to it. Sponges eat by filtering the carbon-rich detritus from passing currents and waves. They have a few different kinds of cells with different jobs, but their division of labor is indistinct, and there are no clear internal organs or layers of distinct specialization. Even their "skeletons" are indistinct branching structures, like a tangled tree, made of either calcium carbonate or silica spikes (and likely shaped by Bejan's tree-making theory).

The most remarkable features of sponges don't come into focus at arm's length but are revealed in their DNA. A surprising number of sponge genes look familiar and line up with animal proteins, including ours—not just a few, but hundreds of these proteins, especially the "kinases" that move phosphates as signals. Sponges may look half-formed, but deep down we have deeply similar phosphate-based signals. Animals have only 17 major cell signaling pathways, and most are shared with plants, fungi, and unicellular organisms, indicating that they existed in some form before the Cambrian explosion, when the family tree diverged.

The problems with the pathways also persist. Cancer happens when this network ignores its instructions, refuses to grow up, and gets stuck in a state of permanent growth eerily similar to that of an embryo. Simple sponges have the same oncogenes and tumor suppressor genes that we do, and even the sponge's cousin *Hydra* gets tumors. One of most important tumor suppressor genes, p53, has been traced all the way back to this root of the multicellular family tree. Cancer is an enemy as old as animals themselves. But there's some hope in this—if cancer is not entirely comfortable with the "new," high levels of oxygen, then oxygen may be part of a therapy against tumors.

Sponges also had new oxygen-based materials for tying cells together. The proteins collagen (in our joints) and keratin (in our hair) are simple, repeated strings of amino acids. Oxidation strengthens the strings in three different ways: (1) with oxygens added onto the chains, (2) with super-strong oxidized links between amino acids formed by a special oxidase, and/or (3) with oxidized sulfur-sulfur links that hold the chains together. All of these bridges are promoted by oxygen.

In humans, the enzyme that puts oxygen onto collagen fibers is under the control of a very old oxygen-sensing network found in the simplest animals. Collagen oxidation is enhanced in low levels of oxygen corresponding to pre-Cambrian levels (2%–5% of today). This may be a relic of the days when oxygen was harder to come by, and its use had to be prioritized for building links. Sponges can live in the same low levels of oxygen.

Oxygen plus sugars may sound more sweet than strong, but it can make rock-hard materials from the decidedly non-rocky CHON elements. Some calcium is also tangled up in these networks, its dense positive charges sticking to the many negative oxygens and carbonates (Figure 9.2).

Consider the squid beak, a material called chitin. Despite being sharp as a knife, chitin has no metal. It's made of CHON sugar and amino acid chains oxidized into a hard, flexible network. The insect eggshell is also hard and made of CHON elements,

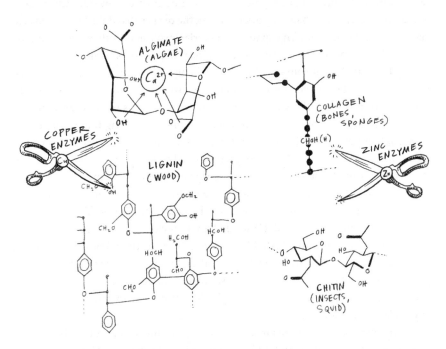

FIG. 9.2 External structures in animals held together by oxygen and calcium. Shown are shell structure, bone structure, alginate structure, keratin, chitin, and cellulose structure. Many of these links are cut by external copper and zinc enzymes.

linked and waterproofed by the oxidizing activity of peroxidases. Today chemists harden plastic strands with the same oxidative cross-linking found outside cells.

These cross-linked networks can be manipulated by chemists into fascinating new materials, starting from watermelons. The red flesh of a watermelon is a network of oxidized sugars that holds the water in the melon. Freeze-drying this network gently flushes out the water, and the resulting material is light and airy, the texture of astronaut ice cream but impossibly light. A lime-sized block of this material can sit on a dandelion without disturbing a single seed. The chemists called this an "aerogel" and magnetized it with iron oxide to make a supercapacitor, but it's just carefully dried watermelon.

Sponge's chemical innovations, oxidized collagen fibers and a spindly calcium/silicon skeleton, clearly predate the Cambrian explosion. Accordingly, sponge fossils have been found in rocks more than 100 million years older than the Cambrian explosion, as well as a specific form of cholesterol only made by sponges. The common factor in forming chemical bridges between cells in these primitive multicelled animals is the unique chemistry of oxygen, which would work even at low levels.

CREATIVE OXIDATIVE DESTRUCTION

The gene networks and multicellularity of sponges were already old when the Cambrian explosion exploded. Something else unlocked the potential of these biological networks, even if, as some propose, that something else was only time. Chemically, oxygen-related elements had the power to break the oxygenated links that joined cells and rocks. In the times leading up to the Cambrian explosion, both biology and geology were broken open, and oxygen enhanced the processes of breaking.

Any body, including yours, that needs to change, grow, and respond to its environment has a set of enzymes used to cut the oxidized networks tying cells together. Collagen strands are built by copper enzymes, mooring cells in place, and they are cut by zinc enzymes, freeing cells to move and migrate (or sag into wrinkles). In animals, these enzymes play a role in the unwanted motion of cells that is cancer metastasis. In the previous chapter, oxygen released copper and possibly zinc; in this chapter, they work together to build and reshape multicellular animals.

To burrow into the tough mats lining the bottom of the ocean, animals had to develop chemical strategies to chew through these mats. To do this, an animal (1) would need to move around, (2) would have to break the mat's links with oxidative chemistry, and (3) would need hard materials for teeth and/or armor. Oxygen provides the energy for #1 through mitochondria, the chemistry for #2 through copper and zinc, and the mineral for #3 (described in the next section). The scientists Peter Ward and Joe Kirschvink, for example, don't think burrowing can pay off until oxygen levels are around 10%.

Having more oxygen in the air would provide fuel for the fires of life by increasing the rate of combustion and ATP generation in mitochondria. Oxygen increases Eric Chaisson's energy rate density—more oxygen means more energy processed per second, which means more watts per kilogram for oxygen-breathing organisms. This supports more complexity and order *inside* the system because much more disorder is created *outside* it. The planetary oxygen increase was a gust of wind that fanned the flames of biochemical combustion.

Even the toughest sugars break down with the right kind of oxidation channeled by the right metal. For example, the bacterium *S. marcescens* breaks insect shell chitin with an oxidizing copper enzyme. This ability to break open sugar armor changes the entire ecosystem. It allows one organism to break open another to prey on the reduced carbon foodstuff inside.

Given enough time, the neighbors evolve better walls that resist the sugar-breaking enzymes—until a stronger wall-breaking enzyme evolves. This sets up a co-evolutionary arms race that can be seen in genes and in the lab. Genetic patterns show a dance of making and breaking in plant-eating fungi and plant cell walls. What is eaten changes, and the eater changes to eat again. Driven by this competition, sugar walls and sugar-wall-breakers constantly change and diversify. Oxygen holds the walls together and oxidation breaks them apart.

Oxygen and oxidation carry out the same creative destruction on geology, as physical and chemical processes crack and transform the Earth. Oceans and magma exert physical forces driving Robert Hazen's "mineral evolution." Within the past billion years, oxygen expanded these processes both directly through chemical reaction and indirectly through biological action.

By the time of the Cambrian explosion, most of the iron had been oxidized to iron plus-three. Minerals made from oxidized iron erode easily. Sulfates dissolve better in water than sulfides, so oxygen-rich water eats away at sulfurous rocks, too. Oxidation changes copper rocks and manganese rocks. What the dry land lost, the oceans gained and life used.

On the beach along the Puget Sound near my house, a round rock the size of a cannonball has been split open. Its outside is an inch-thick rusty red ring, but its inside is olive green. This "reverse watermelon" coloration shows that it was once grey-green iron-plus-two, but oxygen turned the outside to red iron-plus-three. (Did the oxidation of the outside layer weaken the rock enough to make it prone to fracture?) This happened all over the Earth as oxygen assisted in geological breakdown.

Geological weathering can also be mediated by microbes. Ever so slowly, life eats away at rocks, atomizing solid crystals down to mush. Microbes are potters that make their own clay, breaking down a silicon-rich feldspar into moldable grains and reshaping the Earth in the process. Clays have huge surface areas, and one of the laws of chemical catalysis is that reactions happen on surfaces, so clays are inherently catalytic. Clays also adhere to organic carbon and provide a suitable home for life.

This adds up to a positive feedback cycle: oxygen feeds life, life breaks down rocks, life fills the cracks and makes more oxygen. Fitting with this story, aluminum-silicon clays like smectite and kaolinite are found increasingly before the Cambrian explosion. As these cracks formed, minerals dissolved in the ocean, especially phosphate, sulfate, and calcium. The first two are food for life and further expanded the algal mats, assisted at each step by oxygen. The last one, calcium, was the opposite of food and was kept on life's outside and put to use there.

CALCIUM: FROM SHELLS TO BONES

The Cambrian fossils adopt countless shapes and sizes, but they are mostly three members of the geological Big Six: calcium, oxygen, and silicon. After the Great Unconformity, microfossils increasingly developed defensive spikes made from calcium, silicon, and oxygen. Later, after the Cambrian Explosion, life used calcium even more. For example, sponges changed their skeletons from spindly calcium lace to thick calcium layers.

R. J. P. Williams points out that several different types of mollusks developed calcium-based shells at the same time, like widely dispersed inventors filing simultaneous, identical patents. Overall, calcification evolved at least 28 times in separate species. Something was in the air (or rather in the water) that allowed this season of chemical innovation. I propose that this "something" was calcium itself, released from the rock by weathering, assisted in binding by oxygen.

The sea change in calcium is found in rocks as well as fossils. Robert Hazen relates how "massive" limestone deposits, made of calcium, coincide with the Cambrian explosion a half-billion years ago. The broken Earth at the Great Unconformity tells us where the calcium could have come from, and chemistry hints that oxygen assisted in the breaking. Hazen also describes how life made new kinds of mineral, seeded by calcified shells and shell fragments (which are practically rocks themselves). Geology changed biology, then biology changed geology.

Usually the shell fragments that make up these rocks are very small, but some are big enough to be seen with the naked eye. The Castillo de San Marcos in St. Augustine, Florida, is made from blocks of coquina, which is tiny periwinkle shells melded together in a natural concrete of calcium, carbon, and oxygen. This fort was built by Spanish settlers protecting themselves, like a clam does, with a hard wall of calcium and oxygen. Coquina provided an extra advantage over more common rocks: when a cannonball hit, the wall would bend, not break, because of the material strength of the tiny shells. This helped the Castillo resist two sieges, and it never fell to enemy attack.

Calcium, like oxygen, was useful for life's fortifications—but, also like oxygen, it would have been toxic at first. This is reflected in one of the major chemical balances for life from Chapter 5: "eject calcium, import magnesium." When the calcium first saturated ocean waters, life kept shoveling it out like so much snow and bid it good riddance lest it crosslink phosphate-rich DNA.

This rejection of calcium placed its chemistry outside the cell, where it became indispensable. Negatively charged oxidized materials were already outside the cell, like alginate, following the "eject oxygen, import electrons" balance. Positively charged calcium stuck to these negatively charged sugars to make a tough, calcified wall that protected the organism and allowed deeper burrowing.

This reaction is recreated in the application of chemistry to cuisine, when adventurous, science-minded chefs cook with "spherification." They turn practically any liquid, from green tea to apple juice, into small spheres that burst on the tongue like caviar. The secret is to mix calcium with an oxidized sugar strand like alginate. The calcium links the oxidized sugar into a round, hollow shape that looks like a cell but tastes like the original liquid. The same reaction at an organism's surface, with more calcium, makes a harder wall that can be shaped into a shell.

The cells with these proto-shells could manipulate the hardness and shape of these shells by changing the negatively charged sugars and/or proteins outside the cell. One kind of sponge doesn't make the charged material itself, but chops up bacteria and lays them out on its surface. The bacterial surfaces attract calcium carbonate and seed skeleton growth. This would be an easy way for sponges to develop their own skeleton.

Careful positioning of negative charges shapes the shell. Now corals have a "toolkit" of 36 such proteins to arrange charges and make shapes, like a sculptor's kit of tools. Different species use completely different proteins but the same chemistry is seen in them all: negative charges on a surface will arrange calcium and other materials into a shell.

Even very different-looking shells use a generic mechanical process to make diverse shapes. Two shells that look very different were secreted by the same machinery, with a few chemical tweaks that give shells different stiffnesses and growth rates. Complex structures like fingerprints and airways may form from the same type of simple, tunable process. Much biological shape diversity is only skin-deep (or shell-deep), using processes that are mechanically similar and chemically identical. Here, an explosion of forms results from a simple physical process operating on calcium, and the shape diversity masks a deeper unity.

A calcifying creature that needs to fill an area of water efficiently may invoke Bejan's tree-making theory from Chapter 4. Small units connect to larger units as the coral builds a skeleton of calcium carbonate with a predictable shape that fills space both efficiently and with minimal use of material. When calcium is the building material, the branches have rectangular, sharp edges. (When carbon is the building material, the same purpose is accomplished with rounded shapes arranged in a tapering cylinder, forming a pine cone.) The general shapes of these organisms can be predicted from Bejan's engineering principles.

As time went on, calcium was rearranged in other shapes with other purposes, building scales, teeth, jaws, and bones. Similar chemical processes form these by arranging negative charges and collecting calcium on a surface.

Researchers recently found evidence that an early animal called Cloudina built calcium reefs in the period immediately preceding the Cambrian explosion, 550 million years ago. These reefs are fascinating clues in two ways:

1. Only the first reefs were built of a calcium mineral called aragonite, which forms in waters with more magnesium. Later reefs during the Cambrian explosion were built from calcite, which forms in waters with more calcium. This fits a pattern in which reef building immediately predated the Cambrian explosion and was influenced by ocean calcium levels.
2. The reefs also have protective structures that seem designed to repel predators, suggesting that the threat of predation caused these first reefs to coalesce, like

villagers building walls to keep out invaders. Predation depends on teeth, which by and large depend on calcium. Calcium could have armed both sides of this conflict, providing both threat and defense and leading to innovation.

Reefs are underwater oases of complexity and life, and they concentrate the elements useful for life. Many species can find niches for protection and opportunity in the calcium-based nooks and crannies. Phosphorus is also concentrated at the reefs, in some cases by microbes that live symbiotically inside the sponges there. These microbes collect phosphorus as granules that then support the rest of the complex web of life.

Today the most advanced calcium biomaterial is bone, which incorporates phosphate rather than carbonate into its calcium network. Studies of sturgeons and elephant sharks show that bone evolved by repurposing old proteins, like shells did. One protein evolved by lining up five negatively charged oxygen groups in a row, which lined up calcium and phosphate into crystals, and the crystals grew into a simple bone. From this simple start, genes were duplicated and tweaked to produce better calcium-phosphate crystals, culminating in the chemical properties of bone.

Since bones and shells are similar materials built from similar elements, they may work together. This logic inspired researchers to invent a new bone cement by mixing calcium sulfate with ground oyster shells. The calcium and the proteins in the oyster shell encourage the surrounding bone to grow and mend a broken gap. One species helps another through calcium chemistry.

Bone is ideal for our use because it has several unique mineral properties. The most surprising may be that bone generates its own electric field. Negatively charged phosphates can hold positively charged protons loosely, so the protons flit from phosphate to phosphate. These moving charges create an electromagnetic field. When a bone is bent or broken, the electrical field shifts, which may attract bone-making cells to the damaged site from far away.

Calcium and oxygen are major players in other external materials encountered in nature beyond shells and bones. Abalone shells, deer antlers, lobster shells, sponge skeletons, bird-of-paradise stems, porcupine quills, toucan beaks, and feathers are all listed in a single review on biomimetic materials, and all are linked with calcium, oxygen, and/or oxidized crosslinks. This is what calcium and oxygen made possible right before the Cambrian explosion.

THE NEUROTRANSMITTER IN THE ALGA

Another clue to the Cambrian explosion is communication. The neurotransmitters in the human brain are one of the most advanced forms of chemical communication, but this communication molecule is surprisingly old. Algae aren't even true multicellular life forms, yet they contain neurotransmitters.

In neurons, dopamine negotiates the brainiest of brainy functions. It is associated with forms of euphoria and pleasure, from sex to drugs to chocolate. Even though algae don't have complex emotions, they use dopamine, too.

The alga that loves to release dopamine is an ordinary, leafy seaweed. When a predator bites into it, internal holding cells are ripped open and spill out dopamine. Air and sunlight oxidize it to form a reactive "quinone" that colors the surrounding solution red or purple. From our perspective, the water turns to blood. This quinone is toxic to the predator (and to the alga itself, so its victory may be Pyrrhic). The algae turn brown and black as the reactive quinone reacts with itself to make an oxidized network of dopa-melanin like the networks in Figure 9.2.

Life from sponges on up has cells that make dopamine. Bananas are one good source (is this why my kids like them so much?). Dopamine is made in two steps: add an oxygen-hydrogen group to the amino acid tyrosine, then clip off its acid group. The added oxygen makes the dopamine reactive and networkable, changing it from an ordinary amino acid into either a colorful toxin or a neurotransmitter, depending on the context. Even stored in compartments in your brain, dopamine is still a little toxic, so neurons maintain a protective sulfur enzyme as an antidote. That dopamine rush is a toxic thrill.

Dopamine's stickiness has other uses. Mussels oxidize dopamine to glue themselves in place. Algae use another oxidative network for another advanced purpose—they bandage their wounds using a double-ring sulfate molecule that looks a bit like dopamine. They tear off the sulfate to make a reactive oxygen group, then tie the oxygens together with peroxides. This crude network forms sticky clots to plug wounds in the protective walls, made possible by the chemistry of oxygen.

Serotonin is another neurotransmitter made in two steps like dopamine, by adding an oxygen-hydrogen group to the amino acid tryptophan, then removing the acid group. Its biological role may be different from dopamine's, but chemically serotonin follows the same formula: amino acid + *oxygen* − acid = neurotransmitter (Figure 9.3). The different chemical shapes that mediate pleasure, sleep, and algal clotting all come from the chemical activity of oxygen and, despite their advanced function, date back to the time of the Cambrian explosion. Oxygen made the shape of serotonin possible, even easy to make.

SHAPING COMMUNICATION WITH OXYGEN

Imagine an algal horror movie: a predator is chewing on some algae. A cloud of red quinones fills the water, the algae turn black, and everything in the immediate vicinity dies. But there is a silver lining—a little farther away, quinone concentrations are lower, and an animal in that area catches a whiff of the toxic dopamine, enough to repel it, not kill it. That animal learns to avoid the toxic form of dopamine and survives. The same dopamine that is toxic at high concentrations becomes a signal or a scent at low concentrations.

This could happen with any distinctly shaped signaling molecule. Many signaling molecules are old molecules like amino acids that have been repurposed with added oxygen groups. Once you can make different shapes, you can make a code using those shapes. The signaling networks from these molecules can communicate within species or between species, for purposes ranging from attractive to repulsive, and from competitive to cooperative.

Wendell Lim has shown how a rudimentary secrete-and-detect cycle with a distinctly shaped molecule can evolve through simple steps to a complex signaling network. Your glands are especially proficient at making new signaling molecules from old ones, helping your various organs get along. Your adrenal gland, for instance, has a host of iron-heme enzymes that make hormones by putting oxygens onto old cholesterol-related molecules.

Small oxidized molecules can help even organisms from different kingdoms get along. Animals cooperate with bacteria against fungi in a case where a shrimp protects its embryo by attracting bacteria that churn out oxidized, toxic molecules that look like serotonin (and also could be made from tryptophan, in an exact parallel to dopamine and tyrosine).

In another example, fungi "farm" bacteria along long, branching fungal "fingers" (hyphae) that provide food for their bacterial herd—which are probably built using Bejan's tree-making theory and probably secrete small oxidized molecules to herd the sheep-like bacteria. These communication networks co-evolved with random steps, and the exact shape of the molecules to be used may be unpredictable, but the fact that signaling networks will form around oxidized forms of preexisting molecules is entirely predictable, perhaps even unavoidable.

Plants even talk to each other (outside the Little Shop of Horrors, in real life) using small chemicals called green leaf volatiles. The fresh smell of a walk outdoors is made of these chemicals carrying signals through the air to your nose. The signals that carry through air are smaller than those that carry through water, but oxygen is found here, too. Many green leaf volatiles are carbon chains made reactive by oxygen-hydrogen groups.

Sometimes these small volatiles are released when cells are broken open, as with the algae. For example, the smell of cut grass is six carbons with an oxygen on one end, which lawnmower blades disperse in summer. Tomato plants release a similar carbon-chain molecule when chewed on by insects. (In a display of thrift, nearby tomato plants absorb this chain and chemically transform it into a defensive pesticide by adding a sugar molecule to it.)

Because humans and plants both add oxygen to old molecules to make new signals, a plant signal can serve as a human signal, and plants make molecules we can use as drugs. The plant hormone salicylic acid is a signaling molecule for many plants and is one group away from being aspirin, a signaling molecule that blocks human pain signals (Figure 9.3). Both molecules are carbon frameworks made useful and unique with oxygen.

You exude volatiles as well. Mosquitoes home in on human blood by smelling a carbon-oxygen chain called sulcatone that we emit. Genetic patterns show that mosquitoes evolved a receptor specifically to recognize this molecule of ours. Once a specific carbon-oxygen chain was associated with a specific species, then evolution molded itself to that chain. If evolution was rerun, the chain may very well have a different shape, but it would still be made of carbon and oxygen, and would be small enough to float through the air.

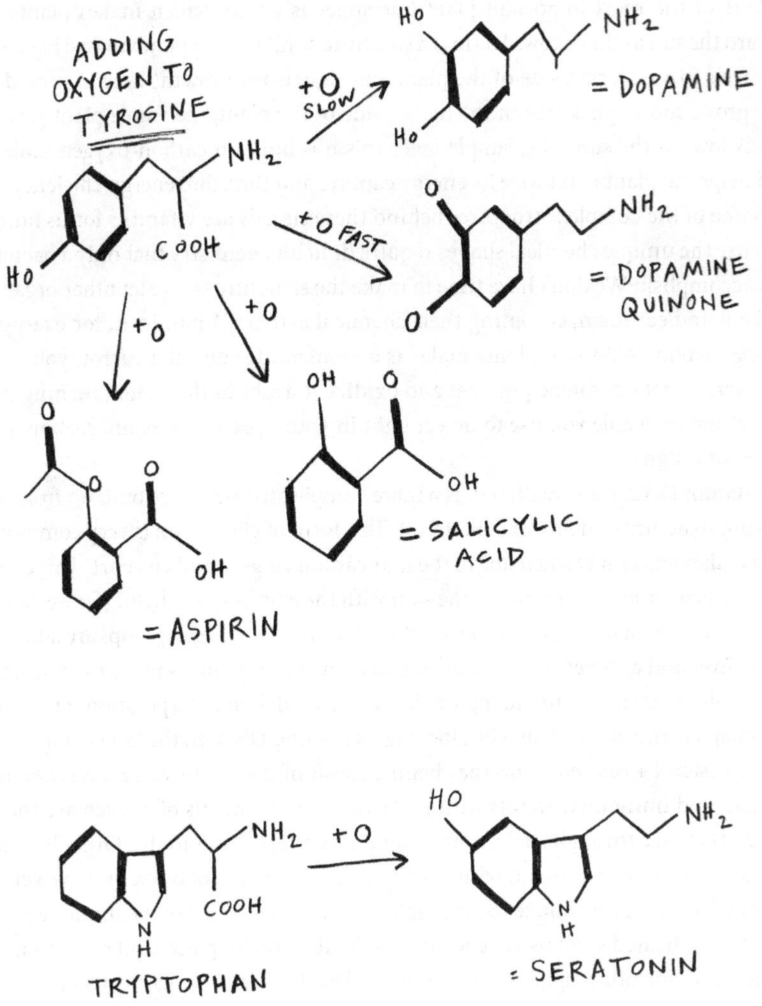

FIG. 9.3 How new signaling molecules are made from old ones by reshaping them with oxygen.

The diversity of smells from a spice rack comes down to small variations in oxygenated carbon molecules, made by the plants as signals, and enjoyed by your tongue as a related signal. Cinnamon gets its flavor from cinnemaldehyde, a carbon ring with a carbon tail and a reactive oxygen at the tip of the tail. Cuminaldehyde, from cumin, is a carbon ring with a shorter oxygen-tipped tail and a "V" of carbons on the other side. Vanillin, on the other hand, is a carbon ring decorated with three short oxygen-carbon tails. These look much the same to the eye, but they are quite different on the tongue. All are variations on a chemical theme, as if Mozart was given carbon rings and oxygen to work with instead of notes and rhythms. Carbon, oxygen, and hydrogen are all plants need to send very different signals.

One of the most important plant hormones is auxin, which makes plants bend toward the sun as they grow. Auxin is a structure similar to tryptophan (see Figure 9.3). It travels to the darker side of the plant and signals for growth there. As the darker side grows more quickly than the lighter side of the plant, the whole plant gradually bends toward the sun. This simple mechanism is built on carbon-oxygen molecules and helps the plant maximize its energy capture and therefore energy efficiency.

Some of the complex structures behind these signals are vitamins for us humans, because the unique chemical shapes require difficult chemistry that only a bacterium can accomplish. We don't have time to make the structure, so we let other organisms make it and eat them, co-opting their chemical artistry. Vitamin A, for example, is a long carbon chain that plants make as a pigment. If you eat a carrot, you take in the carrot's beta-carotene pigment and oxidize it, right in the center, turning it into the retinal molecule you use to detect light in your eyes. Once again, carbon + oxygen = vital signal.

Vitamin D, on the other hand, is a more complicated structure built up from a sterol ring structure (7-dehydrocholesterol). This form of cholesterol is very common but only makes vitamin D when one of the four carbon rings is broken apart. This chemistry is difficult and is done only in the skin with the energy of sunlight. (Or we can have a microbe make it for us and we can eat it.) After this, two oxygen groups are added, one in the liver and another in the kidneys, and then the molecule is ready to send signals. Its signals are related to recent topics: they involve calcium incorporation into bone, in this chapter, and are mediated by zinc fingers binding DNA, in the last chapter.

Cholesterol + oxygen forms the chemical basis of the advanced processes of development and immunity. Sterols with particular arrangements of oxygen are the hormones that rage through adolescents, changing children into adults. Other hormones that look much the same, but with a different arrangement of oxygens, have very different effects that coordinate the multiple arms and functions of the immune system. These complicated systems are coordinated by the precise placement of oxygen.

Because hormone signals are easy to make, they are also easy to forge and are easily co-opted in cancer. Some tumors survive by making oxysterol hormones that send false signals to the immune system, convincing the body to build new blood vessels to the tumor, delivering fresh oxygen. All the tumor has to do is add oxygen to sterols in the right pattern and it survives, to your detriment.

These signals can also explain previously mysterious links. For example, obesity and breast cancer have been correlated in hundreds of studies, but no one knew why. Now we know that the connection runs through an oxygen-sterol. Because diet-induced obesity puts excess cholesterol in the blood, the enzymes that make oxysterols have extra starting material and make extra oxysterols. One of these products is 27-hydroxycholesterol, which mimics estrogen, sending a signal that cells—especially breast cells—should grow. So too much cholesterol makes too many oxysterols, which send too many signals causing tumor growth. The wires in the body are crossed, and the wires are made of oxysterols.

By comparing hormones and their receptors across species, a team including Caetano-Anollés reconstructed a timeline of when new oxidized hormones appeared. The first one appeared in the organisms that burrowed through Cambrian mats. When oxygen filled the cracks of the planet, new signaling molecules were born. Moving forward from the origin of hormones, new arrangements of oxygen on the sterol backbone co-evolved with new receptors that triggered new signals, one day making testosterone and estrogen signals.

Through the changing shapes of signals and receptors, oxygen greatly increased the chemical possibilities of life. This team calculated that oxygen developed 130 new molecular scaffolds for signaling and metabolism. Another analysis concluded that 97.5% of the molecules that send signals to the nucleus were made possible by oxygen and oxygen-based metabolism. The explosion on the biochemical level mirrored the explosion on the biological level, and it used oxygen in new ways. Higher levels of oxygen may have allowed it in the first place.

EVOLUTION FLOWS INTO THE NEW SPACE

The fossil record of the Cambrian explosion is marked by an exuberant experimentation of shapes, as if life itself was trying on new outfits and trying to get this multicellular thing right. This large-scale diversity of shapes comes from an underlying unity of developmental signals provided by oxygenated molecules. As hormones and other signals guided cell growth, cells in one organism would grow with a twist, while in another organism, the cells would grow straight.

Inside the different shapes, the give and take of evolution allowed cells to specialize in particular chemical jobs. Some cells changed shape, stretching and contracting in response to calcium signals, and these became muscle cells. Some cells responded when light or sound hit them, and these became eye and ear cells. Some cells on the surface toughened into epidermal cells. Some specialized in doing tricky chemistry, and these became liver cells. From signals and specialization, organs evolved.

From this profusion of shape and form, one branch led to animals with a backbone and four limbs and a head, like us. Gould concluded in *Wonderful Life* that this meant that the nature of history was contingent, random, undirected, and even drunk. I think Gould focused too closely on the dazzling shapes of this period. My chemical perspective sees a predictable pattern of new elements (oxygen, calcium, copper, and zinc), which opened up the possibility of space that life filled with seeming joy.

At the species level, Gould saw disordered contingency, with species flowing like water. At the planetary level, I see an ordered experimental process that optimized new chemistry, built on the same old chemistry, drawn toward the potential of oxygen like a river by gravity.

In any explosion of new forms, efficient forms would gain an advantage, especially after life over-expanded and ate up the food. Paleontologists have identified one branching body plan that went extinct at this stage because it wasn't efficient enough. Replaying the tape of life would repeatedly reject this body plan, because at the end of the day, this form couldn't compete. If the forms that were better at processing food and oxygen survived, these forms would have a higher energy rate density by

Eric Chaisson's measure, and so the energy rate density of life forms would increase, ratcheted up by thousands of similar steps.

This menagerie of shapes was created through slow changes of growth and development. The shape of an organ can have subtle but important effects on the survival of an organism. The intestinal crypts below the villi have a particular curved shape. This shape is particularly good at *slowing down* evolution for cells at the bottom of the curve—because here, evolution causes cancer. The randomness of evolution is held back by the very shape of your intestinal crypts, which protects you from colon cancer. (Long-lived plants protect themselves with a similar structural damper on mutation and evolution, so this shape may be predictable and universal.)

If one arrangement of cells slows down evolution, then another may speed it up. In particular, plants speed up evolution when they need to. In plant genomes, genes are doubled, shuffled, recombined, and passed around at a rate impossible for animal genomes. Often a plant's genome has a few extra genomes from other organisms, thrown in for no obvious reason. As a result, the nuclei of plants like *P. japonica* swell to sizes several times human nuclei, and hold 50 times the genetic material.

For plants, many evolutionary steps forward coincided with a "whole genome duplication" event that made a second copy of every single gene. One set of genes can change drastically, while the second provides a backup for when things go wrong. Whole genome duplications happened when plants learned to produce flowers 200 million years ago (by sending a signal with a specially shaped oxygen-carbon-phosphate molecule).

In plants, whole genome duplication can happen in response to stress, such as being chewed on by a herbivore. Such damaged plants can duplicate their genome in response, providing more genes for more experimentation, effectively speeding up evolution so that the plant can adapt to whatever is damaging it. This kind of mechanism can speed up variation when an environmental stressor like a new toxic chemical appears.

The flipside of speeding up evolution in response to stress is that it can slow down in the absence of stress. Fossils of ferns found in Sweden from the early Jurassic period, about 200 million years old, are preserved well enough to see the species' genome size. As far as we can tell, it is exactly the same now as it was then. These ferns are truly "living fossils."

Plants can transfer genes "horizontally" from one plant to another more easily than us animals. Even the simple act of grafting can create a genetically distinct species. If a plant were a Facebook user, it would open all the privacy settings without a second thought.

In the face of all this genome fluidity, you might expect that plants would be essentially fluid and random, showing constantly high rates of evolution. But you would be wrong. Plants may develop some structures randomly, but at two levels, the randomness is shaped by order.

At the first, molecular level, the new shapes are not chemically random but are chemically similar. They are carbon backbones (sterols, sugars, or chains) with oxygens added around the edges. At the second, functional level, these diverse signals reshape the organism by fitting it into its environmental niche. Because environments are organized by predictable chemistry, the plants fit into their environments in hierarchical, even organized ways. The species may differ, but the ecological networks have similar shapes. To emphasize the biological and genetic randomness is to minimize the chemical and environmental order, while both interplay to make evolution work.

One study of plant immunity clearly shows this order. Again, the scientists used *A. thaliana* (the other plants must be green with envy). They mapped almost 10,000 of its proteins to see which proteins are targeted by a bacterial infection and by a fungal infection. Despite the fact that 2 billion years of evolution separate the fungus from the bacterium, the two organisms converged in their choices, picking the same targets. Both infections attack the most central, highly connected "hub" proteins in the plant's protein network. Like hackers trying to get through a computer network, they focus on the same points because those are the logical and predictable points of attack.

This ability of plants to change their genomes rapidly to adapt to new environments makes even dramatic evolutionary changes repeatable and predictable. One of the biggest recent biochemical innovations in plants is the transition from the old C3 photosynthesis to a new C4 photosynthesis. The new C4 photosynthesis works in hotter, dryer climates by collecting CO_2 in a special cell type before linking it together as sugar. This new cell type and new C4 process evolved independently more than 60 times from the old C3 photosynthesis, allowing diverse plants and grades to expand into new, dry environments, breaking up the earth with their roots.

C4 photosynthesis also is an extraordinarily efficient process and gives the plants that employ it a high energy rate density (ERD). Primitive plant-like algae and plankton have an ERD of 0.01 watts per kilogram, and plants took steps forward in ERD several times: 350 million years ago with evergreen trees (0.5 W/kg), and 125 million years ago with deciduous trees (0.7 W/kg) and energy-storing plants like wheat and tomatoes (1.4 W/kg).

Eric Chaisson argues that each step processes more energy and is more complex than the one that came before. Thirty million years ago, C4 photosynthesis provided maize and sugarcane, with ERDs of about two watts per kilogram or more. This is so much energy that the plant could afford to save it in the form of starch or sucrose. Is it accurate to say this sweetened the planet?

From extensive study of plant genes, scientists figured out how C4 evolves from C3 (yes, present tense, because it's happening somewhere, *right now*). The pattern is the same as when multiple genes change together in discrete "modules": first a pump, then a shift in where electrons are moved, then a move of the main photosynthetic enzyme RuBisCO. These modules even look like computer sub-routines, and this new, complex process was assembled with steps so predictable that a computer could retrace them. No wonder it evolved so many times.

THE CAMBRIAN EXPLOSION II: THE
CARBONIFEROUS EXPLOSION

Plants' ability to evolve and build new structures with oxygen almost drove the planet out of chemical balance 200 million years after the Cambrian explosion. This later period of time is the Carboniferous period. It is as distinct in the fossil record as the Cambrian, and it coincided with a rise in oxygen levels.

Instead of an explosion of shelled forms as in the Cambrian, Carboniferous rocks contain an explosion of plants and fish, and an increase in size of all sorts of life. Carboniferous forests and wetlands resounded with the sounds of huge insects and amphibians, like dragonflies with two-foot wingspans. Even the *microbes* were big in the Carboniferous.

The rocks from this period are filled with rich seams of a new kind of thick black coal made of carbon, which gave the *Carbon*-iferous its name. The black coal is another indication that there were giants in the Earth in those days: the coal is carbon collected by giant trees and plants. (Arthur's Seat, looming above Edinburgh, is a prominent example of dark Carboniferous rock.)

If trees collected carbon in huge quantities, they must also have consumed carbon dioxide, and therefore released *oxygen* in huge quantities. We might as easily call it the "Oxygeniferous" period. This oxygen rise can be seen in Figure 9.1 around 400–350 million years ago. By some estimates, including direct measurements of the oxygen content of bubbles in amber, oxygen levels were about half again as high as today.

The increased oxygen allowed insects and amphibians to grow to huge proportions before the age of dinosaurs. Some theorize that it constituted a Third Great Oxidation Event and corresponding explosion of life forms. Large predatory fish appear in the fossil record at this point, which require lots of oxygen to move quickly and hunt their prey. Ward and Kirschvink see two separate expansions during this period and conclude that both were driven by rising oxygen. These two scientists go so far as to connect specific biological innovations such as egg-laying to high levels of oxygen.

Oxygen also formed the physical links in the new giant plants. At the very beginning of the Carboniferous, some enterprising plant invented the carbon-oxygen armor lignin (shown in Figure 9.2). Lignin is special because its oxygen links are concentrated and extra-hard to break. In fact, for millions of years after its invention, nothing could break it. For a brief period, plants had won the arms race between building and breaking, and they flooded the air with oxygen.

Chemically, the collagen strands holding your joints together and the lignin strands holding that tree bark together are made from quite different starting components, but they end up with similar properties. Both are tough, oxidized linked carbon strands between cells, and both are laid down by similar-looking rings of copper-containing oxidative enzymes. Formation of each is so oxidative that reactive oxygen species (ROS) are constantly flying off these, so they are built outside cells, where ROS do less damage.

An impervious lignin shield meant that other organisms couldn't steal plants' sugar. Your own stomach still struggles to break down tough oxidized strands like lignin and fiber. More than 10% of the calories listed for high-fiber foods like almonds or seeds are locked up in tough, fibrous strands that pass right through you. As a result, foods like almonds are more low-calorie than the label claims.

Lignin gave plants immunity from the cycle of eating and being eaten, and it resulted in an imbalanced planet with too much carbon on the ground and too much oxygen in the air, instigating global cooling. This pattern of growth and cooling probably happened several times. (The first may have been when moss first grew tough enough to grow on land 470 million years ago was followed by global cooling 440 million years ago.) Like grass breaking through a sidewalk, land plants broke up the Earth's surface. Roots weathered the rocks, sending dissolved silicon, phosphorus, and other elements into the ocean. This is turn caused more ocean life to grow and pull carbon out of the air, depleting CO_2 and cooling the planet.

The cycle would have repeated as the invention of lignin was followed by the onset of the Carboniferous and global cooling. It may have recurred 30 million years ago when the invention of tough grass cracked open the Earth, released silicon, and fed silicon-covered diatoms in the ocean, which again pulled down CO_2 and pushed up O_2.

During the Carboniferous, the imbalance between making and breaking, and between oxygen and carbon, may have lasted longer because lignin itself is such a tough nut to crack. It is good that lignin was finally broken because oxygen levels were approaching a hellish point where any fires that started would be impossible to extinguish. You can have too much oxygen.

Finally, a protein evolved that was up to the task of lignin decomposition and could unlock the carbon from its polymer prison. Fungus gene comparisons show that one type of fungus called "white rot" evolved a manganese binding site in an existing peroxidase enzyme, which gave it the power to break lignin's thicket of oxygen bonds. Like Delilah approaching Samson with a blade in her hand, this protein was able to cut the source of tree bark's strength. The white rot fungus used manganese to oxidize solid lignin into CO_2 gas.

The Cambrian explosion and the Carboniferous period followed the same pattern of making and breaking, but a different type of biological expansion appears in each. In the Carboniferous, organisms grew *big*, while in the Cambrian, they grew different shapes. The Carboniferous explosion was more quantitative, the Cambrian more qualitative. I think this was because the Cambrian explosion had an additional chemical boost that promoted biological invention: an increase in dissolved calcium.

WHEN IS A CHEMICAL CASE CLOSED?

Oxygen provided not only the energy but also the chemical capabilities and even the metal catalysts for advanced life. Oxidative weathering scraped clean the continents, releasing phosphorus for growth, and new life subtracted CO_2 while adding O_2. The same weathering released toxic amounts of calcium, an element that was rejected but then became the cornerstone of shells outside and bones inside. The newly energized and toughened life forms burrowed through the oxygen-linked mats and built new niches for themselves along with new access points for oxygen. Finally, oxygen repurposed old molecules into new shapes that sent new signals from cell to cell, from species to species, and from organ to organ. Oxygen was both reactant and fuel, providing the chemical impetus for multicellular life.

After the Cambrian explosion, life experienced several crises as it adapted to the new normal. Waves of anoxia swept through the oceans and species went extinct at alarming rates (although, because every extinction is different, one mass extinction coincided with a *rise* in oxygen, but it is the exception to the rule). Some scientists speculate that a high-oxygen environment is a high-risk/high-reward system in which the reward is that many new things can be done with the oxygen-fueled energy and chemistry, but the risk is that separated organisms are centralized complex networks prone to crashes and fluctuations. Life grew more unstable as it grew more complex, like a computer's operating system.

I have argued that oxygen is behind each of these changes, but not everyone is convinced. In particular, a Cambridge scientist named Nicholas Butterfield has taken the role of "Oxygen Contrarian" in the literature on the both the Cambrian and Carboniferous explosions. Butterfield notes how we see transitions from anoxic to oxic environments today when turbid lakes are cleared by photosynthesizing plankton that eat up the tasty turbid particles. In these and other systems, "keystone" organisms create their own niche, burrowing into their environment with oxygen following behind, as effect rather than cause. Butterfield suggests that something of this sort happened at the Cambrian explosion and that advanced organisms caused the increase in oxygen, rather than vice versa.

Butterfield may be right to be skeptical about the Carboniferous explosion, which corresponds to a smaller relative rise in oxygen levels as well as less new chemistry in the biological and genetic record. The Carboniferous explosion did not have as many lasting effects as the Cambrian explosion. He's also right that these expansions tend to be followed by contractions, and that the line of oxygen moved forward and back in the short term as it inched upward in the long term.

But recent geological data put oxygen first, allowing us to trace pre-Cambrian oxygen levels with unprecedented accuracy. These reinforce the story shown in Figure 9.1. One study shows that oxygen levels were very low right before the Cambrian explosion, and much higher right after. Another suggests that it wasn't as much about reaching a new high level of oxygen as it was avoiding previous low levels of oxygen, which still retains oxygen's status as a prime suspect for causing this event. Looking at the sequence as a chemist, oxygen may not have acted alone, and it may have been aided and abetted by increased levels of calcium or other elements from newly dissolved rocks. An emphasis on multiple important element levels in the ocean may help us completely untangle the cause of the Cambrian Explosion.

Moreover, in another study, the burrowing activity described by Butterfield appears to have caused a short-lived oxygen spike and drop 520 million years ago as a clear *result* of the Cambrian explosion. This leads to two interesting implications.

First, burrowing appears to have stabilized the planet's chemistry. It connected the oxygen and phosphorous cycles so that feedbacks between them would stabilize each and dampen the magnitude of later booms and busts. As a result, later events like the supposed third Great Oxidation Event were relatively less drastic.

Second, the burrowing ultimately is a mixing of the planet that exposes more material to oxygen's effect (through branching patterns that look like Bejan's river-trees). This provides more space for life. I think more space for life, in the end, leads to more oxygen released and energy processed, with more chances to increase energy efficiency and energy rate density through evolution. In the long term, then, burrowing may cause a temporary increase or decrease in oxygen, but oxygen trends upward as long as the sun is shining on oxidizing life and the planet cools to release less reduced material.

My argument is the story of this chapter. It is both an ordered series of events (with oxygen coming first) and a network of diverse pieces of evidence, such as how signaling molecules were made possible by oxygen. This could be investigated with experimental evolution of signaling and intercellular links in different levels of oxygen.

I agree with Nick Lane that mitochondria allowed the energy for genome expansion a billion years before oxygen. If Lane is right, life had the extra "disk space" for genetic programs a long time before, and developed those gene networks before the Cambrian explosion. The crucial point is that, despite this preparation, the diverse shapes and species did not start until *after* oxygen levels rose.

Most important is the coincidence of all these events. Several changes happened practically at once, after oxygen increased, which can each be related to energy and oxygen. Does burrowing happen without calcium release, and do the continents weather enough without oxygen's oxidative reactions? As a chemical historian, I follow this chain of events back, and I keep coming back to oxygen both standing as the first domino and participating in the later steps.

To the extent that oxygen reaction ordered and enabled the course of evolution, this force was made possible by oxygen's special chemistry coming from its special place on the periodic table. When the systems of complex life emerged in the chemical context of this planet, evolution allowed life to fill the planet and respond to changes. Similar ordered functions resulted from dissimilar biological arrangements that themselves rested on the same ordered periodic table. Randomness played a role in this system, but it was constrained from the bottom up by chemical reactions and from the top down by the chemistry of the environment.

Here, evolution is like water flowing through a riverbed. The water molecules flow randomly, as biological species change randomly. But the random change is shaped by the environment, whether rocky or smooth. When observed from a distance, the major events become predictable, and even beautiful, as the water flows over the constant stones and flickers in the changing patterns of sunlight. The river is never the same river twice, but the waters always follow the same course, driven by gravity, hemmed in on the right bank and the left, as evolution is hemmed in by the dual forces of geology and chemistry.

10

THE RETURN OF THE EXILED ELEMENTS

the shape-shifting wave element // calcium trumps magnesium // sodium and chloride return // ramping up by shutting down // from vampire bats to pandas // the fast energy of bird wings and human thoughts// neurons vs. nucleotides // evolution's predictable convergences // ruins come to life

THE TWIST YOU COULD HAVE SEEN COMING

The last page in a comic book is often a cliffhanger, so you'll be more inclined to buy the next issue. It happens so regularly that as I read through the comic (yes, I still read a comic or two), I find myself trying to anticipate what kind of twist will be on the last page. The best twists are the ones you could have seen coming, but didn't.

The story in this book also has a chemical twist here, near the end. This twist is innovative, expensive, and predictable from chemistry. For this twist, the periodic table plays spoiler.

Before the Cambrian explosion, hidden in the nets of signaling proteins within cells and signaling molecules outside cells, the cells held a secret chemical potential that could send a much faster signal, built from four elements involved in two of the balances set up in Chapter 5. This form of signaling would be incredibly expensive, but also incredibly fast. It would be electric in its nature and in its effects, the basis of both muscles and brains.

Like water flowing randomly down a rocky slope, this fast signaling built from fast chemistry spread out in many different ways in life. At certain points, evolution came together and converged, repeatedly finding that a particular shape or signal was the best solution to a particular problem. Because the liquid flow of life was increased, it could diverge and converge more quickly, while predictably fitting into the shape of its landscape and efficiently moving downhill.

The fast chemistry that forms the basis of fast muscles and faster neurons developed with the Cambrian explosion, along with oxygen and calcium use. The explosion of life provided predators that ate and prey that was eaten. Oxygen's energy (resulting from its place on the periodic table) allowed more complex food chains, with more predators and more prey. For example, some calculate that more oxygen in the late Cambrian made more predators evolve.

In response to this oxygen, certain species moved onto dry land, where they had more contact with that element. Plants and insects colonized dry land at the same time. The extra oxygen in the air allowed land insects breathing that air to enjoy a 10-fold metabolic increase when moving quickly (running or flying). Crustaceans can gain only a 3-fold metabolic increase in the same situations, so most of them stayed in the ocean as insects crawled onto land. A network of eating and being eaten soon developed.

Nick Lane explains in *Life Ascending*:

> oxygen respiration is about 40 per cent efficient, while most other forms of respiration (using iron or sulfur instead of oxygen, for example) are less than 10 per cent efficient. This means that, without oxygen, the energy available dwindles to 1 per cent of the initial input in only two levels, whereas with oxygen it takes six levels to arrive at the same point. That in turn means that long food chains are only feasible with oxygen respiration. (p. 62)

Such relationships shape the overall look of the food chain network in predictable patterns. The old Cambrian predator–prey network was modeled in a computer, and it looks like a modern network, taking the same predictable shape with different organisms forming the links—different species, different time, same network shape. This regularity all depends on oxidative, metal-based chemistry that gave these newly evolved animals excess energy to move around quickly and grow new shapes over time.

This process of building an ecosystem can be seen as a different kind of selection driving evolution up rather than out. The process of species spreading out into cracks and niches of an ecosystem, each diversifying according to its kind and forming stable wheels of energy and material cycling, can be called *horizontal selection*. But when one type of organism makes a different type of organism possible, that would be a different kind of selection called *vertical selection*.

In vertical selection, the "higher" organisms depend for energy and chemicals on the "lower" organisms. As the network grows, many more "lower" organisms allow for more "higher" organisms to develop. This results in a vertical pyramid rising up like a volcanic island from the ocean, with plants holding up the bottom, herbivores above them, and increasingly few (but increasingly complex) layers of carnivores surviving on the energy provided by the lower levels. Resource consumption increases throughout, ultimately spreading out sunlight to higher-entropy energy, and complexity increases at the top. The different layers of organisms may align with R. J. P. Williams's "chemotypes" from Chapter 2.

The organisms at the top of the pile of vertical selection require energy for fast motion. This energy was latent in the oxygen-rich atmosphere and the sugar-rich plants, but the chemical for quickly channeling this motion required a different kind of cell, using an element that was strong enough to change a protein's shape but weak enough to let go quickly when the process was done. One element in the periodic table was both abundant (especially after the pre-Cambrian chemistry released it into the oceans) and able to walk the line between strength and speed: calcium.

CALCIUM AS PROTEIN GLUE

Calcium sits below magnesium on the periodic table, and like magnesium with RNA, calcium glues things together. Calcium is too big to fit between phosphates like magnesium. Instead, calcium turns DNA into flakes of calcium-phosphate snow. Remember that its sticky incompatibility with phosphate is why calcium is ejected from inside every cell. Calcium has a different role to play.

Calcium and magnesium, being in the second row of the periodic table, have two positive charges in water. This means that calcium and magnesium have a high "charge density" relative to the single-charged metals to their left. (For example, calcium glues a gel together about a thousand times more tightly than its immediate neighbor sodium.) On top of this all, calcium is one of the abundant Big Six elements found in Earth, and so a protein can use many calciums at once without fear of running out.

High charge density helped calcium stick to carbonate and phosphate to make shells and bones in the previous chapter. Calcium is also found tangled up and stuck tight in arterial plaques. These show how the default position for calcium is outside the cell. Today, in advanced organisms, that ancient pattern can be reversed if the calcium is let in as a wave—but only fleetingly, lest it cross-link the insides of the cell into sticky solid gunk.

Calcium is strong enough to bend a protein out of shape (Figure 10.1). A protein forms a "nest" for calcium's positive charge with four to six negatively charged oxygens on its surface, and calcium will fly into this binding site. Calcium binds so intensely that if a few of the oxygens are a little out of alignment, the calcium will tug on the rest of the protein and pull the oxygens against itself, like it's snuggling up in a blanket of protein.

Calcium binds and bends the shape of a whole family of protein-cutting enzymes called calpains. These proteins are loose and unformed until calciums bind, drawing oxygens to themselves and reshaping the protein to open a canyon—really, a sort of mouth—that accepts protein chains and breaks them apart. Calcium opens up calpains and turns them on.

Once turned on, the calpains do a lot of important things. For example, we may sleep because of the mess left behind by calpains. Several lines of evidence suggest that complex brains require sleep so they can be cleaned up. Neurons are so energetically active that for every minute of wakefulness detritus accumulates, like fragments from proteases and leaked reactive oxygen species. Calpains in particular leave behind a lot of protein fragments, like pieces of paper that didn't quite make the trash basket.

One theory about sleep's purpose is that all these fragments would eventually gum up the intricate workings of the neuron. Brains solve this problem by shutting down everything during a period of sleep. Spaces swell between neurons, and the cells shovel their trash out there, where it washes away as if down a gutter. Ultimately, this trash is caused by the chemical activity of calcium.

(a)

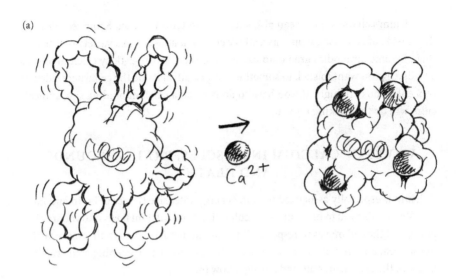

(b)

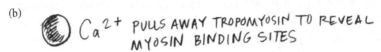

Ca^{2+} PULLS AWAY TROPOMYOSIN TO REVEAL
MYOSIN BINDING SITES

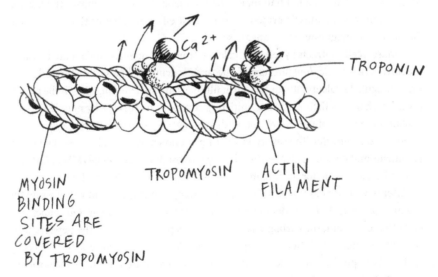

Ca^{2+}

TROPONIN

TROPOMYOSIN   ACTIN
FILAMENT

MYOSIN
BINDING
SITES ARE
COVERED
BY TROPOMYOSIN

FIG. 10.1 Calcium's charge density can move proteins. *Top*: Calcium tightens a protein's shape (calcium shown in exaggerated scale).
Bottom: In muscles, calcium movement pulls a protein strand, opening up binding sites for myosin to move along, causing muscle contraction.

Signaling protein S100B, from B. R. Dempsey et al. "S100 Proteins" in 2012, *Encyclopedia of Signaling Molecules*. Springer, p. 1711.

From J. L. Krans. "The Sliding Filament Theory of Muscle Contraction." 2010. *Nature Ed.* 3(9), p. 66.

Calcium's charge density can also stick two proteins together. Some receptors on the outside of cells use calcium as a sticky claw that attaches to a variety of different shapes. Some antibodies grab onto bacterial molecules using sticky calcium ions as a bridge. If you want to stick to something, especially outside the cell where calcium concentrations are high, all you have to do is make a calcium binding site, and calcium will do the sticking for you.

WAVES OF CALCIUM IN MUSCLES, BRAINS, WOUNDS, AND DEATH

Calcium is especially associated with the energy-intensive tissues of brain and muscle. When cells need to move in a particular direction, calcium flows into the area in a pulse, and the cell moves in response. The calcium is sticky enough to move proteins in the area, and in many cases a wave can be created just by opening some calcium doors in the cell membrane and letting it flow in.

Calcium provides the chemical force for a pull-up. When a muscle needs to contract, calcium ions stick to the muscle proteins and change their shape, opening up a "road" for the muscle protein myosin to walk along seen in Figure 10.1. Thousands of tiny shape changes add up to make one big shape change that moves the muscle. Calcium moves the muscle protein myosin most efficiently, better than the similar plus-two ions magnesium or manganese.

A muscle cell contracts by releasing calcium from a set of intracellular bubbles called the *sarcoplasmic reticulum* (presumably because "calcium bubbles" didn't sound scientific enough). As calcium floods the cell, it fills oxygen-lined binding sites in the muscle proteins, changing their shape and initiating contraction. This is why muscle proteins contain high levels of the two amino acids that have negatively charged oxygens (*glutamate* and *aspartate*). The savory taste of meat known as umami is associated with glutamate, and it is mimicked by the additive monosodium *glutamate* (MSG). Calcium is the ultimate reason for the umami taste of soy sauce and grilled steaks.

After the muscle contracts, a pump in the sarcoplasmic reticulum vacuums up the calcium, burning ATP in order to pack the calcium back in its bubble. One paper suggests that capsaicin, the carbon chain from hot peppers that gives the peppers their spicy burn, interferes with this pump and flips it "on" so that it continually breaks up ATP without pushing any calcium across. Capsaicin may literally burn off calories by short-circuiting calcium waves.

From a distance, the cycles of calcium release and return look like waves of calcium washing through the cell. Heart muscle beats steadily by timing the calcium waves from its sarcoplasmic reticulum in a rhythmic "calcium clock" pattern. Other cells pick up on the beat set by one calcium clock, coordinating in a "pacemaker symphony." Because the calcium ions are charged, this chemical wave is electrical, too, and can be coordinated with an electrical pacemaker.

In the lungs, the intercellular symphony can fall out of tune. During an asthma attack, lung muscle cells are contracting too much out of sync from each other. One class of asthma drug fixes this by sending a signal to close the calcium channels that bring calcium in. If the calcium is stopped, the cells stop contracting and the constriction in the chest loosens. (In an odd connection, the drugs activate bitter *taste receptors* on the lung cells, which are sense proteins we will discuss later in this chapter.)

When a wound punctures a fly's cells, calcium pours into the wounded cells. A hurt cell senses the high levels of calcium and activates calcium channels in the surrounding cells, which also fill with calcium in apparent sympathy. There, calcium-activated proteins make hydrogen peroxide, which attracts immune system and wound-patching cells to the area. The initial trigger of this complex healing response is a calcium wave.

Even a plant may communicate with long-distance calcium waves. A wounded plant cell sends an electrical signal that moves from cell to cell at a speed of 9 centimeters per minute. Calcium pores initiate this signal by letting a wave of calcium ions into the cell. Salt stress also sends rapid calcium-based signals from the root to the shoot of a plant. We think of brains as electrical and plants as decidedly non-brainy, but this finding blurs the lines a bit.

The eeriest calcium wave is found in the roundworm *C. elegans*. When worms die, their guts (under UV light) start to glow a ghostly blue called "death fluorescence." As this glow spreads from cell to cell, the cells acidify and fall apart. Calcium is the angel of death carrying this signal. The blue glow happens when the calcium wave breaks down vesicle bubbles, which happen to include glowing blue stuff.

Since calcium communicates death, a superficial reading of this finding suggests that if you stop calcium signals you'll live forever. But that experiment has already been tried, and it didn't work out too well. If calcium signaling is impaired in zebrafish, colon cells start to grow, and grow, and grow into tumor cells. Here the problem is too *little* calcium signaling. The cancer cells may technically be immortal, but that's not the kind of eternal life anyone wants.

Many brain diseases have a link to the calcium channels that create calcium waves. A huge study searched the genes of 60,000 people suffering from five mental illnesses—schizophrenia, bipolar disorder, autism, major depression, and attention-deficit/hyperactivity disorder—looking for disease genes common to all five. They found five candidate genes were shared by all diseases, and two of these genes were calcium channels. The authors speculated that, despite all the differences between diseases and additional signals coursing through the brain, a drug that blocks a calcium channel might be able to treat some of the mental illnesses on that list.

Finally, the importance of calcium is shown in new fluorescent sensors that track where and when neurons activate. These sensors turn on when they bind calcium, because if you know where the calcium wave is, then you know where the neural circuit is turned on. The chemical element and the biological event coincide.

WHY CALCIUM IS THE BEST ELEMENT
FOR THESE JOBS

Calcium has these distinct roles in signaling because it occupies a unique spot on the periodic table, giving it a distinct size, charge, and abundance. Substituting another element risks a toxic reaction. Two other elements play similar roles, but their chemistry is not quite as good for signaling as calcium's, so their roles will always be limited. Calcium need not fear a coup.

The column of the periodic table that calcium lives in (two from the left side) appears to contain several chemical alternatives at first. Above calcium sits beryllium and magnesium; below it, strontium and barium. In a lab, they might substitute, but we aren't in a lab. We are in a living cell connected to the environment. If this molecule is going to be pulling muscle fibers along after it, we're going to need a lot of it.

The criterion of abundance knocks out all the alternatives but magnesium. Strontium and barium suffer the same fate as most of the bottom half of the periodic table, being too large to be abundant in the universe. In the transition metals, manganese has a size and shape similar to calcium, but it is also too rare to be used in quantity.

Usually beryllium, with only four protons in its nucleus, would be the most abundant element and the favored son in this contest. But beryllium's nucleus is too unstable. On Figure 3.3 in Chapter 3, beryllium is as rare as strontium and barium.

Magnesium is already spoken for because of its association with phosphate. That said, there are lines of evidence that magnesium is used as a secondary signal that can fit into calcium-binding sites. In one blood-clotting protein, seven sites in a row bind calcium, and magnesium can fit into these as well (although it occupies the outside sites, leaving the most central sites to calcium).

Immune system cells called T cells can pump in a wave of magnesium to send a signal. The authors of the paper that reported this finding suggest that magnesium supplements may augment immunity. Nose nerves may use waves of zinc to communicate, too. Yet these examples are scattered and calcium is the default for this kind of signal.

Calcium is used more than magnesium, zinc, and all the others for this kind of signaling, not only because of how *much* calcium is available for binding, but how *fast* it binds and how fast it unbinds. After all, when a muscle moves, it needs to move quickly, then stop moving quickly. Only calcium's timeframe of binding fits muscle's needs (Figure 10.2).

First, calcium is the fastest ion with two charges. Sodium, potassium, and chloride ions bind their targets more quickly, but they only have one charge, so they have little staying power and fall off quickly. Magnesium, zinc, and phosphate all bind too slowly for muscles, wasting precious milliseconds. Figure 10.2 shows a sort of ionic race, and none of the other bars match calcium.

Copper binds about as quickly as calcium, but the bottom of the graph shows copper's problem. Copper sticks so tightly that it never falls off. Zinc is only a little less clingy. On the cell's time scale, only calcium has the right combination of fast-on with medium-off that allows for a firm but temporary signal.

All advanced muscles use calcium, but some invertebrate muscles use phosphate in the same places. Phosphate binds with a more permanent bond that must be deliberately cut, while calcium falls off by itself. In a world without as much calcium, phosphate could have caused suboptimal protein movements that would later develop into calcium-triggered movements.

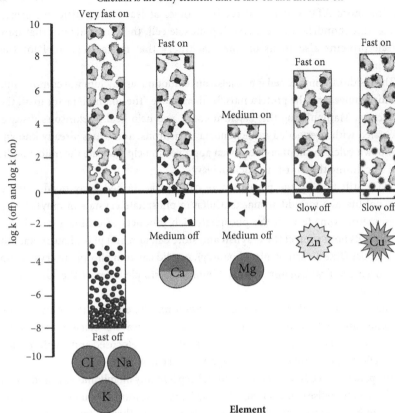

FIG. 10.2 Binding speeds of different elements. Calcium is the only available element that can bind proteins with a fast on-rate and medium off-rate.
Data from Table 2.12, R. J. P. Williams and J. J. R. Frausto da Silva, *The Chemistry of Evolution*, p. 72.

In the Olympics of temporary chemical signaling, calcium wins a gold medal, magnesium takes silver, and zinc would be a possible bronze. Any organism using the periodic table in water-based systems would come to the same conclusion, so I expect that calcium signaling is a universal feature of life.

MITOCHONDRIA UNDERSTAND THE CALCIUM CODE

Calcium is so good for signaling that life sent signals with it long before muscles and neurons developed, even before the global calcium wave that precipitated the Cambrian explosion. Calcium signaling was around long before animals and fungi evolved apart. One way to trace this is through seeing who has calcium channels. Animals have many calcium channels and fungi have fewer, but only a few calcium channels have been found in bacteria. Animals and fungi also have mitochondria while bacteria don't, and they use calcium to talk to their mitochondria. Considering that mitochondria are 1.5–2 billion years old, this implies that calcium signaling is about as old.

Mitochondria use calcium as an "on-switch" for burning more carbon and making more ATP. Calcium directly changes at least seven different proteins inside mitochondria. In a contracting muscle cell, the same calcium that moves muscle proteins also turns on the machinery that makes ATP fuel for those proteins.

The inside of a muscle cell is a maelstrom of motion, as calcium waves move, mitochondria move, and the protein muscle fibers move. The common factor to all these is calcium. Manipulating this calcium could also help heal. Parkinson's disease is associated with overloaded mitochondria in neurons, and drugs already known to slow down calcium channels have been approved to help alleviate the mitochondrial load by calming the churning calcium waves.

We're still finding the places where calcium is stashed away inside different forms of life like forbidden snacks. Calcium phosphate crystals in crayfish shells are sometimes made with the phosphate-rich products of ordinary metabolism. These shells both protect the crayfish and store calcium and phosphate metabolites for later use (like an armor-plated pantry). A solid calcium phosphate crystal could have been the first calcium storage "bubble," a simple form of the sarcoplasmic reticulum.

Calcium's chemical ability to move things is multiplied when the things it moves are other storage bubbles inside the cell (vesicles). A wave of calcium can release a whole bundle of other molecules and its signal can be refocused and amplified, leading to effects that change the entire organism, like insulin release.

The pancreatic cells that release insulin keep it in tiny membrane-enclosed bubbles until a calcium pulse comes along. The vesicles move toward the outer membrane like a fleet of boats approaching a dock. Protein ropes are thrown from vesicle-boat to membrane-dock. Calcium reorganizes and stabilizes the proteins, with movements like contracting muscles, until the boat and dock merge. The insulin inside the vesicle is pushed out of the cell, out of the pancreas and into the blood.

The same thing happens at the gaps between neurons (synapses). A calcium pulse moves vesicles containing neurotransmitters inside the neurons until they merge with the membrane and spill out into the synapse. These molecules are hardy enough to persist across the gap and signal the next neuron.

Because they release so many neurotransmitters, neurons need many calcium channels, which also show up in many neuron-related diseases. Only calcium has the strength to fuse proteins and membranes this dramatically. A variety of molecules can be used as neurotransmitters, but calcium's chemical strength is irreplaceable.

Calcium does have its limitations. Its timeframe of binding is fast, but not the fastest. Other elements are faster but weaker. They can move more quickly as long as we don't ask them to move heavy loads. Long ago, two of these elements were ejected from the cell in the name of electrical balance, but now, like exiled leaders, they return to play a crucial role.

THE OTHER OUTSIDERS RETURN:
SODIUM, POTASSIUM, AND CHLORIDE

It's common to talk about neurons with electrical words, such as "firing" and "flickering," and of getting one's wires crossed. We also talk about them with chemical words, whether as a specific dopamine rush or a general chemical imbalance. Both languages are right. The best language for talking about neurons would be simultaneously electrical and chemical, because neurons' *electrical* voltages come from *chemical* ions. Neurons fire with three chemical elements: sodium, potassium, and chloride (the dissolved, negative form of chlorine).

The final chemical balance of the five found in all cells provided electrical balance with the rule "potassiums inside, sodiums/chlorides outside." Neurons excel in maintaining this balance, with a top-of-the-line sodium/potassium pump that moves two potassiums in for every three sodiums it pumps out. The cell pays for this pump by burning through stacks of ATP.

This sodium/potassium pump is important both medically and energetically. Medically, mutations in this pump that slow the passage of the ions cause neurological disease and hypertension. Energetically, neurons devote their ample mitochondrial resources to this one protein type. Sodium/potassium pumps use a full 25% of your energy when you're resting. They use even more if you're thinking hard. In active neurons, this one pump uses two-thirds of the neuron's energy, because every time a neuron fires, the pump must rebalance (or re-un-balance) sodium and potassium concentrations.

The sodium/potassium pump creates a stark difference between sodium and potassium concentrations across the cell membrane. In this context, if the cell opens up a door for sodium ions (a sodium channel), the ions flow in freely and create a wave of fast positive charge. Potassium freely flows out, so an open potassium channel has the opposite effect and erases the positively charged wave. Sodium channels make positive waves inside the cell and potassium channels take those waves away. The charged wave sends a signal that is as fast as sodium and potassium, and Figure 10.2 shows that sodium and potassium are the two fastest elements in this race.

These channels are sensitive to the electrical charge around them. When one channel opens and lets in charge, its neighbors sense this and open in turn. First, a row of sodium channels opens to let charge in, and later a second row of potassium channels opens to let charge out, each responding to its neighbor like a chorus line doing a ripple maneuver, or a crowd doing the wave. Then the cloud of positive charge moves down the neuron in a single direction (Figure 10.3).

A pulse of positive charge moves across the neuron, injected by sodium coming in and erased by potassium going out. At the right end of the neuron, a different process takes over. Calcium channels sense the wave of positive charge and introduce a wave of the strong, slower calcium ion. The calcium wave moves vesicles full of neurotransmitters to the synapse, and the electrochemical signal moves to the next neuron.

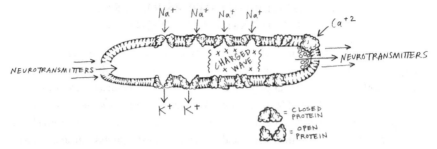

FIG. 10.3 The chemistry of thinking. First, a neuron fires in response to chemical neurotransmitters like dopamine from outside the cell. This starts an influx of sodium ions into the cell through sodium channels, causing a wave of positive charges that travels down the length of the cell. Potassium channels open an instant later, ejecting positive charge from the cell and ending the charged wave. At the end of the cell, calcium ions enter, causing movement of small bubbles that spill neurotransmitters into the synapse.

A mess of sodium inside the cell remains for the sodium/potassium pump to clean up. All is quiet, with the pump constantly pumping and burning ATP, until another neurotransmitter arrives from the left side of the page and the chemical wave begins again.

These sodium and potassium cells methodically upset the electrical balance to create a fast wave of positive charge. At one level, it's just salt in water creating a voltage, but thanks to the coordination of the channels, it moves like a thing alive. After its ancient rejection, sodium returns through its respective channels to send a fast, directional signal.

Problems with sodium and potassium levels are problems for the health of the whole organism. Small changes in sodium and potassium channels can have big effects, which doctors diagnose by measuring sodium and potassium levels. For example, a broken sodium channel causes erythromelalgia, an inherited pain syndrome where the arms burn in pain from misfiring nerves. Blocking pain-causing channels with a specific drug may stop the pain.

Problems with sodium and potassium lead to problems in neurons and muscles for all complex forms of life. Some evidence suggests that roadside salt introduces so much sodium into the ecosystems that it alters butterfly development, both of neurons and of their version of muscle cells. In the sea, oil spills hurt ocean animals when a component in crude oil plugs the potassium channels in animal hearts.

The other rejected element, chloride, should not be forgotten. It is let back in by chloride channels, making waves of negative chloride charge that interact with the positive waves in complicated ways. Chloride must be important because malfunctions in its related proteins are implicated in diverse brain disorders. To give two examples, chloride channels are implicated in how tumors cause brain seizures, and also in a mouse model of autism. In this model, mice treated with drugs that target chloride channels in the womb developed less severe forms of that disease. Chloride has its own areas of importance, but we know less about it.

SENSING THE ENVIRONMENT AND PRUNING THE BRAIN

These pulsing waves of charge explain how a single neuron works, but that's only the beginning of understanding. The human brain has so many neurons (*almost* as many as the stars in the Milky Way) that the real complexity comes when billions of these flickering cells, crackling with activity, are placed in close proximity and then are woven together with branches and support cells and multiple kinds of neurotransmitters. This complexity makes a human brain the most complex known object in the universe.

As a child's brain grows, neurons connect and reconnect, forming networks guided by proteins. Even before birth, patterned waves fire from the retinas through the brain that look like the waves seen after birth when the eyes look around. It's as if the neuron symphony is tuning up and practicing for the post-birth experience.

After birth, when a neuron fires repeatedly, that connection grows stronger. I imagine it thickening when it fires. Maintaining this dynamic network costs an extraordinary amount of energy, and so I tell my students that if they feel tired after studying, that means they're doing it right, thinking hard enough to burn off some calories.

The brain is physically reshaped by the day's events, in "rewiring" events that can be reproduced in the lab. When whiskers are removed from mice, their ability to sense the environment is reduced. Their brains compensate by remodeling neuron connections in specific patterns that are predictable and repeatable.

The brain makes a map from sense data in order to survey and understand its environment. In a stable environment, the brain detects regular patterns and remembers them. Our planetary environment has been remarkably stable for billions of years (Chapter 4), and our galactic environment is stable enough for us to see patterns that extend back to the birth of the universe. It's easy to take this stability for granted (like water's fluidity), but without it (as without water's fluidity), brains could not even function. Brains also need chemicals that work faster than the environment changes, making sodium, potassium, and calcium necessary.

All five senses feed into these brain patterns that serve as inner maps of the environment. As different as the five senses seem, they are all built on the same sodium/potassium signaling chemistry. Once a sensation is triggered, its information is carried through the body by similar neurons and then is encoded and combined in the brain. If the information is useful, its brain area can be expanded and enhanced, and more sensory cells detect and discriminate other patterns.

For example, herbivores that eat more plants have more taste receptors for bitter molecules (and toxic molecules are usually bitter because that sends a clear signal: "don't eat me!"). Because plants are so good at experimenting with evolution, plants are always making creative new toxins, more so than animals. A herbivore with enhanced taste could better discriminate food from poison.

The fast neurons that run on fast elements allow fast change, even evolution fast enough to be measured in decades, not millennia. In the 1980s, cockroach poisons started to include glucose sugar as bait to attract the cockroaches, a spoonful of sugar to help the poison go down. But soon even young cockroaches who never had the chance to learn for themselves wouldn't touch the stuff. Somehow, in the time between Presidents Reagan and Obama, cockroaches evolved.

Specifically, they evolved their neurons. Non-exposed cockroaches lit up sweetness neurons when they tasted glucose, but exposed cockroaches tasting glucose lit up bitter neurons and suppressed their sweet neurons. In a few decades, the cockroach genes changed, and glucose receptors moved from sweet neurons to bitter neurons. Their children avoided the bait and survived. The plastic, adaptive nature of the fast signals sent by neurons allowed this rapid response, to the benefit of the cockroach population (and the chagrin of the human population).

Neurons shape the physical shape of the organism as well. Fruit flies sated with a diet of yeast will grow versions of blood vessels called trachea in their guts, which are branching roads that take in more food, built according to Bejan's tree-making theory. The cells that start the process of building these food-roads are oxygen-sensing *neurons*. (Does this show that fruit flies think with their stomachs?) Food or oxygen deficiency drives neuron signaling through calcium, which drives the growth of trachea branches through the fruit fly gut. The fast signaling of the neurons causes the gut to grow branches in one direction rather than another. Is this a physical or mental process?

In the complex context of the brain, the mechanisms that remove connections can be as important as the mechanisms that make them. Immune cells in the brain called microglia don't attack pathogens, but attack their neighbors the neurons, plucking off stray neuron branches and removing weak connections. This strengthens the connections left behind, so that a well-pruned brain is a strong brain.

Your neurons are pruned and your sensory information attenuated so that what's stored in the brain is less than what is experienced, and what's experienced is less than what is detected. Only the simplest organisms avoid this editing process. Annie Dillard wrote that "[t]his is philosophically interesting in a rather mournful way, since it means that only the simplest animals perceive the universe as it really is" (p. 20). I prefer to think of it from a different angle. Our focusing and pruning lets us see through the superficial chaos, the sound and fury of everyday experience, to the ordered foundation where law and flow interplay.

BRAINS, MUSCLES, AND BIRDS NEED FAST ENERGY

Once these neurons are tangled together in a brain, they form something of incredible complexity. There are 100 trillion connections between neurons (synapses) in the brain. To put this in perspective, there are "only" 300 billion stars in the Milky Way. By measures like this, the brain could be considered the most complex known object in the universe. Marilynne Robinson writes, "If we are to consider the heavens, how much more are we to consider the magnificent energies of consciousness that makes whomever we pass on the street a far grander marvel than our galaxy?" (p. 8).

With Eric Chaisson's energy rate density (ERD), the contest between brains and galaxies turns out to be no contest at all. The Milky Way processes 50 *microwatts* per kilogram of energy, while your brain processes 15 *watts* per kilogram, so by this measure, the brain is thousands of times more complex than our galaxy. If the Milky Way's ERD was a millimeter long, your brain's ERD would be 300 meters.

This happens because stars are very bright but also lightweight. The sun's vast bulk consists of hydrogen and helium, the lightest elements possible, so it is closer to "fluffy" than "heavy" on any cosmic scale. A piece of the sun placed in a full bathtub would almost float on the water.

On the other end of the energy rate density and complexity scale, the human brain is set in a body that processes 2 watts per kilogram overall, a completely ordinary number for animals. Altogether it adds up to say that you (and your dog) glow with heat energy like a high-wattage bulb, and that this energy glows most inside your skull. (When a light bulb blinks on above a cartoon character's head, how true is that?)

You and your dog differ in that you devote a greater percentage of your watts to your brain's galaxy of neuron connections. Your brain is an energy hog, taking 2% of your body's mass but 20%–25% of your resting energy. The neurons in your brain consume energy (ATP and sugar) up to 10 times faster than average cells. To do this, neurons buzz with motion, even down to their mitochondria. Recent imaging techniques show the mitochondria in mouse neurons moving around to where their energy is needed most like support trucks on an airport tarmac.

Neurons are specialized and highly adapted cells, yet their mitochondria run the same combustion reaction as a candle on a birthday cake—they just run more of it. Muscles have even *more* mitochondria for making the ATP energy that drives muscular stretching and squeezing. By providing energy, mitochondria keep the brain thinking and the muscles moving, two traits that distinguish animals from plants. (Plants have mitochondria, too, but they do not use as many of them as intensively as animals do.)

All of this mitochondrial activity gives the brain a special relationship with oxygen. It is the first organ to balk when oxygen levels drop. This has many unintended consequences. For one, submarines must keep oxygen levels low to reduce uncontrolled combustion reactions (fires and explosions, which would be extra-bad on a sub). But this safety measure also impairs the *controlled* combustion reactions of mitochondria in the neurons. On a long submarine voyage, reaction times slow and decision-making falters. Most submarine officers have stories of the amazingly illogical things that can happen when your brain is combusting less oxygen than it should.

Animals with more or more-advanced neurons and muscles have higher energy rate densities in general, giving rise to a hierarchy in ERD that parallels evolutionary history, like that for plants mentioned in the previous chapter, but higher. Fish, amphibians, and reptiles are the earliest animals and have ERDs of about half a watt per kilogram, while mammals process 4–5 watts per kilogram. But mammals do not win the ERD crown, not individually at least—birds process energy at a rate of about 10 watts per kilogram. This fits with structural complexity, because birds pack more complexity into less weight, minimizing mass so they can take flight. It also fits with the evidence that reptiles evolved before mammals, and mammals before birds.

So by the measure of ERD, we are a little more complex than the reptiles, but birds have the most complex bodies, because flight makes so many energetic and structural demands. This also appears to apply to flying insects and presumably would apply to flying dinosaurs if any powered their own flight (an experiment I'd love to try if I could just figure out how).

Bejan's tree-making theory also applies here because air is a flowing fluid that supports the mass of the flying object (whether animate or inanimate). Bejan predicts that all designs of powered flight find similar structures that maximize the efficiency of flight, or to put it another way, that minimize the work spent to travel through the air. Because they have started from many different points but are optimized toward the same efficiency, a graph of maximal flight speed versus mass makes a straight line for insects on the left (with the least mass), birds in the middle, and airplanes on the right (with the most mass). Whether engineered by evolution or by Boeing, a flying object with a certain mass has a particular optimal speed.

The high energy expenditure of keeping yourself in the air means that birds must have more intricate structural complexity than mammals, but when it comes to meta-bolic *focus*, we mammals are indeed more complex than birds. Humans have focused both energy flow and structural complexity in our heads, not our limbs or wings. We developed neurons so complex that our brains process 20 watts using about one and a half kilograms of matter. Ultimately this internal complexity resulted in external complexities of culture and society, which Chaisson also expresses in energy units (more on that in the next chapter).

All of the animals' ERDs listed here are resting rates, so a running horse or jump-ing alligator momentarily increases its ERD. Chaisson writes that "fishing leisurely, playing a violin, cutting a tree, and riding a bicycle" require about 3, 5, 8, and 20 watts per kilogram, respectively, because of the brain power and muscle movement (p. 44).

The chemical basis of these activities is listed on food labels. Calories are energetic units, like watts and ergs. After you eat calorie-rich food, its stored energy is released by oxygen and then put to use in these activities. A violinist turns carbon into a melody, spreading out energy in the form of heat and sound waves to the tune of 5 watts per kilogram, making the world a little more beautiful and complex as a result.

Beyond neurons and muscles, a third energy-burning trait separates mammals and birds from other animals: endothermy, which is the ability to control internal temperature. Reptiles have to bask in the sun to raise their temperature, but we have a biochemical thermostat keeping our insides at a predictable, toasty 98.6°F. Birds and some insects can do this, too.

The energy for this internal heat is also supplied by mitochondria. Some of your heat-producing fat tissue is brown, not white, from all the iron-containing mito-chondria in it. Newborns have deposits of brown fat over their heart and kidneys to keep these vital organs warm. In this tissue, iron chemistry breaks down the fat and pumps protons out of the mitochondria—but instead of using those protons to make ATP, these mitochondria have holes punched in their membranes that let all the protons back in. The energy from the protons is scattered randomly as heat. Not a particularly elegant solution for generating heat, but it works.

In general, animals produce their own heat with more, larger, and/or more wrinkled mitochondria. Some fish build a sac full of warm mitochondria next to their eye and brain tissue, making them "regionally" warm-blooded. The biochemistry inside these mitochondria is unchanged, and because mitochondria used to be bacteria, they even reproduce on their own, making it relatively easy to multiply their numbers.

It is expensive to generate heat this way, and to make the proteins that turn these kludgy reactions on and off. One paper calculates that the precise heat regulation in your body requires tens of thousands of molecules, an onslaught that the authors call a display of "brute force." No price is too high when it comes to keeping a warm-blooded animal's blood warm. The energy is spent to support complex biology.

Once the energy is made, the flow of the energy arranges itself through a complex network of blood vessels that branch throughout the organism. Adrian Bejan's tree-making theory shows that the blood vessels have very efficient (possibly optimal) cross-sections and lengths for carrying the flow of warm blood to the farthest reaches of the organism. This pattern of heat flow predicts that a mammal's mass must be supported by a specific, optimal metabolic rate. Observation matches Bejan's theory, because despite the diversity of shapes and species, a warm-blooded animal's metabolic rate can be calculated from its body size alone.

The structures that take in the oxygen also appear optimized for efficiency. Figure 4.3 in Chapter 4 shows the branching pattern inside your lungs that brings oxygen into contact with your bloodstream with near-maximal transfer of oxygen, and how that pattern matches that of trees and rivers. All these are patterns that spread energy or matter throughout a volume.

Oxygen efficiency causes predictability. Fish with very different shapes and sizes all have the same spacings between certain structures inside their gills, because these structures have been spaced to maximize the rate of oxygen intake. Scientists tried different spacings on an artificial "gill on a chip." The best spacing was the one fish already used. Instead of "the best answer is in the back of the book," we might say, "the best answer is evolved in the fish."

Mitochondria are at the heart of these patterns of profligate energy use in neurons, muscle cells (especially for flight), and warm-blooded heat. Such vast amounts of energy can only be accounted for by the abundance and reactivity of oxygen, as shaped by the iron and copper in mitochondria. These three areas of seeming overinvestment by the animal must be of crucial importance.

WHY BRAINS ARE BETTER THAN DNA

As organisms evolved, resources poured into complex systems: systems to keep blood warm, muscles to move animals across the ground and birds through the air, and, of course, brains. But why are brains so much better than DNA, the previous information-processing system? Both neuron connections and DNA undergo patterns of expansion and loss, respond to the environment, and record information, but in all these things brains are more complex, faster, and just plain better than DNA.

For example, brains win the numbers game because humans have more neurons than DNA bases. Each DNA base can achieve four states, while each neuron can achieve a near-infinity of states, responding to many different neurotransmitters, connecting to many other neurons, and even firing with differently shaped charged waves.

Because DNA is changed through survival and selection, in order to change it, somewhere something has to die. David Krakauer writes that when using DNA, "[t]he amount of information that is acquired for performing a preferred function is directly proportional to the loss of life" (p. 232). But brains dynamically change and record information throughout an organism's lifetime. Brains remove a burden from genomes by processing information more quickly.

Individuals who care very much about their individual fate would much rather record information in brains and live than use DNA and die. Of course, brains have high ERDs due to all that sodium and potassium pumping, which costs enormous amounts of ATP. But if you can afford it, the brain is the best way to fit an organism into its environment.

As brains go, ours may even be optimal. Our neurons use the optimal chemistry for fast signaling out of the available elements (sodium and potassium). Human brains may also make optimal connections. Human brains are special not because of their size (elephants and whales are bigger) or the nature of their neurons (there's a few weird ones but not enough to alter the entire organ), but because of the sheer number of connections between neurons.

Neuron size may be optimized. Neurons can only grow so big before running slow or leaking ions, and so they can't increase complexity by growing bigger. What they can do is connect to more of their neighbors. One hypothesis is that "tether cells" tied primate neurons together, but the expansion of the evolving human brain tore these tethers apart, which allowed more and more distant connections. Neurons can only form so many connections before they max out, so it's possible that they can't get much more connected than they are now.

As more information is taken in through the senses, the brain has more to process, integrate, and edit. As new senses developed and converged, the brain's burden increased proportionately. The current human brain may be maxed out in regard to the amount of sensory information it can process. Of course, considering that our brain can distinguish more than 1 trillion smells, it's not clear what advantage another trillion would give.

In *The Mind's Eye*, Oliver Sacks tells the story of John Hull, who went blind in his mid-forties. Sacks describes how Hull lost even his visual thinking over time. Yet Hull's brain compensated for the loss by processing the information from his other senses in new and surprising ways of nonvisual thinking. The calculus of loss and gain is impossible to fully integrate, but Hull's experience shows that the brain is fluid and adaptive—it can re-edit its information processing and re-optimize the data flow from fewer senses.

FAST ELEMENTS CAUSING FAST ANIMALS AND FAST EVOLUTION

Brains are made of fast neurons, using fast elements, responding quickly to changes in the environment. The ion waves in neurons come from altering old balances inside and outside the cell—even bacteria can create faint sodium-ion waves. Perhaps it shouldn't be surprising that evidence suggests the brain itself evolved quickly, as early as the Cambrian era, into a complex structure.

In 2012, a 500-million-year-old fossil of an animal from the crab and lobster family was found preserved well enough to see the details of its brain. These details had the same multilevel organization as modern brains from this family. (This early complexity is also supported by a study of zebrafish brains.) Simpler does not equal earlier. Some simpler brains may be later, streamlined versions of an original, complex, quickly evolving brain.

A genetic analysis shows that brain proteins have been consistently evolving ever since we split from the fungi (the two other most innovative tissues were both important for immunity, the spleen and the thymus). It would be helpful to find the gas pedal for evolution and to discover mechanisms for speeding up biological change, especially during periods of environmental change, such as times when a new toxic element challenged the ecosystem. Because it is so energy-intensive, the brain is a stressful place, and stress may have enhanced brain evolution.

One strong possibility is that organisms have mechanisms for manipulating their own genes in response to stress and damage, These would speed up evolution when risk becomes worthwhile, when the alternative to change is death. The easiest changes to detect are the largest, so we should look for cases where plants undergo large-scale DNA duplications, even copying their entire genomes, triggered by environmental stress.

This may be seen in genes related to potassium. For plants, potassium is the element of summer and of stress. When summer heat and droughts stress the plant, it needs extra potassium to balance the shifts in osmotic pressure. Plants that make extra chromosomes have more "genetic space" to evolve genes that move the potassium around. Extra DNA gives the plant a better ability to control its potassium and respond to its environment.

This won't work for animals because we can't handle extra chromosomes, but we have other mechanisms on a smaller scale to accelerate change in particular genes and gene regions. Animals may duplicate just one gene, but like plants, one copy is used as a functional backup and the other for rapid, experimental change. When duplicate genes fail miserably or are no longer needed, they are overwritten, like old computer files. Gene duplication uses random change, but if it is turned on by environmental stress, that stress can be chemically predictable.

Advanced genetic detail shows different speeds of evolution in different organisms encountering different environments and stressors. It probably doesn't surprise you to read that alligators evolve slowly while mice evolve quickly, or that bacteria evolve more quickly when they first infect a new organism (which may be triggered by a stressful new environment). But it did surprise scientists to see, for example, just how recently polar bears evolved. These surprises tend to be on the side of underestimating the power and speed of evolution, even among evolutionary biologists who are very familiar with their subject.

Environmental novelty spurs evolution, which spurs more novelty. Evolution speeds up in symbiosis and in more complex environments. It slows down when environments are filled and during the battle for iron. The transcription factors that turn on DNA organize into a robust network that is itself more evolvable. Evolution evolves, and the most common factors for speeding up evolution are novelty and stress. In this story, oxygen provides both.

Evolution can be seen when comparing sodium and other ion channel genes, which show telling patterns of duplication and loss. Older genes have genetic signatures that betray their age, and we can trace the patterns of duplication from species to species. Scientists conclude from this that sodium channels are the youngest child, having evolved more recently than potassium channels (which regulated osmosis long ago) and calcium channels (which created calcium waves long ago to communicate with mitochondria). Of course, in this case "more recently" means about 800 million years ago, a few hundred million years before the Cambrian explosion. In the time since then, the sodium channel made up for lost time by becoming one of the most frequently duplicated genes.

The change from channel to channel is surprisingly easy, and it could have happened both quickly and multiple times. It takes only a handful of changes to turn a particular sodium-potassium channel into a chloride channel. Likewise, a calcium channel can be changed into a sodium channel with a few tweaks, and evidence shows that at least once, when the sodium channel gene was disrupted, a calcium channel substituted and kept this vital function going. For one class of protein pumps, evolution can be traced backward from sodium-potassium pump to copper pump to zinc pump to proton pump.

Sodium and calcium channels are practically interconvertible, and sodium channels probably came from mutated calcium channels. Some one-celled organisms have a sodium channel that conducts both sodium *and* calcium. This could be a half-changed calcium channel, stuck in the place between old calcium channels and new sodium channels. A few bacterial sodium channels (including one from Mono Lake) can be changed into a calcium channel with a single mutation. Such a mutation could have accomplished the same change 800 million years ago when sodium channels were born.

The transition from calcium channel to sodium channel is easy enough that it could have happened multiple times—and it looks like it did. A sea anemone sodium channel looks suspiciously like an older bacterial calcium channel, with some distinct features that suggest that this type of channel turned from calcium to sodium by itself, rather than with the others.

This ease of change may explain some remarkable similarities across continents. Sodium channels rapidly duplicated to make electric fish, twice. African electric fish and South American electric fish are separated by half a planet, but each developed an electric mechanism for sensing the environment through muddy water. They even developed it in the same way, by duplicating sodium channel genes, placing the extra version in a special organ, and then mutating that version 10 times faster than normal to make super-strong changed waves.

When the process was completed, the modified channels were changed in the same way in the same places, making them identical twins separated by a hemisphere. The investigating scientists call it an example of "striking parallelism." The flow of genes ran the same way on two separate continents, from gene duplication to fully functional electric organ. Not just genes but broad genetic structures were re-created by evolution. The end result was a new electric sense based on an old sodium channel. Such parallel evolution is potentially predictable.

HOW VAMPIRE BATS SEE IN THE DARK, AND HOW THE PANDA LOST ITS TASTE FOR MEAT

Through fast and fluid evolution, the sensory apparatus connected to neurons quickly developed sensitivity to different aspects of the environment. At the creepy end of the spectrum of senses, vampire bats can sense hot blood in the dark because of ion channel evolution. Just like electric fish duplicated sodium channels to gain an electric sense, vampire bats duplicated ion channels to gain an infrared "heat vision" sense.

Vampire bats started the process with a normal heat-sensitive calcium channel. You have this channel in your fingertips, when you touch something well above 100°F, opening the door to a calcium-wave signal. This channel is also found in tongue nerve cells where it responds to, say, hot soup. (The same channel opens when the chili-pepper molecule capsaicin binds to it, which is why these peppers taste "hot"—they are literally sending the "hot" signal.)

Vampire bats repurposed this hot-sensing channel into a warm-sensing channel by cutting off a part of the gene so that the channel protein is incomplete and unstable. Because it is slightly broken, the vampire bat's channel opens more easily, at temperatures down to about 85°F. This channel activates when it is pointed toward the heat of a warm-blooded creature, like a human at 98.6°F. The bat turns a broken protein into a sixth sense.

Senses can be lost as well as gained, as seen in the genetic history of pandas. Pandas are essentially herbivores, consuming huge amounts of plants every day (especially bamboo) in order to meet the energy demands of their muscles and brains. Pandas evolved from large bears that ate meat, but pandas have gone vegetarian all the way down to their genes and have lost their meat-tasting gene.

Recall that the muscles in meat use the oxygen-rich amino acid glutamate to bind calcium, so that glutamate signals the savory umami taste. The bamboo that pandas eat has little glutamate and no umami taste, so a panda used to eating plants could lose its umami taste receptor and not even miss it. Change that "could" to "did"—scientists found the genetic footprint where the umami-tasting gene used to be in pandas.

Pandas aren't the only ones. Hummingbirds have also lost their umami taste. Penguins, not to be outdone, have lost three distinct tastes. In general, it appears that a few general bitter taste receptors can do the job adequately, but specialized diets evolve specialized receptors. Tastes literally change.

This rapid movement of genes is seen in nearly every sensory gene, including the opsin proteins that sense light in your eyes. Opsins are found all over the animal kingdom, often in places that are definitely not eyes, suggesting that they originally had a more general signaling role. Jellyfish have stinging cells with opsins in them. In bright light, the opsins make those cells less prone to fire their stingers. In the hydra microbe, the light-sensitive stinging cells also contain *taste* receptors. The researcher who found this speculates that these cells were all-in-one sensing cells that responded to all sorts of stimuli, from white light to bitter tastes, sending a non-specialized signal back to the brain that only later diverged into separate senses.

The story of gain and loss runs throughout natural history, especially with the fast sensory neurons built on the fast elements sodium and potassium. The closer we look the faster (and stranger) evolution seems, and the more parallels and rules we find.

Both genes and neurons grow quickly and are quickly cut back, their branches constrained by energetic demands, the chemistry of ion speeds, the biology of the organism, and the geology of the environment. Brains and gene trees are both pruned by their physical context, like a topiary sheared by a gardener guided by natural law. They fit into their context like a river into its canyon.

IF YOU'VE SEEN ONE METAL-BINDING SITE, YOU'VE SEEN THEM ALL

In the face of all this biological change, chemistry remained constant. Whenever a protein bound a metal ion, evolution adapted a changing organism to an unchanging chemical. The constant chemistry of the ion shaped the configuration of the ion-binding protein. As a result, metal-binding sites in proteins all look the same, no matter how oddly shaped the protein.

All zinc-binding sites have a shape that fits zinc's binding chemistry, such as four sulfur atoms arranged in a tetrahedron will do. This precise arrangement is seen binding zinc in very different proteins, clearly showing that these sites were brought to this shape by the need to bind zinc, and that these proteins converged on this arrangement, driven by the need to bind zinc.

Calcium prefers six or seven oxygens. Similar calcium sites can be found in many different proteins, showing significant convergence on a single solution. Calcium-binding proteins called annexins have completely different overall structures, but they all show a distinctive pattern of amino acids that binds calcium (by their the one-letter abbreviations, it's K/R/HGD). Evolving independently, these very different proteins all found that this sequence of amino acids would give them calcium-binding ability and calcium's shape-changing chemistry.

In the same way, sodium channels and sodium-potassium pumps match the chemical shapes of the sodium and potassium ions passing through. This chemical requirement is no secret—it leaves the channels open to attack from carefully designed toxins.

If one organism wants to kill another, it can make a chemical that will fit tightly into the target organism's channels, blocking them and altering the charged wave. In extreme cases, a milligram of this channel-blocker can kill the target organism by killing its neurons, and the chemical is called a neurotoxin. Because the toxin-making organism also has its own ion channels, it must somehow protect its own channels from the toxin, or it has just gone through a lot of trouble to invent a complicated suicide mechanism.

This is why the scorpion genome shows a huge expansion in genes that make poison molecules, mutating constantly to experiment on new chemical toxins for this kind of purpose. To the observing biologist, scorpions look much the same as always, but to the geneticist, their toxic genes change and rechange over time.

Scorpion toxin and target channel are tied together in evolutionary flow. As the scorpion toxins change shape over time to bind the target sodium channels, the sodium channels change shape to try to stop the binding, and the toxins change again, and so on. In the long run, the toxin shapes mirror the channel shapes.

Since these neurotoxins target channels with similar shapes, the neurotoxins themselves look similar. Animals like amphibians defend themselves from predators with tetrodotoxin, a small molecule that blocks predator's sodium channels, leaving them with quite a bad taste in their mouths. Then the snakes that eat amphibians remolded their sodium channels so the tetrodotoxin can't fit snugly. Six species of snakes at three places around the world all reshaped their sodium channels in the exact same way, making them worse at moving sodium but better at resisting the toxin (they sacrificed sodium channel efficiency so they could eat toxic amphibians). The shape that resists the channel-blocking toxin is as predictable as the shape that channels the sodium.

Another poison, called cardenolide, follows the same pattern. This chemical is made by some insects and plants, like the toxic flower foxglove. Chemically, cardenolide is a four-ring steroid that looks like a hormone with a strange oxygen-containing pentagon and a sugar. This toxin blocks the sodium-potassium pumps of the animals that try to eat the plants or butterflies, causing an unpleasant death.

Birds or other predators resist cardenolide's toxicity by changing the shapes of their sodium-potassium pumps. Although it seems like there should be many ways to reshape these pumps, 14 different species surveyed in one study and 18 different species in another were all found to make the same shape changes in the same places. These shape changes are predictable, although the mechanisms that change the shapes involve randomness. The path is unpredictable, but the end result—the shape of the toxin-resistant pump—can be written down ahead of time.

Finally, you have no doubt noticed that coffee and tea contain caffeine. Both drinks come from very different plants that make the same small chemical, because both evolved the same shape *independently* as a chemical defense. Caffeine mimics the shape of adenosine (the "A" component of the ATP energy molecule). Coffee and tea plants found that caffeine kills insects because it binds in the place of adenosine (as did the cocoa plant). In us more complex animals, that insecticide chemical blocks just enough brain receptors to give us a buzz of energy in the morning. There may be other plants out there that have independently converged on the same solution as well, just waiting to be turned into energy drinks.

PREDICTING THE PATTERNS OF EVOLUTION

The biological level converges as much as the biochemical level does, as organs and organisms converge on similar shapes for similar functions. Evolution eventually repeats itself and finds convergent solutions to the problems posed by the environment. The question posed to each organism is simple: "How can you use this limited stuff (resources, energy, and volumes) most efficiently?" In each context, those that answer that question best survive and multiply. After a while, the answers developed by the organisms start to look suspiciously similar.

One extensive study of mollusk genes found that organisms that look the same on the surface can have very different genes, and vice versa. Consider a snail, a clam, and an octopus. Which two do you think are most similar? If you said snail and octopus, then congratulations, you drew the same conclusion as biologists, and we all can be wrong together. Snails and clams have similar genes but octopi have very different genes, despite the physical similarities.

If they are so genetically different, why do snails and octopi have heads, while clams do not? The genes show that the nerves gathered together in a head twice, independently. Shells also emerged many times, independently. The more we look at genes, the more we see that complex traits can develop multiple times and can be thrown away as species evolve. Whether the pattern is one of convergence or fast evolution with later gene loss, evolution happened faster than we thought.

Simon Conway Morris is a prominent scientist who has spent much of his career looking for examples of convergence (after he started his career by uncovering evidence of the Cambrian Explosion). In his book *Life's Solution*, Conway Morris lists the areas which he thinks have converged in the same way that snails and octopi convergently developed heads: "the emergence of larger and more complex brains, sophisticated vocalizations, echolocation, electrical perception, advanced social systems including eusociality, viviparity, warm-bloodedness, and agriculture" (p. 307). Many of these depend on neurons, which in turn depend on fast elements.

The senses provide many examples of odd but unmistakable convergence. Katydids are so different from us that they put their ears by their knees. But katydid leg-ears hear with the same *shapes* as human ears: an eardrum, a plate amplifier, and a crest of sound-wave-sensing nerves. The same three-part mechanism with similar shapes developed from very different starting materials.

Other senses also converge. We see convergence in camera eyes, which must have developed independently at least six times. Octopus eyes look like mammal eyes, although the two organisms are far apart on the tree of life. Far from being a difficulty to evolution like Darwin thought, eyes developed repeatedly throughout history.

Convergence can be seen by comparing the ecosystems that evolved on different continents, where different species fill similar roles. Remember that electric fish in Africa and South America independently converged on an electric sense. Australian mammals and North American mammals also look the same, as if they have reshaped themselves to fill similar roles in the ecosystem network. Both continents even have a flying squirrel (Australians call theirs a flying phalanger). But there's one big difference: Australian mammals are marsupials that carry their young in pouches, while the North American mammals are placentals, like us, that don't. The simplest explanation for this pattern is that the first Australian mammal was a marsupial, while the first North American mammal was placental, and then the two ecosystems converged independently.

Islands separated by water can show the same kind of convergence as continents, but on a smaller scale. A group studying lizard evolution on four Carribean islands published a paper with a title that says it all: "Exceptional Convergence on the Macroevolutionary Landscape in Island Lizard Radiations." Scientists don't put "exceptional" in their title unless they really mean it. All four islands developed similar lizard features in parallel, showing "that many features of large-scale radiations may be surprisingly predictable."

Each year the scientific literature shows more examples of the predictability of evolution. For the past decade I have collected papers on the topic for a class exercise, and I typically collect a few dozen without even trying.

Several groups have taken proteins at the beginning and end of an evolutionary path, and made all the variants in-between to see how the protein changed from Function A to Function B while functioning at each step. One group analyzed these experiments and found that the proteins' paths were smooth and predictable—so predictable that they want to modify the old concept of "evolutionary landscapes," because sometimes the paths within the landscape are so well defined that the rest of the "landscape" doesn't change things much. Here, evolution is more a narrow river than a broad wetland of possibility.

Other groups run evolution in the lab, by challenging bacteria or yeast and watching how their genes respond. More often than not, the experiments are repeatable and the evolution is predictable. (One paper even predicts the predictability of bacterial evolution in the lab!) Many times, even the DNA undergoes the same changes in different populations. The figures in these papers show genes ebbing and flowing like flowing rivers through time, with random changes adding up through selection to converge on the same solutions.

This is good news for doctors, because we may be able to anticipate how viruses and other diseases evolve. One group is anticipating the path of malaria's evolution and is designing new drugs that will outflank malaria as it develops, ambushing future resistant strains. Another group found that some cancer patients who relapsed after chemotherapy had all developed resistance in the same way, by stopping production of a gene called PTEN. This new gene may be targeted by a second therapy. We can anticipate that tumors will evolve in that direction.

These experiments usually study populations, not individuals, because convergence is a big-picture conclusion and a group effort. Individual genes may change randomly, but the environment sets the rules of gene selection, and the environment is shaped by constant chemical rules that change predictably.

Microbial evolution is fast, so microbes give the most examples of how in new environments, evolution can quickly converge with predictable paths:

1. An enzyme that adds electrons to mercury changed from a normal-temperature to heat-stable form with only 10 mutations, a small, predictable number.
2. Two very different microbes from different kingdoms of life, which both live in high-salt environments, have genes that look alike and have converged on similar chemical solutions to survive the environmental saltiness.
3. Bacteria in hot and normal environments show genetic signs that they can quickly adapt to a hot environment and then re-adapt to a temperate environment. Once they evolve (or re-evolve), their genes match their environment, whether hot or lukewarm.

This even works in humans. It's a little scary, but it's true that the environment can give your DNA new genes—all we have to do is redefine exactly what we mean by "your DNA." Not the well-protected DNA in your nucleus, but the DNA in the microbes that live in your gut. Every time your diet changes, they change, too.

Asians who eat seaweed have more seaweed-breaking microbes than North Americans who don't. The gut bacteria got the DNA for seaweed-breaking enzymes from ocean-dwelling bacteria, so this DNA sequence moved from the ocean to the human gut. There's also evidence that this type of microbe can be passed from mother to daughter through the mother's milk.

Evolution that runs this quickly causes gene loss as well as gain. Microbes will throw out the kitchen sink if it's not needed in a particular environment. If yeast are grown in a constant environment, they will jettison *all* their signaling proteins. If there are no changes to signal, the cells burn their signaling proteins and throw away the genes. Also, bacteria at some point threw out at least two protein-moving systems. If a network of proteins doesn't fit the environment, that network can be excised from the genome.

Loss and convergence can be difficult to detect, which is one of the reasons that it's taken so long to see some of these patterns. It was originally assumed that "adipose fins" in fish evolved once and were lost multiple times, but biologists re-examining the data now suggest that it's simpler to conclude that these fins evolved multiple times.

Another example with fish involves internal and external fertilization. Originally, we assumed that the simpler external fertilization evolved first and the more complex internal fertilization evolved later. But a species was found that "devolved" from an internal-fertilizing ancestor to fertilize its eggs externally again. That seems as unthinkable as humans reverting to laying eggs, but it happened. Complexity was lost when the new fish converged on an old form of fertilization. In each of these cases, evolution is more creative than we thought, although its creativity is distinctly repetitive.

Evolution is both faster and more comprehensible than we expected. In the lab, we can watch as species evolved their way back from what seem to be huge losses. One bacterium lost its entire flagellum when its flagellum-making gene was destroyed. This would be like someone stealing your car (if you had to build your own car). But in the lab, it took *less than four days* for the bacterium to re-evolve its flagellum-making ability. The bacterium rewired its internal pathways in a two-step process and adapted a protein to flip the switch for the right genes, regaining its flagellum in about the same time as a long weekend.

All this adds up to give a picture in which evolution is more like a fast-flowing river than a dice-rolling casino. The chemical flow of water so important to the history of life fostered a biological flow of genes. As water flows predictably downhill, evolution flows predictably to fill the environment with a diverse network of creatures. Individual species fluctuate and flicker back and forth, like water molecules turning in the turbulent flow, and the overall flow of evolution fits its environment as predictably as a river fills a valley.

WHEN EXTINCTION BROUGHT LIFE

The patterns of gain and loss in the flow of life are writ especially large in the all-too-frequent extinctions that punctuate the fossil record. The most prominent (and recent) of the five major extinction events is the event that ended the dinosaurs, but the Permian extinction 250 million years ago was even bigger. Huge percentages of species disappeared during these events, never to walk the Earth again. In the wake of each extinction, life re-converged through evolution and rebuilt the planet. The stress of extinction actually may have sped up the process.

These extinction events usually correlate with the introduction of rocks from outside the oxidized zone, whether from below (volcanoes) or above (meteors). The third and biggest extinction, the Permian extinction, was preceded by a huge chain of volcanic eruptions. Lava is the least of the worries here—more troubling is the ash that blots out the sun, plus the toxic carbon dioxide and sulfur dioxide gases that warm the atmosphere, choke out life, and acidify rain. Volcanism is an intrusion of the old, reduced Earth into the new, oxidized biosphere, and it shows just how inhospitable those old days before oxygen were to us complex organisms. But it seems chemically inevitable that an active planet will periodically bring its reduced insides in contact with the oxidized outside. The timing of the event is random, but the nature of the event is a chemical consequence of the fluidity of the planet.

During the Permian extinction, the oceans warmed to 104°F. Volcanic gases reacted with and depleted the protective ozone layer. As a result, 70% of the species on the land and 96% of the species in the sea vanished. A "reef gap" appears in the fossil record because the reefs were destroyed. Elements from previous chapters were injected into this chapter, including nickel, which supported methanogens with its hydrogen-moving power in Chapter 6. One group suggests that enough volcanic nickel returned to the oceans that these old organisms grew and filled the atmosphere with methane, suffocating oxygen-dependent life. It temporarily reversed the nickel famine. Dangerous chemical nostalgia engulfed the world.

But perhaps 5 million years after the storm had passed, life began to recover. New reefs formed, different from the previous species but still recognizably "reefy." After the Permian extinction, complex communities survived better and outnumbered simpler communities three to one, when before the extinction they had been about equal in number. Twenty million years after the extinction, dinosaurs and giant amphibians proliferated to fill the Earth.

Today, effects that kill or threaten a branch of an ecosystem can help the ecosystem as a whole. Take modern rainforests: these are diverse hot spots of life because of pathogens. Fungal diseases thrive in the rainforest, and if any one plant species dominates, the pathogens evolve to attack that species. They eventually succeed, creating an ecological void that other species fill. Extinctions created similar voids on a planetary scale that convergent evolution filled with life.

A later, minor extinction 94 million years ago (Extinction #4.5?) shows how "reduced" chemicals can cause widespread death by reducing oxygen. Sulfur from inside the Earth flooded the waters and competed with oxygen, creating low-oxygen zones. At first, the extent of the damage to the fossil record caused geologists to conclude that the low-oxygen zones filled the oceans, but more precise measurements showed that this event extended through only 5% of the oceans. Even this was enough to upset the oxygen levels and cause a minor planetwide extinction. Ocean life is even more sensitive than we thought to sulfur-rich waters. Good thing that today sulfurous waters have shrunk to 0.15% of the oceans, and that the planet itself has moved on from sulfur chemistry.

The last great extinction 66 million years ago is ensconced in every future paleontologist's imagination as the end of his or her beloved dinosaurs. A meteor impact blotted out the sun, destroying the plants and the animals that ate the plants, up to and including the dinosaurs. Obviously, this was bad for the dinosaurs and those who love them. But the extent to which this random event steered the course of Earth's natural history is not so obvious.

First, it wasn't just, or even primarily, a meteor. Increasingly, studies are finding that geological unrest increased well before that fatal star appeared in the Cretaceous skies. The Deccan province, now in India, spewed more than a million cubic kilometers of basalt into the biosphere in about a million years of eruptions that began 250,000 years before the asteroid impact. The eruptions are inevitable stress with unpredictable timing.

Then there's the question of how the survivors responded to the stress. Plants, mammals, and birds each rebounded from the devastation. Plants shifted toward a "fast" growth strategy in order to collect more sunlight more quickly. This could have led to "higher rates of ecosystem functioning" according to one reference, which may correlate with a higher ERD and greater internal structural complexity—roughly, faster evolution.

Although they were already around before the impact, mammals diversified afterward. This would represent an increase in ERD over the heavier, more inefficient dinosaurs. Simon Conway Morris is among those who argue that dinosaurs were already on their way out, and that the asteroid impact only accelerated the process by removing the less-evolvable dinosaurs from the competition for resources. Conway Morris thinks each mass extinction advanced the evolutionary clock by 50 million years. Disaster accelerated change but did not necessarily redirect it. As Conway Morris puts it, extinctions are "paradoxically . . . creative."

This creativity may be seen with birds. The branch of dinosaurs that eventually became birds started to reduce in body size as far back as 200 million years ago, according to careful measurements of fossil leg bones. (Smaller bodies have greater ERDs, so this would represent an increase in ERD.) Structures that would become feathers evolved convergently, using old genes that predated even dinosaurs. The avian body plan was gradually assembled before the meteor impact, but its results would not be seen until after.

This evolution in shape was reflected at deeper levels, in some genes but not in others. Genes specifically from this chapter evolved especially quickly (e.g., brain genes, spinal cord development genes, and bone resorption genes). In contrast, genes from other chapters, such as hemoglobin genes, remained remarkably static. These increased rates of evolution in these advanced-function genes may have contributed to this branch's survival when the sky fell.

A full-genome analysis of 48 different birds dates a huge burst of evolutionary innovation at 60–65 million years ago, immediately after the impact 66 million years ago. This has been described as "an evolutionary Big Bang." Before the impact, we had the ancestors of ostriches. Five million years later, we had distinct ancestors of penguins, owls, cuckoos, doves, and falcons. We may owe the existence of birds to the absence of the dinosaurs. When dinosaurs vacated, birds filled in the space. (Insects may have had their own "Big Bang" right after the Permian extinction.)

What is at stake is the predictability of evolution. Would the mammals with their high ERD be able to usurp their tyrannical dinosaur overlords with their low ERD? Would a naturally cooling planet doom dinosaurs without warm blood? Does intelligence eventually emerge in birds or dolphins if it doesn't emerge in primates? Right now, we just don't know, but given the trends in increasing ERD over time, I side with Conway Morris in suspecting that the mammals would have slowly won out due to their enhanced energy processing and increased complexity, impact or no impact.

If this is right, then the extinction that killed the dinosaurs was a surprising twist followed by a more surprising counter-twist: new life filled the vacated space and innovated so that what came after was more complex than what came before. Like a flowing liquid filling a basin, life filled and refilled the Earth as it combined the energy of sunlight with the resources of air, sea, and land.

THE END OF (CHEMICAL) EVOLUTION

Sometime in this chapter, oxygen-driven chemical evolution ended. Since the Carboniferous era, oxygen levels have changed much less dramatically, dropping to about 15%, then rising to today's 21% level. Some scientists tie the development of large animals and a placenta to this rise, but I'm not convinced. The change is a rise by less than 50% at most, which is not on par with the more than 90% scale of previous increases. This scale of change may allow more energy consumption by quicker predators, but it is not the scale of change that would bring new chemicals out from the Earth. The next agent that would bring new chemicals out from the Earth would be the human race.

Precisely because evolution is so good at its job of increasing efficiency, at some point, it maximizes efficiency and therefore reaches a point where further change is pointless. Once you're at the mountaintop, there's nowhere to go but down. Despite what football coaches say, one cannot give 110% effort, so if evolution if effective, it may find the mechanism for 100% effort repeatedly and then maintain it from that point.

This can be seen in the patterns by which diverse animals forage for food. Math dictates that the most efficient pattern for foraging is to follow a set of rules called a "Levy walk," resulting in many small steps mixed with rare, random long steps. Fossilized tracks from 50-million-year old sea urchins show a near-optimal Levy walk pattern, and older fossils show similar shapes, although they are harder to read. The authors of this study suggest that Levy walks were figured out multiple times by multiple animals. Extinctions allowed the organisms that figured out Levy walks to survive, while those without such efficient rules perished.

Eventually the world changed so that evolution itself changed. Most human brains, including yours, are now set in a biological context where intelligence and brain function are *not* life-and-death matters. Survival of the fittest does not apply in the same way to a society where every human life is valued and helped to survive. Natural selection does not function without death's Sword of Damocles hanging over everything. In the next chapter we will see that genes continue to change in humans today, but their changes are far exceeded by the learning of a single brain.

It goes without saying that this is a good thing. Now we have other yardsticks for human achievement than counting offspring. If we and our brains are near-optimal, human life is equally gifted with potential, no master race or superman is imminent, and evolution has ended in this sense as well.

And yet the world changes and even accelerates. Energy rate density increases at a greater rate today, and complexity has grown, but not primarily from biological change. What happened after human brains evolved is not biological evolution built from DNA, but a pattern of change built from the energy and power of human brains. In the midst of accelerating change, the periodic table still holds in the human era. From the right perspective, we can see how chemistry continues to shape biology, psychology, sociology, and even civilization.

11

HOW CHEMISTRY SHAPED HISTORY

held together like a hive of bees // the creation of nature // the evolution of speech and music // energy rate density rises again // chemistry inside and out // the chemical rules that shaped the Iron Age // layers of metal in rocks encode history

THE EVOLUTION OF COMMUNITY

Just as music must be like a hive of bees, with each note that strains to go its own way gently held to a thriving plan, a great empire depends for its driving force upon the elements that will eventually tear it apart. So with a city, which if it is to make its mark must be spirited, slippery, and ungovernable.
Mark Helprin, *Winter's Tale*

Evolution accomplished its final major innovation 125 million years ago. Not brains—complex brains had already been around for a few hundred million years. Not flight—animals with wings can be found as far back as brains, although the birds that really *flew* developed during this time. It was something that is so commonplace that it seems more annoying than astounding. Evolution invented the hive.

Hives are annoying because of their success. Only 2% of insect species form hives, but all hive-makers weigh as much as the other 98%. The most complex hive-making species exhibit "obligatory eusociality": their genes encode physically different levels, or "castes," each with a specialized job, led by a queen that reproduces for the entire hive. The physical hive itself is easy to see, but the true innovation is the network of life protected inside. The hive acts as a skull for the hive mind.

Eusociality was a step forward, on par with the inventions of mitochondria and multicellularity. These other two great evolutionary innovations made specialized organelles inside cells and specialized organs inside organisms. Eusociality made specialized roles *outside* organisms.

Eusocial species changed externally in protection, specialization, and communication. A hive protects; the queen specializes in reproducing (like muscles specialize in movement, and the brain specializes in fast sensing); and external chemical signals like pheromones form parallels to internal chemical signals like hormones.

The most telling sign of eusociality may seem mundane: organized, communal child care. One extraordinary day, a termite tended a child not its own. Such au pair termites cannot directly pass on genes to the next generation. But loss is gain—the individual took one step back while the species took two steps forward. Ants and termites succeeded by losing the ability to reproduce.

This specialization is coordinated through a chemical communication, mediated by small molecules messaging from bee to bee and termite to termite. Queen bees make a carbon-chain "perfume" molecule that suppresses reproduction in the surrounding bees, which communicate back with molecules of their own. This chemical language can be mimicked with organic chemistry: researchers dabbed carbon-chain molecules onto different insects and saw them change behavior as if they heard the voice of the queen.

The same molecule dabbed on very different insects (bees, wasps, and desert ants) produced the same effect, meaning that the same chemical messengers send the same messages in very different insects. These insects must have developed eusocial organization independently. The same molecules are fertility signals in their non-eusocial cousin species. As eusociality evolved, insects predictably co-opted the same signals to send the same hive-making message.

Chemically communicated specialization may include other species. Leaf-cutting ants keep special fungus gardens that pull inert nitrogen from the air and break it apart into ammonia fertilizer. (Some scientists think we should call them "leaf-cutting/fungus-growing ants.") Chemicals form the bridges of communication between the species. Termites keep similar farms, and they control their fungus gardens so carefully that even the fungus genes are kept the same. In genetic language, the gardens are cloned.

Another species can join the network and increase complexity. The leaf-cutting ants keep their fungus gardens pristine by coating them with antimicrobial chemicals made by a *third* species, a bacterium that lives on the ants. The ants feed the microbes, the fungi make fertilizer, and the bacteria make pesticide. This love triangle of species communicates with chemistry.

The ants have gotten so used to this arrangement that it is written in their genes. The ant has lost all its genes for making arginine, a nitrogen-containing amino acid, because its fungus garden gives it all the nitrogen it needs. The ants can't live without arginine, but they have entrusted its production to their fungal gardens. Other genes have changed, too. The ants have gained genes that cut up proteins, which they use to mulch their gardens.

Ant nerves and brains also changed. Leaf-cutting ants have many new neurotransmitter-making genes that are internal chemical messengers to interpret the external chemical messengers in their symbiotic network. Different castes have different patterns of receptors and will respond to chemical signals differently. As the genes reshape themselves around the chemical messengers, genetic losses and gains interplay, and complexity evolves.

Interestingly, most eusocial species became eusocial not so much by changing their DNA, but by changing the organizing molecules next to the DNA, or by adding a few atoms to the DNA, like icing on a cake (these are what geneticists call "epigenetic" changes)—or even by changing the DNA of their symbiotic bacteria. This may be because DNA is too slow for complex back-and-forth networks demanded by eusociality. In a sense, the DNA is computer hardware, but next-to-DNA molecules are software that can be installed and updated.

These changes can be also be directly caused by the environment as part of an animal's stress response and then inherited by later generations without needing to wait for the DNA to randomly mutate into a favorable configuration. Many epigenetic changes are found in human brain development and human evolution as well. In this way, evolution may have evolved to operate at a different, more tunable level.

The hives built by eusocial species have a stabilizing effect on the local environment. Termite mounds help buffer the surroundings against wild swings of climate. As islands of productivity and stability in resource-limited areas, termite mounds are oases, like coral reefs located above ground. While the coral reefs host a diversity of species, eusocial hives protect fewer species but more complexity within the species. (Coral reefs, too, evolve faster than expected after climate stress, with a mechanism that seems almost Lamarckian, a term I will soon explain.)

Over history, eusocial complexity developed multiple times. Eusociality in the form of different castes and protective hive-like structures evolved at least two dozen times in very different animals like aphids, parasitic shrimp, and two separate species of mole rats. Again and again, a group of animals has self-organized into different castes and has developed a chemical code to keep everyone on the same page. It's reasonable to predict that somewhere, another species will re-evolve eusociality soon.

THE CREATION OF NATURE,
THROUGH CONSCIOUSNESS

What but the energies of the universe could be expressed in the Great Wall of China, the St. Matthew Passion? . . . Yet language that would have been fully adequate to describe the ages before the appearance of the first artifact would have had to be enlarged by concepts like agency and intention, words like creation, that would query the great universe itself. Might not the human brain, that most complex object known to exist in the universe, have undergone a qualitative change as well?
Marilynne Robinson, *Absence of Mind*

The communities of the leaf-cutting ant are almost *too* familiar. I can empathize with an ant that mulches its garden. The ant is better at gardening than I am. Yet if I put myself in the place of the ant, I object. The idea of an odor sterilizing me for the good of the community is an inhuman conceit reminiscent of twentieth-century eugenics. The termite isn't bothered by its lot in life, but we humans would be. We invent hive-like enemies to scare ourselves around the flickering campfire of the TV screen (whether the Borg in Star Trek, the Silence in Doctor Who, or the Formics in *Ender's Game*).

I am not so chemically determined—am I? Are humans eusocial? Eminent proponents of eusociality like E. O. Wilson say we are, but others disagree, and it all hinges on the question of agency.

The wild card is the human brain. This organ of galactic complexity confounds all previous biological paradigms. Even if my actions are completely physically determined, any unalterable fate is buried beneath so many layers of complexity and at least the persistent illusion of agency that, for now, I am as unpredictable as if I had true free will. (At any rate, I choose to act as if I have free will, and you can't stop me.) This complexity means that the emergence of social humans was a planet-transforming innovation.

Perhaps we are eusocial but can opt out. Our eusociality is communicated by words, not chemicals, and our specialization is determined by our brains more than our genes. Human eusociality is complex and fluid, as different from termite eusociality as water is from ice. Humans navigate an external, complex social structure with the internal, complex sense-memory of the human brain.

What's fascinating is that the interior complexity of brains is itself a complex, networked, even *social* web. After all, it's not the number of neurons but the social connections between neurons that exceeds the number of stars in the Milky Way. As much as hive minds creep us out, every thought is the act of a community of neurons, and therefore every mind is a hive mind. The unity of consciousness comes from many neurons working together.

This is seen by scanning brains as they wake from sleep. Neurons are always active, but to be awake and self-aware, separate regions of the brain must talk to each other. When an unconscious person hears a sound, neurons fire in the middle of the brain and pass the information to the front. But the person is aware of the sound only if the information is passed backward again to the middle of the brain, closing the loop. Awareness is a two-way conversation.

Another clue about awareness comes from studies of brain activity with the anesthetic propofol. When this chemical saturates the brain and knocks you unconscious, all brain areas participate in a synchronized undulation. At the moment when awareness is lost, a low-frequency wave of firing neurons sweeps across the brain. Most brain areas continue to connect *within* their own area, but the connections *between* separate areas are lost. As consciousness fragments, so does the connectivity of the brain.

Consciousness integrates all the information from the brain into a single viewpoint, which is the self. A common test to see if other species have this same sense is the mirror self-recognition test, in which you put a dot of rouge on an animal's face and see if they notice it in a mirror as being "on me" as opposed to "on that other animal in the mirror." Not many animals pass this test, but magpies, dolphins, elephants, and great apes have recognized themselves. Some monkeys pass; others do not. Like eusociality, the sense of the self is scattered among species and must have developed independently several times. Our species has both eusociality and self.

Magpies, dolphins, and great apes, especially humans, have hugely enlarged frontal lobes relative to similar species. This focus on the brain is costly. As children develop, they shunt metabolic resources away from the body and toward the brain, literally sacrificing brawn for brains. We are relatively half as strong as our primate cousins (correcting for size), but pour much more energy into the complex tissue inside our skulls. This allocation of resources away from the body is seen even on short times-cales: our species' bone density has decreased measurably in the 7,000 years since we invented agriculture.

The path that evolution has taken through those past 5 million years was extremely complex. The tree of life in this area has been erased and rewritten so many times that the tree metaphor itself seems wrong. Our genome is a pastiche from different sources, including a few Neanderthal genes mixed in and other branches like the Denisovans. Our family tree is more a braid or a thicket. The genes are like a flowing river, branching and rejoining.

What emerges from the dust is us: a single, well-defined species with a single, well-defined brain. We use the same genetic code as all life down to bacteria, and we are so similar to other animals that we catch their colds and flus. Our brains light up the same way as animals with the primal emotions of fear, hunger, and desire.

But there is something different with us, something as distinct as the phase change of water to steam. We can change our own perspective, mentally stepping outside Nature for a moment to study it and ourselves. This moment was when Nature was created, when the mind perceived Nature as something other, something "not me," and yet something that could be predicted and even controlled.

When consciousness emerged, evolution changed. Before consciousness, Darwin's biological mechanisms of competition and cooperation dominated the only paths for change in life. Afterward, increased memory and agency added new possibili-ties. We could learn and change and carve our own hives out of the Earth on an unprecedented scale.

This is the mode of evolution proposed for the natural world by Lamarck and disproved by Darwin. Now, with humans, Lamarck's evolution returns, through a brain built by Darwin's rules. This new type of evolution is built on top of the old, constrained by the old, and faster than the old.

From this, culture evolved, along with the first science and technology. Archaeological finds in Africa may place the first instance of humans performing what can be called chemistry 150,000 years ago. The most fundamental chemical reaction is heating, whether to make food, paint, or metal. Both tools and pigments made by chemistry have been found at Pinnacle Point on the south coast of Africa dating back to this time. Advanced stone tools were chemically tempered by fire, and nearby ochre pigments from the same period—the first art?—must have been puri-fied by rudimentary chemistry and heating. Scattered around are shellfish fragments showing that these people liked their seafood, and that was probably cooked, too.

Pinnacle Point is a focal point of biodiversity and is therefore an ideal place to wait out an ice age. Thousands of species of plants grow in the region, including starchy, filling tubers. The sea provides shellfish protein, but to harvest them, one has to *remember* tides. The area is not big enough to support many large mammals, but this is a good thing for medium-sized mammals like us, especially ones that can map the complex and shifting landscape with complex and fluid brains.

This tiny niche sheltered a small band of humans through an ice age, and when it ended, they had invented fire-worked tools and pigments. This echoes earlier movements in the development of life. First, oxygen stressed the planet as a toxin before it was used to burn fuel. Then geological weathering, pushed along by the oxygen and oxygen-caused glaciations, stressed the planet, cracking it open and bringing calcium into the ocean before the Cambrian Explosion. This time, the scarcity associated with ice ages stressed the populations of primates and cracked open the possibilities of the brain, producing chemistry and culture.

WORDS AND MUSIC SHAPED INTO CULTURE

For Tolkien, as for Barfield, language is not a set of abstract concepts applied at will to phenomena, like a coat of paint; rather, it is the expression of perception of living reality in all its manifestations.
Verlyn Flieger, *Splintered Light*

After a single brain mapped the inlets around Pinnacle Point, it communicated that internal process to an external audience using words. Internal organs "communicate" with specially shaped hormones; outside, brains communicate by reshaping sounds instead of carbon atoms. Sound and word are connected with an inner image through metaphor. When people communicate through words, the brain activity of the listener takes on some of the shape of the brain activity of the speaker. This replication is much faster than any strictly chemical or biological process, and it deserves its own level built on top of the others: the social.

J. R. R. Tolkien intuited that cultures are built from words, because he built his imaginary world, Middle Earth, only after he invented imaginary languages. Tolkien's languages were so detailed that they came together, almost of their own accord, into the stories of the peoples that spoke them, and the geography of the planet that formed the stories. Tolkien created backward, from word to world, and inadvertently showed that languages and maps have a strong and mutual overlap.

Speech must have co-evolved with the brain, because our brains are shaped by the advantages that speech gave. The closest analogue to speech is probably bird song, and bird song shaped bird brains in the same way that speech shaped ours. The speech centers in human brains and the song centers in bird brains have converged on similar shapes, structures, and genetic patterns. The large-scale genome study that showed the birds' "Big Bang" of evolution also showed that genes in songbirds have triply accelerated rates of evolution, which could have caused this neural similarity. Some evidence even suggests that the hermit thrush constructs a musical scale that matches ours (and prefers consonant to dissonant intervals). Have both species discovered a universal music?

A complex conversation evolved between tone, rhythm, and melody that walked the line between music and speech, and helped connect brain to brain. Our sense of rhythm is rare. Only a few other species appear to share it, such as parrots and horses. This rhythm shaped the tones and consonants of speech. Think of how, when you talk to an infant, you draw out the vowels and emphasize the tones. When you are teaching someone else to speak, you subconsciously make your words more musical.

In this way, music restricted the possibilities of speech. Physics shaped speech in other ways, too. Sound waves are shaped by the physics of the atmosphere, so high-altitude languages have more "ejective consonants" that click loudly in dry air. (This is either to make the sound carry farther or to minimize water loss.) When the sound waves hit another human's ear and are transmitted to the brain, the shapes of the brain waves match the shapes of the sound waves. Even the letters used to represent the words, in all languages, match natural shapes in the environment, which are themselves formed by chemistry, geology, and biology.

As more dynamic forms of communication, music and speech foster a more dynamic and faster form of culture. Musical sounds superimpose into a chord, carrying huge amounts of information both consonant and dissonant, as they change over time, setting up dynamic harmony and melody. Anticipation and resolution, expectation and surprise, draw you into the musical phrase, and the music moves you.

You can feel this time-based complexity when you listen to music. Music propels your expectation forward, carrying your perception forward through the progression of chords. Music requires the flow of time. It is a conversation between your brain (shaped by your past) and the sounds you hear (in the present), taking place as you move into the future. Over history, in different cultures, music cycled from periods of simplicity to periods of complexity, and its progression is anything but Darwinian in nature.

The string of notes on a page is one dimension of a multidimensional experience, and a brain must be trained to appreciate it. In the nineteenth century Richard Wagner wrote a four-hour opera, *Tristan und Isolde*, in which no chords resolve fully until the very last note. When I saw *Tristan und Isolde*, all that anticipation both drew me into Wagner's music and wore me out, even physically (much more than you'd think from sitting still for that long).

I had to learn how to listen to opera, reshaping my own brain's perspective on music, until I could step into Wagner's conversation, the dynamic "call and response" of his tense, unresolved chords. I tuned my neurons' fast chemistry in order to appreciate the jagged edge of the dissonant "Tristan chord." And in opera, the words, the music, the sets, and the actors all work together in an overlapping, social pattern to create one unified experience in the audience member's consciousness, like the brain's neurons work together to make your singular consciousness.

The experience of opera is a new thing that is built by reshaping and combining older things (music, plays, and poetry). The brain did the same thing at this time in past history, when it repurposed primitive reward pathways to respond to the complex physical waves of how sounds fit together and change over time to tell a melodic story. In other animal brains, these same pathway loops provide rewards for the simple stimuli of hunger, thirst, and sex. These pathways were co-opted by the musical reward pathways. This may explain all that hunger, thirst, and sex in operas—it's vestigial.

Just like music changed over time, words also changed over time. Because words don't superimpose into chords, they are simpler than music and can be represented as one-dimensional strings of information. Once software was written that could find similarities between DNA genes (which are strings of four nucleotide letters) and trace their common history, it was a relatively simple conversion to apply that software to language words (which are strings of about 20 alphabet letters, depending on the language).

For centuries, philologists have reconstructed the evolution of languages through academic debate and logic. So far, the computer analyses have confirmed the general language trees deduced by the philologists, showing that despite their different methods, philologists and bioinformatics specialists agree. The bioinformatics methods date modern language to an ancestor language 15,000 years old. This date also coincides with the end of an ice age, again suggesting that scarce resources foster innovation.

A few debates deeply split philologists, such as whether the root of the tree of languages sprouted from Anatolia or the Pontic steppes. A massive computer analysis has favored a root in Anatolia. If that holds up, then computers may help resolve that debate about the shape of the flow of words through history. (Although my guess is that before long the Pontic steppe hypothesis will discover its own, competing computer model, bringing the two hypotheses back into a static cold war of ideas.)

We can then move from reading words like genes to reading genes like words. Genes tell their own historical story, which harmonizes with what we know from history's words and archaeology's ruins. Most genetic findings reinforce what historians thought. The story of migration told in the genes of UK residents mirrors what we know about migration from history. Genes confirm national identities: the Scots and the Irish are very genetically similar, despite living on different islands, and the Ashkenazic and Sephardic Jews are genetically similar despite their dispersal. (In their case, Hebrew words shaped culture and kept genetic words very much the same.)

Even our pets have legible genetic histories: the DNA for dog breeds backs up the stories of how and where dogs were domesticated. We can read the same story now for horses, tomatoes, rice, and more. In most cases, the domestication recorded in the genes changed the shape of different animal species in particular ways. We are part of that same process—the human shape has changed according to the "domesticated" pattern, as we tamed ourselves.

Old, diseased bones contain genes for ancient plagues, and their genes are familiar. Medieval leprosy genes look like modern ones, and the Justinian plague in the sixth century was caused by the same *Yersinia* microbes that caused the Black Death in the fourteenth century. Plague has stayed the same, and so has the immune response to plague. Different human populations develop similar immune systems to fight off the same plagues, so here is another way in which evolution converges, predictably reproducing the same result (similar immune system responses) at different times (during separate plagues).

WORDS ACCELERATED ENERGY RATE DENSITY

. . . the ground zero of innovation was not the microscope. It was the conference table.
Steven Johnson, *Where Good Ideas Come From*

The physical complexity brought about by words may be captured by Eric Chaisson's energy rate density (one last time). Chaisson's equation can be applied to any system if you measure the energy processed by the system and divide by its weight. This works for plants, animals, brains, and for collections of plants, animals, and brains, for added-up societies, and for their energy-using devices. If it gets hot and has mass, it has an energy rate density (ERD). Every technology, from the prehistoric cooking fire to the laptop on which I type these words, gets hot and can be weighed. Technologies increase in ERD over time.

Fire was the first of these technologies to be harnessed. We saw it forge stone tools at Pinnacle Point, and it also helped with the advanced human activity that is cooking. Fire breaks down tough parts of food and makes it easier to digest as it kills microbes. This means that more calories can be extracted from cooked food to support more energy-hungry neurons in bigger brains (and the little brains perched around the campfire). Some have even linked the development of fire to the evolution of language, because the extra time provided by fire at night is used to string words together in stories.

Society controlled this energy with its collective brainpower, through words, stories, and music. The fire-using culture increasingly burned plants to give heat at greater rates, increasing ERD. Chaisson estimates that foragers use 2,000 calories a day and fire-using hunters use about 4,000 (including the energy of the fire). This comes out to 2 watts/kg for a human body without fire versus 4 watts/kg with fire.

The next step up in ERD happened with the invention of agriculture, when farmers controlled calorie production with irrigated, plowed rows of plants and ordered pens of animals, 11,000 to 7,000 years ago. Agricultural society directed energy with controlled burns and powerful oxen to break up soil and grind grains, both of which increased ERD. Overall, more energy was released as heat and sweat.

All this brought agricultural ERD up to 10 watts/kg (Figure 11.1). As this happened, words increased in complexity, too, bringing about the *Iliad*, the Torah, the New Testament, and the Koran, although the impact of these is not quite quantifiable.

The next big step forward in ERD was the Industrial Revolution. The list of things that get hot and move expanded to include steam engines and house heating, mostly driven by fossil fuels. Defining more complex societies is a more complex calculation, but Chaisson insists that "what matters is the flow of energy through the aggregated social network." He estimates that ERD for industrial societies increased to 50 watts/kg.

Today, seemingly everything up to and including my phone gets hot and spreads energy. Even refrigerators chill their insides by spreading out heat on the outside. Cars, computers, and planes are now the major energy processors, which process more energy and are more complex than in previous devices. Over time, new devices are smaller and/or lighter, which increases ERD by shrinking the denominator of the fraction. The Boeing Dreamliner is the same size as before but is more energy-efficient and has reduced mass, so it has an increased ERD.

All these calculations are about complexity, not value. If your iPhone has a higher ERD than you, all that means is that the iPhone processes more energy with less

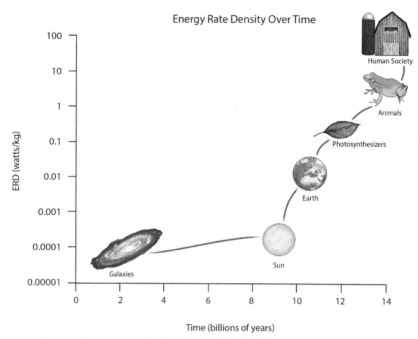

FIG. 11.1 Increase in energy rate density over time, from the Big Bang to now.

Data from Figure 1 in E. J. Chaisson. "Energy rate density as a complexity metric and evolutionary driver." 2011. *Complexity*. 16 (3), p. 27. DOI: 10.1002/cplx.20323.

weight, not that it is smarter or more aware than you. Also recall that birds have higher ERDs than mammals, because they must stay light to stay aloft. Yet aren't you worth more than many sparrows?

Consciousness is more than information, and wisdom is more than energy. ERD is a physical quantity for a system that puts a number on how many moving parts are working together, and it increases over time—as long as the system can be maintained in a cyclic, energy-processing state. Death reduces ERD to zero.

Within these limits, ERD's virtue is its simplicity. It provides a number with which we can compare systems from dwarf galaxies to blue whales to agricultural societies, and over time this number increases. It matches the nagging sense that the world is getting more complex. By connecting complexity to energy, it parallels the second law of thermodynamics, because increasing ERD goes along with the second law's statement that the entropy of the universe must increase. A stable system that spreads more energy out generates more entropy outside itself, even as it lowers entropy inside itself.

Bejan also has ideas in this realm, although as usual, his focus is on the structure of the societal flows. Bejan sees the transportation networks that move people around cities and nations as trees. The tree-like qualities of a map of a city are easier to see if we establish two new things about them: (1) transportation trees are built from road *loops* rather than one-way branches; and (2) a city will have multiple destinations in it, each of which has its own transportation tree that uses the same roads.

(Bejan's theory is about the flows that move on the roads, not the roads themselves.) If so, city maps are actually multiple trees with multiple roots, superimposed.

Bejan theorizes that economic distribution networks, too, establish flows of material with a branching structure built up from smaller parts, just like a branching tree or river. Each of these is an example of efficient, bottom-up complexity. As society increased in complexity, so did its structure, by maximizing its local flow. Bejan correlates this with what economists call the law of parsimony, meaning that his theory even sees economic law as a branching structure of small, improving parts.

Obviously, there is more going on than Bejan and Chaisson capture with their ideas—but it's fascinating that their ideas are predictive at all. For both, the process increasing the complexity is not entirely Darwinian. Survival of the fittest and descent with modification work in the realm of biology, but brains that learn and empathize change by patterns that are better described as Lamarckian. Humans are not coded to form hives, but they choose to build cities, no longer bound to define the world in terms of predator and prey. The development of history becomes less predictable. Perhaps Darwinian evolution is so predictable that it's boring, and our unpredictable brains help keep life interesting.

WE'VE ALWAYS BEEN EATING CHEMICALS

No one is a good historian of the patent, visible, striking, and public life of peoples, if he is not, at the same time, in a certain measure, the historian of their deep and hidden life; and no one is a good historian of the interior unless he understands how, at need, to be the historian of the exterior also.
Victor Hugo, Les Misérables

A man strikes the lyre, and says, "Life is real, life is earnest," and then goes into a room and stuffs alien substances into a hole in his head. I think Nature was indeed a little broad in her humour in these matters. . . . Nature has her farces, like the act of eating or the shape of the kangaroo, for the more brutal appetite. She keeps her stars and mountains for those who can appreciate something more subtly ridiculous.
G. K. Chesterton, The Napoleon of Notting Hill

Chaisson's ERD works because the energy-producing and energy-consuming networks on the inside of your body are connected to the energy-producing and energy-consuming networks on the outside of your body, using flows structured by Bejan's tree-making theory. An outside energy measurement sums up the complexity inside the organism or device. Chemically, inside and outside are connected.

The increasingly efficient energy-processing networks have changed gene-words in the human genome. One such story comes from when agriculture introduced milk into the human diet. Milk brings lactose, a specific sugar that must be broken down with a particular enzyme, helpfully called lact-*ase*. Since we are mammals, we were born with an active lactase gene for digesting milk, so we could take in its calcium, protein, fat, and sugar. After weaning, the lactase gene is turned off by the body's waste-not want-not mechanisms. An adult mammal should have lactase turned off and therefore be lactose intolerant.

When society transitioned from hunting to farming, domesticated animals provided milk for adults (and dairy products, with a little bacterial help). This new source of calories increased ERD and helped humans survive. Any adult with a lactase gene that was accidentally left on could take in more calories and would pass along their turned-on gene to their children.

We can read this in the genes of humans and cows. In different societies, the genes for lactose tolerance and mammal domestication started together multiple times. These genes were written according to Darwinian survival, but survival happened because people chose to domesticate cows and drink milk. Brains directed the gene flow and caused human and cow genes to change together, several times.

Lactase was turned on in the Middle East 10,000 years ago, then it moved northeast to central Europe 6,500 years ago. The ages of the genes match the dates of the archaeological evidence of milk usage (like traces of dairy products in old bowls). This story requires a broad spectrum of efforts from the fields of genetics, biochemistry, geography, and archaeology. This collaboration is its own social network of specialized scientists, working together in a hive of activity.

Society can shape human interior chemistry in negative ways. Paul Tough in *How Children Succeed* describes the chemistry behind educational challenges for struggling students. A few decades ago this would mean a focus on external toxins like lead paint, but Tough instead focuses on internal "toxins" like cortisol.

Cortisol is a stress hormone, a four-carbon-rings-plus-oxygens structure like other sterol hormones. Its signal puts the body in a state of "red alert" in which resources are diverted from long-term building toward short-term breakdown and energy use. But this hormone is only meant to be experienced in episodes. If the stress is chronic, cortisol's protracted alert eats away at the body.

In low-income environments, the stress of being in want and the words and experiences from outside elevate cortisol levels inside, so that a social, intangible network of ideas erodes physical health. Stressed children with high cortisol levels grow up to become physically sick adults who contract a range of ailments. Tough thinks that elevated cortisol takes its toll on the body's complex energy-consuming systems like immunity, so that cortisol connects outer stress with inner health. Tough does deliver some good news: the damage of cortisol can be counteracted with the invisible social power that is parental love and involvement. (Another study found similar results for a different stress marker called CRP.)

The effects of cortisol and CRP show that we are chemical inside and out. This is a great advantage, because chemicals put in from the outside can manipulate, fix, and augment our insides. Even eating is a chemical process by which we break down food chemicals and harness their energy. A river of chemicals flows through us, good and bad, for better and for worse.

Reducing social analysis to chemical analysis will always run the risk of oversimplification, but social status and chemistry do correlate at some levels. For example, how much money you make correlates with the heavy metal toxins in your blood. The poor have high levels of cadmium and lead from toxins like lead paint, but the rich are poisoned with mercury and arsenic from more expensive foods like seafood.

Of course, not all metals are bad at all concentrations. Many of the elderly need zinc, and Chapter 8 described how the chemistry of zinc is associated with late-evolving, complex biochemical processes such as aging. The chemical solution is to supplement the diet with zinc-rich foods or even a spoonful of zinc sulfate powder. The hard part is not the chemistry, it's the *society*. One article suggests that we supplement many common chemical deficiencies in the developing world not with vitamin pills but with "Sprinkles," a cheap powdery salt that contains iron, zinc, iodine, and vitamin A. The article acknowledges that the biggest problem is not paying for the mixture, but getting people to use it. (I myself have an as-yet-unopened bottle of vitamin D pills that I fully intend to take starting on Monday.) Chemistry's effects are not immediate, and the human tendency is to throw away habits without immediate benefits, much like our bodies threw away the lactase gene after weaning.

The changes in the chemicals we eat can be traced through history. Bones and especially teeth preserve chemicals for analysis. Because barium levels in teeth change when switching from mother's milk to other food, they tell us that Neanderthal children were weaned when they were a year old—yet another way in which Neanderthals behaved like us.

A little later in history, during the time when agriculture spread, strontium levels in teeth show that Middle Eastern farmers physically moved to Europe and brought farming with them. Waves of migrating Middle Eastern farmers actively spread agriculture, not Europeans who traveled and then returned.

Instead of teeth, we can look at ancient tartar, which still contains intact bacterial DNA. (Personally, this makes me grateful for the invention of the toothbrush.) Neolithic tartar, medieval tartar, and Industrial Age tartar were compared. The Neolithic mouth and the medieval mouth were environments in which a thousand microbes bloomed, living off the diverse sugars and fats that passed through the mouth. This changed in the Industrial Age, as industrially refined sugar and processed starches simplified the chemical diversity of the diet. The microbes in the mouth co-evolved and simplified, too.

This sounds like a good thing until we ask *which* microbes thrived most in the simplified Industrial Age mouth. The bacterium best at eating simple sugars is *S. mutans*, an acid-producing microbe that is the bane of dentists everywhere. Its genes started changing 10,000 years ago with the advent of agriculture, especially its genes for acid tolerance and sugar metabolism. *S. mutans* evolved to eat sugar and to produce acid that eats holes in teeth.

Together, these examples show that the metals, sugars, and microbes in our bodies and food interact with chemical logic. Because the laws of chemistry have been constant over time, even human history has been ordered by chemistry in predictable and understandable patterns.

CHANGING THE WORLD WHILE FOLLOWING CHEMICAL RULES

Man is a centaur, a tangle of flesh and mind, divine inspiration and dust.
Primo Levi, *The Periodic Table*

I suspect that every chemist, deep down, loves the Star Trek episode "Arena" (I assume here that all chemists are Star Trek fans, and feel free to challenge me if you can find one who isn't.) In "Arena," Captain Kirk has to fight a big lizard-creature called a Gorn using only his wits. Kirk subdues the Gorn with chemistry, by making his own gunpowder from yellow sulfur, black coal, and bluish saltpeter, and using it to fire diamonds at the Gorn out of a bamboo bazooka. (Never mind that the Mythbusters showed this can't work. Kirk found some titanium space bamboo or something.)

Chemists aspire to change the world by channeling energy with clever chemistry, like Captain Kirk. We are still connected to and shaped by the environment, but now we shape it back. First with fire and then with other types of chemistry, ancient chemists changed the colors of Earth and extracted new metals with new properties.

To a certain extent, the order in which new metals were extracted can be predicted from chemistry. Gold, being nonreactive and yellow, would have been the easiest to pull from the Earth. Silver was a bit more reactive than gold and was found mixed up with other things which chemistry had to remove, so it was second in this sequence. As foremost among metals, gold was associated with the life-giving Sun, and silver with its counterpart the moon.

Five metals followed gold and silver, which are found as sulfide ores. The chemistry to remove the sulfur would have been similar: heat ore, remove sulfur, and cool. Tin, lead, and copper were mined and purified, followed by mercury/quicksilver and zinc. These five metals were associated with five of the visible planets in Greek mythology and early alchemy.

These seven metals represent the first time that humans, using the collective brainpower and energy of society, ventured into other parts of the periodic table. The rough order in which they were purified can be read from their chemistry. If the metals are laid out by redox potential, gold is the farthest to the left, followed by the next five classical metals, with abundant iron as the only exception (Figure 11.2). Since humans live in an oxidized environment, they have to *reverse* the oxidation and add reducing electrons to the metal ores to reach these new elements. This is easier to do for the metals with the more positive redox potential.

As in Figure 8.4 in Chapter 8, metals were released into the environment according to their redox potential, but this time it was humans, not oxygen, driving the release. The metals were released in society's quest for more energy and lighter, stronger materials. As miners cracked open the Earth and melted it in forges, the smelting of ores began to pollute the Earth, unbalancing the environment with new chemicals.

Iron's affinity for oxygen means it is found as the oxide, not the sulfide, and a different sort of chemistry was needed to extract it from ore. In the Middle East, the Hittites discovered the secret of making iron, and made war chariots with it.

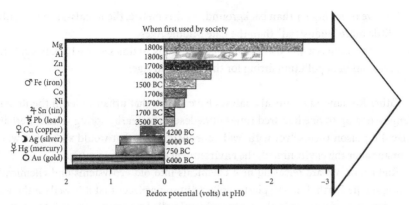

FIG. 11.2 Humans first used metals with the higher redox potentials, because those metals were easier to isolate from ores. Those metals are placed near the bottom of the graph. The oldest metals were associated with the visible plants and have alchemical symbols as shown.

Data from Table 5.18 from R. J. P. Williams and J. J. R. Frausto da Silva, *The Natural Selection of the Chemical Elements*. 1997. Clarendon Press (Oxford).

Humankind turned from the Bronze Age to the Iron Age when their secret was leaked and iron use became widespread.

Farthest to the right (the "reduced" end) sit magnesium and aluminum, which are commonplace now but were not unlocked from the Earth until the 1800s. Both metals are stuck in their ores so tightly that only Industrial Age electricity could break the bonds. Once this happened, magnesium and aluminum were the modern wonder metals. Napoleon III made utensils of aluminum that were valued more than solid gold. Now you can own your own aluminumware, as electrical power has made aluminum commonplace. If you have the chance to time-travel, bring a roll of aluminum foil and you'll be rich.

The chemical redox potentials determined the order of the chemical advances that shaped history. On any replicate Earth, gold and silver should be the currencies of trade, the Bronze Age should come before the Iron Age, and magnesium and aluminum should not be found by the early empires. Chemistry determines the order in which metals will be discovered and used.

Some of these chemical steps forward can be corroborated by physical evidence in the strata of rocks or the layers of microbial mats. One group of scientists dug down through layers of microbial mats off France's Mediterranean coast and found that mercury levels in each layer fit history:

1. The first mercury layers are dated to 500 BC, when Roman mines first opened the Earth in Spain and spread the mercury into the atmosphere and oceans. At the height of the Empire (100 BC–200 AD), mat mercury levels were double the "natural" background levels, and then they fell back as the Empire crumbled.
2. A second pulse of mercury is seen in layers dated 1000–1350 AD, and it is four times higher than background. A third is centered on the sixteenth century, and

it is five times higher than background. By this metric, the Renaissance was only a little more "industrial" than the Late Middle Ages.

3. Mats from the last 250 years show mercury levels 10 times above background, an acceleration of pollution fitting for the Industrial Age.

Other Roman-era chemical analyses have shown that urban water in the Roman Empire had up to one hundred times more lead than nearby spring waters. Perhaps they did poison themselves with lead aqueducts. If so, that would show that when humans alter the environment, the environment can alter us in return.

Right now, we are releasing new chemicals into old ecosystems and chemically changing the Earth. Enough plastic collects in some places that it is well on the way to being metamorphosed into a new mineral called "plastiglomerate." Most cities have high levels of carbon-oxygen drugs in their wastewaster, with unknown consequences for the ecosystems absorbing these drugs. Our fertilizers have put so much nitrate in the ocean that it may change from being limited by nitrogen to being limited by phosphorous. If cyanobacteria changed the world's chemical balance, then it stands to reason that we could, too.

THE UNPREDICTABLE CHAOS OF BEING COMPLEX

[Continued from the first quote in this chapter] . . . *So with a city, which if it is to make its mark must be spirited, slippery, and ungovernable. A tranquil city of good laws, fine architecture, and clean streets is like a classroom of obedient dullards, or a field of gelded bulls—whereas a city of anarchy is a city of promise.*
Mark Helprin, *Winter's Tale*

The predictability of chemistry is a broad, not a specific, predictability. Some culture would have learned how to smelt iron, but why the Hittites and not the Israelites? (Was Israel too bookish?) A culture fosters innovation by providing resources and protecting risk-takers, but the timing of any single advance will always be contingent on complex events. Chemistry predicts no more than that a Bronze Age will precede an Iron Age.

This is true for biology as well as history. Chemistry predicts what *type* of species will survive, but it only places odds on whether a *particular* species will survive. Also, complex systems are more fragile than simple ones. The Cambrian Explosion led to periodic mass extinctions. Life walks the line between order and chaos. Complex systems fall off that line more easily.

Life adapts and carries on. The construction of the US Interstate highway system made countless overpasses that inadvertently shaped the evolution of cliff swallows. When these birds abandoned their old homes to nest under overpasses, the new artificial habitat resulted in new shapes for swallow wings and bodies. Their wings grew shorter to avoid fast-moving cars, for example. The source of this evolutionary change was human society.

Human society has split several species in two, producing urban and rural versions of the same animal, as those animals have adapted to the new urban environments. One example is a kind of fox that has adapted to live in London. The gene flow between the country fox and the city fox may be reduced so much that, soon, they may no longer interbreed—one definition of a new species.

City life provides certain advantages, as shown by the Oxford ragwort. This plant can endure the harsh environment along railroad tracks. By paying that price, Oxford ragworts gain the benefit of dispersing their pollen long distances on the moving trains, and have spread across Britain as a result. No longer is it just the birds and the bees—now it's the birds, the bees, and the trains.

All species do not adapt as quickly. A large-scale study mapped animal species in Thailand before and after forests were fragmented by roads and cities. A scant 25 years after the human activities broke up the forest, almost all small animals died out. Hopefully "almost all" does not become "all" in the next 25 years, but it may.

History is littered with human-induced extinctions. In the 1200s, Crusaders moved north through what is now Poland and brought with them a host of cultural and biological changes. The majestic aurochs died out, large wild species like wolves and bison were decimated, and new domesticated species were introduced. (Notice that the largest, most complex mammals were most vulnerable.) Some of these changes were deliberate, and some were inadvertent consequences of environmental changes like clear-cutting.

Some evidence says that humans changed the environment even in the Stone Age. On the south and west coasts of Africa, near Pinnacle Point, shellfish fossils from the late stone age are distinctly smaller, suggesting that as humans became more human, they "overfished" the larger shellfish out of the population in their quest for calories. Also, human activity 8,000 and 5,000 years ago may have added carbon dioxide and methane, respectively, to the air in quantities that warmed the planet by an extra degree Celsius. If so, global warming got an early start as a direct consequence of increasing ERD.

If technology got us into this mess, we can redirect technology to clean it up. One step is to objectively compare options with quantitative measures. Another is to provide enough resources and energy to help promising ideas succeed. Humility reminds us that previous societies had the best of intentions when they inadvertently poisoned the Earth, and we may be no different. We should direct our steps carefully and objectively, but we have to try to fix our mess.

The chemical story of this planet says that widespread chemical change has happened and can happen again. When that change hit, many good things were lost, until new biochemistry turned toxins into vitamins (as in Chapters 8 and 9). We can retrace this path, finding chemical solutions to chemical problems, and constraining the chaos in our biological, chemical, and social networks.

12

A FAMILIAR REFRAIN

are tapes (of life) obsolete? // Gould vs. Conway Morris vs. Kauffman // paging Charles Darwin // Rothko and the power of perspective // mistaken mistakes and accidental accidents // the melody of chemical evolution // home after a long day

SIX ISLANDS NORTH OF THE ROCKETS

Play the tape a few more times, though. We see similar melodic elements appearing in each, and the overall structure may be quite similar. . . . Look at the tape as a whole. It resembles in some ways a symphony, although its orchestration is internal and caused largely by the interactions of many melodic strands.
Leigh Van Valen, in a review of Gould's *Wonderful Life*

When I was a child, I lived near Kennedy Space Center in Florida. To me, science was all about rocket science: big rockets that flew where no one has gone before, powered by fiery oxidative chemistry and guided by precise measurements. I could watch the space shuttle go up, and trust it would come down, connecting Earth's orbit to the ground at my feet. That we could send hollow metal capsules on such immense journeys was a wonder I never quite got over, not even when two of the shuttles failed to return home.

I had a different view of evolution. It seemed messy, haphazard, cruel, and wasteful, nothing like the sleek engineered rockets at the Space Center. Most of all, I think it just seemed boring. Why study the unpredictable? To be sure about something, you need to see it work multiple times, right?

I missed the chance to see evolution work multiple times, right in my home county. In the next lagoon over from where I once dredged up sulfurous muck for my science projects, on six small islands, the tape of life was being replayed six separate times. All six times, it gained the same result. Evolution might be messy, haphazard, cruel, and wasteful, but it is also predictable.

In 1995, scientists brought Cuban brown anole lizards to six small islands in Mosquito Lagoon, which only had green anole lizards at the time. Each time, scientists watched what happened and compared the results to five nearby islands that remained green-lizard-only zones.

After three years of competition, the green lizards changed their behavior and perched twice as high in the trees. In 2010, the scientists returned and examined the lizards' feet. On the invaded islands, the green lizards had larger toepads with more wrinkly lamellae than on the non-invaded islands. This difference in feet was mirrored by differences in the lizards' genes. In a 15-year span, the lizards evolved. (During the same time span, my opinion of evolution also evolved, as I learned about a chemical order behind the biological messiness.)

The predictable divergence of evolution on small time scales has been glimpsed in many other biologically complicated systems. Predictability is found in the genes of Californian stick bugs perching on different plants, and the genes of laboratory-grown yeast cultures that shuffle genes in a very basic form of sex. These experiments involve the co-evolution of different organisms that interact with each other, or to put it broadly, complexity in the *biology* surrounding a species.

The *geology* and *chemistry* surrounding a species can also cause regular, pre-dictable patterns in evolution through short-term parallel divergence or long-term convergence. These patterns can run deep, as when the type of bedrock far below determines the soil chemistry that shapes which species will grow above it. Giant sequoias grow on phosphorus-rich bedrock, for example, because they need a rich diet. (And don't forget the trees with gold in their leaves from Chapter 2.) In this way, you could map the bedrock from a sophisticated map of trees.

As these trees evolve, the amount of rainfall can set their paths, and sometimes gives them only a few options. When conifers adapt to a dry climate, they need to close pores called stomata in order to hold in the water. They do so with what Brodribb et al. call "surprising simplicity," taking one of only two options to adapt quickly and effectively to the external stress.

One group even found that a new, stressful environment may have caused fish to walk long ago. A long, thin fish called *Polypterous* can survive on land, so they raised one group of fish on land and another in the water. The fish raised on land experi-enced a more stressful lifestyle, being literally fish out of water.

The fish raised on land learned to move in a way that was more like walking than swimming. They also evolved different shapes in the lab. Some of those shapes match shapes in the fossils of the first walking fish. These scientists rewound and replayed the tape of life, and its tune sounded familiar.

HUCK AND JIM, CONTINUED

We had the sky up there, all speckled with stars, and we used to lay on our backs and look up at them, and discuss about whether they was made or only just happened. Jim he allowed they was made, but I allowed they happened; I judged it would have took too long to MAKE so many.

Mark Twain, *Huckleberry Finn*

I first encountered this quote in a comic strip by Berkeley Breathed, a century after Twain wrote the passage. It's an echo of a much longer conversation, not too different from the debates of the Stoics versus the Epicureans in ancient Athens. What kind of universe is this? Is it fundamentally ordered or disordered, driven by necessity or chance?

Through this book I have been in conversation with Stephen Jay Gould's *Wonderful Life*. Gould took Huck's side of the old debate. Gould emphasized the bizarre and alien nature of the Cambrian life forms and inferred that only a tiny number of paths lead to anything like us. Let's return to Gould's quote from the first chapter, expanded: "Wind back the tape of life to the early days of the Burgess Shale; let it play again from an identical starting point, and the chance becomes vanishingly small that anything like human intelligence would grace the replay" (p. 14).

But more evidence has come to light in the past quarter-century. We have found more Cambrian fossils and have seen that the explosion was *longer* in time than the Burgess Shale alone indicated, and the longer it took, the more time evolution had to select the most efficient forms. We have seen stress accelerate evolution in Chapter 11, again making the most efficient form more likely. More thorough comparison of Burgess Shale fossil shapes revealed more patterns, showing that many of them do in fact follow familiar body patterns. Then the debate becomes a he said/she said (or Gould said/Conway Morris said) about biological shapes.

A different angle has come out in the lab, where scientists run the tape of life repeatedly, with microbes in flasks. This field, founded by Richard Lenski, is called experimental evolution. Lenski's experiments have run continuously since *Wonderful Life* was published, producing over 60,000 generations of bacteria and finding rules that predict the direction of evolution in specific cases. Gould's "tape of life" hangs over these experiments. A recent profile of Lenski claims, "Gould was mistaken when he claimed that, given a second chance, evolution would likely take a completely different course" (p. 790).

Lenski's experiments show that evolution presses forward relentlessly, even in the constant conditions of a shaking lab flask. It is easy to see evolution when it diverges to a new function, but it is harder to discern when it converges onto a previously discovered function, or normalizes a metabolic network. The convergence shown in Chapter 10 means that there is more evolution around us than we thought 25 years ago, and it is faster and more effective than originally assumed.

This story repeats when the context changes from microbes to larger creatures. The authors of a study (Mahler et al.) that found extreme convergence for lizards on Caribbean islands (which were briefly mentioned in Chapter 10) come right out and say it: "Gould famously argued that evolution over long time scales is 'utterly unpredictable and quite unrepeatable' (32, p. 14) due to historical contingency. Widespread convergence among entire faunas of Greater Antillean Anolis refutes Gould's claim and shows that adaptation can overcome the influence of chance events on the course of evolution" (p. 292). In science, words like "refutes" are what Huck and Jim would call "fightin' words."

PREDICTABLE CHEMISTRY CONSTRAINS BIOLOGY

Another scientist with "fightin' words" for Gould is the principal exponent of convergence, Simon Conway Morris. In *Wonderful Life*, Gould extolled the scientific virtues of Conway Morris, but later Conway Morris contested Gould's "tape of life" conclusion. Many articles by Conway Morris have long lists of traits that have evolved multiple times to converge on a common structure or function. He has even written his own book on the Burgess Shale called *The Crucible of Creation*, which opposes Gould's interpretation.

In a recent collection of essays, Conway Morris has a typically rich entry on how complexity came from evolutionary convergence. But in the very next essay, Stuart A. Kauffman plays Huck to Conway Morris's Jim. Kauffman is an origin-of-life researcher who first detailed self-catalyzing peptide cycles such as may have occurred at the origin of life in Chapter 5. I think that Kaufman's work on how complex cycles emerge naturally from chaos may be our best bet for moving forward on the origin of life. This emphasis on how cycles may emerge from many possible paths in a complex mixture may have influenced his Huck-like views in this essay.

Like Gould, Kauffman points out the huge areas that life could occupy (but doesn't) and the unforeseeable numbers of paths it could take through that space. A huge number of amino acids could be strung together into a single protein. A biological structure could be purposed and later repurposed in a huge number of ways. Because these possibilities cannot be listed beforehand, Kauffman argues, we cannot know how a species will use a protein and we cannot predict how it will evolve.

Ironically, Kauffman's arguments echo those heard from Intelligent Design advocates like Stephen Meyer. Meyer, too, focuses on the huge numbers of possible amino acids in a protein, but Meyer does so to say that it must have been rationally designed, while Kauffman does so to say that there must have been no design.

Kauffman focuses on examples such as exaptation, when a swim bladder became a lung, and says, in essence, who could have predicted that? I, for one, could not have predicted that—but that's not the kind of predictability we have. Chemical evolution is not a specific prediction, but a narrowing of options and a tilting of the field. I *could* predict that somewhere some species would evolve the ability to collect oxygen-rich air in an open cavity (whether it comes from a swim bladder or some other air sac). The cells exposed to oxygen's chemical potential will evolve mechanisms to withstand its toxicity, and then to use its carbon-burning power.

Evolution reshapes the oxygen-gathering cavity predictably toward greater energetic efficiency. A better shape with more oxygen contacting more cells processes more energy and conveys a selective advantage. The shape of this pouch can start from multiple points, and it can vary and flow over time like water molecules in a river. But just as water molecules are pulled down by gravity, cavities and life forms are pulled toward an increased use of oxygen.

John Torday has proposed another series of logically ordered events that may have turned a swim bladder into a lung, driven by oxygen stress. That argument uses the other side of oxygen's double-edged sword. Whichever road it used, it appears to have happened repeatedly, because swim bladder genes evolved and converged four times in teleost fish, providing many structures from which a lung could develop. Some estimate that fish overall evolved air breathing 68 independent times. These fish took evolutionary paths that differed in the details, but they reached the same destination dozens of times, predictably.

Kauffman's questions focus too tightly on points of origin, arguing about a particular species and cavity, when a pattern can be seen at the point of destination, as embedded in the larger scale of the ecosystem and constrained by chemistry. The originating structure may not be predictable, but the functional chemistry is, which arises convergently from multiple origins. Since oxygen is the point of highest energetic efficiency, life gravitates toward its use with lawlike inexorability. All the rest is glorious detail.

When Kauffman writes, "Not only do we not know what *will* happen, we do not even know what *can* happen," he is too tightly focused on the individual, when patterns can be seen in the chemistry of the ecosystem. Why is it that so many things that chemically *can* happen are never seen? Chemically, strontium can substitute for calcium—why don't we see that? Channels bringing bromide in could erase a positive wave as easily as channels bringing potassium out—why don't we see that?

Even when some scientists thought arsenic could substitute for phosphate in DNA at Mono Lake, other scientists knew that that was not a chemical possibility. The chemists knew that, despite its similarities with phosphate, arsenic cannot form long-lasting bonds in water. Chemistry predicted that phosphate (but not arsenic) could shape the biology of the Mono Lake microbes, which narrowed the possibilities for how those microbes could build DNA. Chemistry narrows down both what *can* happen and what *will* happen.

LEVELS OF FREEDOM

CÉLINE: Well, the past is the past. It was meant to be that way.

JESSE: What, you really believe that? That everything is fated?

CÉLINE: Well, you know, the world might be less free than we think.

JESSE: Yeah?

CÉLINE: Yeah, when given these exact circumstances, that's what will happen every time. Two parts hydrogen, one part oxygen, you'll get water every time.

JESSE: No, no, no, I mean, what if your grandmother had lived a week longer, you know? Or passed away a week earlier, days even, you know? Things might have been different, I believe that!

CÉLINE: No, you can't think like that.

Richard Linklater, Julie Delpy, and Ethan Hawke, *Before Sunset*

If all this talk of predictability feels stifling, it may come as a relief to allow that, at several levels of life and history, contingency and unpredictability rule. The geneticist Michael Lynch has convinced many (including me) that genes drift about randomly inside a genome. Lynch's emphasis on the non-adaptive nature of change applies less well to other levels of life, such as symmetry in proteins, or in other higher-level processes that can be ordered by chemistry and physics. Lynch's genetic randomness allows life to flow into the predictable chemistry of its environment and its ecosystems.

Both environmental/chemical selection and random genetic drift shape the path of evolution. The broader the timescale, the more chemical selection determines the path, as life flows through a chemical sequence set by the periodic table.

On a smaller timescale, random flow can accelerate evolution by altering the genes that serve as on/off switches for the cell. These can be rewired with great fluidity to produce a large-scale change, and they are implicated in some steps of brain evolution. But even these show predictable patterns once you step back and look at the network. These patterns can result in different speeds: mammals evolve a type of "wiring" called "enhancers" quickly, but evolve "wiring" called "promoters" slowly. Despite the small-scale differences in specific wires between mouse and human, a comparison of mouse "wiring" with human "wiring" found that the large-scale features of the two networks were, in the words of the authors, "strikingly similar." Close up, gene expression patterns change; from a distance, however, the core regulatory programming stays the same.

Above the DNA and RNA levels, at the protein level, a certain unpredictability persists. Joseph Thornton and colleagues have carried out a series of experiments reconstructing the evolutionary path of hormone-binding proteins. Some proteins evolve through a specific sequence in which two or three rare changes that have no effect must come before a final change that has a large effect on binding. These first changes are rare and give no benefit to the organism, so they must come about randomly. Thornton interprets this as saying that proteins evolve contingently, because without those rare initial mutations, that hormone would never work at all.

Thornton is right at the close-up level, for a *particular* hormone—but one that follows the *predictable* chemical pattern described in Chapter 9. Like the other sterols, this hormone is made from cholesterol plus oxygens in various places. The path is narrow for a protein to develop binding to this particular arrangement of oxygens, but clearly other arrangements can be bound by other proteins with other mutations. The general chemical pattern is what is predictable, not the specific oxygen arrangement. Focusing too closely on the unpredictability of the specific hormone is like focusing too closely on the unpredictability of the evolution of a particular species. It is like saying that since individual water molecules move randomly, we cannot predict that water will flow downhill.

Other studies at the level of protein interactions have found a certain predictability. What is difficult about these studies is that you have to account for every single possible path a protein can take, but some have managed to do that. One study of RNA-protein interactions found that 81% of the time, a protein will take paths in the top 30% of the possible paths. Essentially, the road is only one-third as wide as a simple undirected theory would predict. Other times, the road is even more predictable, as when a single change will open up a brand new activity for an enzyme.

DARWIN'S STRUGGLE AND POPE'S WONDER

I am inclined to look at everything as resulting from designed laws, with the details, whether good or bad, left to the working out of what we may call chance. Not that this notion at all satisfies me.
Charles Darwin

Darwin contained the equivalent of both Huck's and Jim's voices inside him, and struggled with each in his letters to Asa Gray. Darwin also struggled with the very concept of progress. Michael Ruse notes that in Darwin's thinking, "[t]here is no absolute necessity for upwards change. Darwin was always clear on that. He wanted to put a firm line between himself and Germanic-style notions of inevitable upwards progress" (p. 285). Still, Darwin could not deny that later forms of life appear to follow some sort of progression of increasing complexity.

Too much emphasis on progress makes for a brittle, unyielding determinism. Too little results in a nebulous, chaotic, freewheeling science with no predictive power. Between is an emphasis on life's flowing but predictable order from a particular perspective. This emphasis sits between the first two like a liquid sits between solid and gas (in Chapter 4), like a river for life. Just as a river optimizes its branches in a predictable pattern as it flows downhill, life optimizes its flowing networks with a predictable tree-like shape, producing greater complexity here, greater entropy there.

The question is not whether randomness and order are both present, but rather which one is leading the other. The answer may lie in perspective. I once visited an art installation for the ears, in a bare room with 40 loudspeakers arranged in a circle. Through each speaker, a voice sang a part of Thomas Tallis's sixteenth-century motet "Spem in Alium." Walking close to each speaker, you could hear the individual line, or you could walk to the middle, where the lines joined together in a single song too complex for the brain to fully process.

Imagine if this music were the tape of life, and each speaker carried a different label for a different level of evolution. One speaker is "Genome Architecture" and produces a low bass note barely distinguishable from static. Next to it is "Protein Structure," which has a perceptible shape but seems to meander. Another is "Oxygen Levels" and sings an ascending line, like the line on the oxygen graph at the Redpath Museum. Another is "Biological Families," which follows the "Oxygen" speaker after a few measures with its own ascending counterpoint (see Figure 9.1 in Chapter 9). One labeled "Energy Rate Density" may also increase in a way that blends with other speakers from across the room (see Figure 11.1 in Chapter 11).

My hypothesis is that the chemistry speakers, especially "Oxygen Levels," are the voices that precede and open the way for the other music. At important points, the chemistry changed before the biology, and the chemical changes could have been predicted from the regular mathematical patterns of the periodic table, for oxygen, calcium, sodium, phosphate, and so on.

Do you hear what I hear? This sounds like something that Jim would intuit and Huck would argue.

The struggle between Huck and Jim, or how to interpret order and disorder, pervades history. In the eighteenth century, the emphasis was on the order, as Owen Barfield writes:

Poets and philosophers alike were delighted by the perfect *order* in which they perceived the cosmos to be arranged. They sought everywhere for examples

of this orderliness. Pope, for instance, praises Windsor Forest on the ground
that it is a place

> Not Chaos-like together crush'd and bruis'd;
> But, as the world, harmoniously confus'd:
> Where order in variety we see,
> And where, tho' all things differ, all agree.

This appreciation of Nature's regularity—from which we do not ourselves so
easily derive poetic inspiration—is now so familiar that it is difficult for us to
realize its freshness at that time. (pp. 178–179)

In the twenty-first century, we cannot regain the freshness of Pope's ideas, but we
can recognize that order and chaos have been differently emphasized over history,
and that both play roles in our exploration of this understandable universe. In the
interpretation of life's history, there is a balance to be restored and a conversation to
be continued. In discussions of natural history, the conversation has been uninten-
tionally tilted to emphasize the layers of flowing randomness. We need to once again
recognize the beauty of lawful predictability.

SO HYPER-ORDERED AS TO SEEM CHANCE-MADE

ROTHKO: Wait. Stand closer. You've got to get close. Let it pulsate. Let it work on you. . . .
Let it spread out.
John Logan, Red

I didn't even admit this to myself, but I used to think that art museums were not
worth the bother. I thought most of the images could be seen online. Then I discovered Mark Rothko through Simon Schama's The Power of Art and I repented of my
error. After that, I drove to the next state just to see an exhibit of Rothko's paintings
at the Portland Museum of Art. I only regret that I didn't go twice.

Rothko created rectangular paintings of colors as big as double doorways.
When these are shrunk to fit on a calendar or computer screen, they're pleasant
and pretty. But in person, close up, they come alive. Rothko knew this and made
specific rules about how his art should be presented. The surrounding lights should
be dimmed and the paintings illuminated with natural light. There shouldn't be
much wall around the painting—which is not usually a problem, because a Rothko
tends to fill up a wall. The viewer must move in and out, letting the painting fill
the eyes. The paintings call you closer and closer, to dive into their pools of color.
A guard told me that every half hour she had to stop someone from touching
the art.

To experience a Rothko, you need to slow down and change your perspective,
moving back, forth, in, and out. Only then do you see the many layers of colors and
washes, the texture of the brushstrokes, the interface where a rectangle fades into the
background. The color speaks as you move.

Rothko's paintings are an experience, a conversation, and an exercise in perspective. They change as you change your view. When you step back they float like clouds or crystals, but up close they flow like water, writhing with colors and tangled brush-stokes. They are as balanced and chaotic as a chemical equation.

Science has a place for perspective, too. A biological perspective that focuses on species or population pays attention to the flow of genes in an environment. A chemical perspective that focuses on ocean and air chemistry over billions of years (and how organisms respond) gives another view of the same process. At the biological level, genes change randomly. At the chemical level, oxygen increases and the processes that use oxygen also increase. Biology's randomness is constrained and channeled by the chemical possibilities: what will be solid and liquid, what will dissolve, and what will give energy.

Chemically, there are limited options, and some species must discover that the best chemical for processing energy is oxygen. Bejan's tree-making theory predicts the shapes of oxygen and heat flow. Chaisson's energy rate density supposes that more stable flow will create more structural complexity at multiple levels. From this, order will emerge at a different level, from a broader perspective, perhaps not as a law but as a predictable *narrative* describing how chemistry changed and biology followed. Lower levels are unpredictable, but higher levels are quite predictable. Physics also follows this pattern: quantum unpredictability at pico scales gives way to classical predictability at human scales. Prediction requires the right perspective.

Darwin himself sometimes saw things from a contingent biological perspective that differs from a predictable chemical perspective. Darwin wrote,

> Let an architect be compelled to build an edifice with uncut stones, fallen from a precipice. The shape of each fragment may be called accidental; yet the shape of each has been determined by the force of gravity, the nature of the rock, and the slope of the precipice—events and circumstances, all of which depend on natural laws; but there is no relation between these laws and the purpose for which each fragment is used by the builder. In the same manner the variations of each creature are determined by fixed and immutable laws; but these bear no relation to the living structure which is slowly built up through the power of selection. (p. 228)

This analogy works for rocks and for many biological structures. Vestigial shapes like the "Panda's thumb" were assembled for another purpose and do not have a purpose right now. Individual features on individual species are levels that exhibit a certain randomness.

But this analogy does not apply as well at the chemical level. The atoms that an organism uses are not a jumble of rocks, because they have shapes set by math and availabilities set by the chemistry of dissolving in water. There is a relation between chemical shapes and chemical uses. There are only a few dozen available elements, and they have predictable jobs. Phosphorus is used for ATP and DNA. Sulfur is used in proteins and for energy. Iron and manganese are used for old chemistry, while copper and zinc are used for new chemistry.

Carbon is the most unpredictable element in a sense, because it can build so many shapes, but even its shapes have clear chemical patterns. Signaling hormones of various shapes and effects tend to have carbon rings on the inside with oxygen decorating their outside. Protective links around the molecule are made of carbon joined by oxygen. These are not arbitrary choices made by the organism; rather, they descend from chemical laws, and the element in most of these structures (oxygen), predictably became more available over time.

In *The Old Ways*, Robert Macfarlane writes, "Mountain landscapes appear chaotic in their jumbledness, but they are in fact ultra-logical landscapes, organized by the climatic extremes and severe expressions of gravity: so hyper-ordered as to seem chance-made" (p. 192). The landscapes are both random and ordered, constrained by laws of physics and geology.

On Darwin's landscape, I cannot predict where each rock will fall, but I can predict how the law of gravity pulls different types of rock, and how the laws of chemistry cause some to fragment with particular shapes. I could also predict that an architect would use bigger, more square-edged rocks as a foundation. The architect would have a range of choices within that, but I can predict that the building would need to stand up.

When something is truly random, that randomness can be used for a purpose nonetheless. Conor Cunningham writes, "Yes, there is randomness. But like selfishness, it is derivative" (p. 148). Science brings out both a hidden randomness and a hidden order underneath nature. Geology shapes biology through chemistry, and biology shapes geology through chemistry. The planet is inherently interdisciplinary and knows no division between fields of science as they change together over time. Yet for all its complexity, it surprises scientists repeatedly with its consistency and predictability.

THE STORY OF CHEMICAL EVOLUTION

The reductionist approach to evolution . . . has proved enormously successful in illuminating the ultimate basis for the transformation of one organ into another, or even one species into another: but these momentous discoveries do not explain the course of biological history. The description of a cast of characters alone does not determine the shape of a drama.
Richard Fortey, in a review of Simon Conway Morris's *The Crucible of Creation*

By contrast, I have always regarded natural history expansively and seamlessly as a long and continuous narrative not only incorporating the origin and evolution of a wide spectrum of ordered structures, but also connecting many of them within an overarching framework of understanding.
Eric Chaisson

Our objective was in large part to challenge this traditional perspective—to reframe miner-alogy as a historical science.
Robert Hazen, *The Story of Earth*

If you want to know what people really think, look at their mistakes. One mistake in science journalism is misuse of the word "mistake." Two examples that came across my browser were "500-Million-Year-Old '*Mistake*' Led to Humans" and "*Accident* of Evolution Allows Fungi to Thrive in Our Bodies." The first article tells how whole-genome duplications preceded the evolution of the backbone. The second is about how a fungus infected humans after it evolved a tough coat and protective pigments.

Each of these "mistakes" is not necessarily a mistake at all, but is part of a rea-sonable and predictable genetic response from a broad perspective. Biologist James A. Shapiro gives several examples in which drastic genetic manipulations, includ-ing whole-genome duplications, are turned on in plants to accelerate evolution and increase the chances of survival. Whole-genome duplications can be a predictable response to environmental stress.

It is a mistake to jump to the conclusion that a genetic change is an undirected "mistake." Even when genes change randomly (as they often do), the survival of the organism is then determined by the environment, which sets limits on the results of the mutation. "Mistake" may be more clickable than "mutation" or "change," but it forces a close-up view of the story that crops out the larger context.

The story of science should include the context of *why* changes happened. Eminent scientists like E. O. Wilson encourage telling science as a story, and a letter to *Science* recently urged scientists to use narrative structures like "And But Therefore" (which was promoted by the creators of *South Park*, proving that all sorts of art can speak to science). We are story-driven creatures, and science can tell a great story.

And yet it doesn't always work. Stories, like music, take time. The universe was not built in a day, and stories about the universe must be edited significantly to fit into human brains via human language. If a crucial part fails, the story may turn out to be wrong. The detritus of failed scientific stories can be communicated with just a few words: phlogiston, ether, the steady-state universe, cold fusion.

But the risk of being wrong is a risk inherent in science. Even the most elegant story is simplified somehow and may fall apart with further testing. This is why every scientific story continues to be tested with each new experiment. Stories may be frag-ile, but when they cohere, they resonate with our story-driven minds.

Gould's tape of life was a story that was too simple. It assumed that genetic changes were largely independent of other events, when in fact they were hemmed in by the biology of other species in the ecological network, by the chemistry available in the environment, and by the physics of energy efficiency. Ironically, considering that Gould showed the power of evolution many times over, the tape of life assumed that evolution was too slow and too ineffective. Gould's assertion that the tape of life is unrepeatable requires a type of evolution that can solve hard problems only once, rather than a type that converges on repeated and efficient solutions. Gould's evolu-tion is weak tea compared to a chemically driven and convergent evolution.

All stories have a finite storyteller who tells the story to other finite minds. All are incomplete and (consciously or not) edited by politics. In the nineteenth century a physician named Samuel George Morton collected and measured a thousand human skulls. From these, he concluded that different races had significantly different cranial dimensions.

A hundred years later, Stephen Jay Gould remeasured the skulls and found no significant differences among them. In *The Mismeasure of Man*, Gould used his measurements to conclude that Morton had either consciously cherry-picked or unconsciously massaged his data. Gould's is a reasonable conclusion, and books like Daniel Kahneman's *Thinking, Fast and Slow* discuss how prior beliefs can shape even seemingly objective measurements.

But the story doesn't end there. In 2011 a group of anthropologists re-remeasured some of Morton's skulls and found evidence of statistically significant differences by race. They concluded that *Gould* was the one who mismeasured based on preconceptions, and Morton's data were more objective. And, of course, a *Nature* editorial quickly (and inevitably) followed, pointing out that the 2011 group had *its own* set of preconceptions!

With each round in this bout, I become a little less certain that we can know anything at all. But there's no need for epistemological despair; this just serves as a reminder that even straightforward measurements are shaped by stories in the mind of the measurer. Part of hearing any story is understanding the background of the author. Biography is part of interpretation.

Gould was motivated by his opposition to racism and his correct conviction that humans are one species springing from a single root. Michael Ruse, in *Monad to Man*, a book about the controversial notion of progress that so beset Darwin, writes that Gould wanted to emphasize the randomness and contingency of evolution because he despised social Darwinism, eugenics, and racism—all bad stories born from notions of evolutionary progress. No doubt this shaped the story of randomness that Gould told in *Wonderful Life*. But now Gould's story has taken too much hold on the scientific imagination and it too must be corrected.

My motives are complex as well. I always loved science and words from a young age, and like other kids, I ate up the stories in my environment: from the famous ones like *Star Wars* and *Indiana Jones*, to more obscure ones like the anime series *Seven Cities of Gold*, to the Hebrew and Greek stories from Sunday School that combine into one interlocking narrative. I make up stories for my boys every night, even when I feel as trapped by the arrangement as Scheherazade.

None of us is guiltless of telling stories. The question is, which story do you tell? This story of a world built from dust begins with an ordinary miracle: we have been assembled out of a consistent and regular universe, by laws so reliable that we can look back 13.7 billion years and reconstruct how it happened.

The core of the story in this book is robust, even if some parts of it may change in the future, as science is prone to do. Much of it has stayed the same for decades now. The chemical sequence that R. J. P. Williams predicted two decades ago remains recognizable as it is reinterpreted by recent data.

As Gould proposed, let's look at this story as music, and ask what the tape of life sounds like. Both music and evolution unfold over time, with a varying speed or tempo. Evolution accelerates with stress; it likely decelerated during the "boring billion" while the planet waited for oxygen. In times of stress, genetic mechanisms change DNA quickly in big chunks, rather than slowly in random single bits. If convergence is the dominant mode of evolution, then evolution is even faster than we thought, as life repeatedly converges on the same solutions.

Both music and evolution are contextual and harmonize parts together into a coherent whole. Life cooperated to change from asexual to sexual and from solitary to social. Complex organisms delegated vital chemistry to microbes, cooperating in the form of mitochondria or gut microbes. All of these cooperations required communication to keep them in sync or, in musical terms, to maintain the harmony.

Finally, impetus drives both music and evolution as they unfold over time. In Western music, this is the interplay of consonance and dissonance. In evolution, it is an interplay of random genetic drift and ordered chemical change. Increasing oxidation over time and increasing energy rate density correspond to an overall movement from left to right on the periodic table in the arrow through the sky (Chapter 4), the arrow through the ocean (Chapter 8), and the arrow of mining metals (Chapter 11). Without foreknowledge, evolution is pushed from behind by waves of chemical change and energetic optimization, as complexity increases. Music pushes toward tonal resolution, and chemistry toward complexity through increasing oxidation and energy flux.

Thousands of experiments, run by hundreds of scientists, tell a story that spans billions of years and yet fits onto a single piece of paper, in the form of the periodic table of the elements. This story is painted in the bright colors of the individual elements and the blended shades of complex biochemical molecules. The tape of life unfolds through catastrophic dissonance and resolves with eucatastrophe after inevitable loss.

To start with an orderly periodic table and end with beautiful complexity is a marvelous story. This narrative thread ties us to the world, making it our home.

I found this once, when we were staying near Fort Casey on a cold December evening. It was the lull between Christmas and New Year's Day, when everyone else was home. I was exhausted when I walked out to the shore, and I wanted to be alone. Going on vacation is supposed to be relaxing, but with four boys under the age of 12, it had required too much from me to temper their chaos. They spread out so much energy that their energy rate density must have been off the charts.

I walked out to the beach and stood at that thin place, between worlds. Something was within reach from each chapter in this story. I stood on the green serpentine rock, the one that may have supplied the chemical hydrogen to the first life on the planet. Waves pounded and surrounded the rock as I stood at a triple point between three phases of matter: liquid ocean in front, gaseous air above, and solid earth below.

At first, I could only see the periodic lines of wave-foam and the lights of Port Townsend across the Sound. After about 15 minutes, a twinkling star caught my eye—below me. As I realized that stars don't reflect in rippling waves, I saw another one, and another. They winked in and out in the black water. I followed one as a wave washed it up on the shore. It was faint green. These were bioluminescent plankton, the type still uncomfortable with oxygen after a billion years, lighting up to burn it off like a gas flare, a beautiful waste of energy poured out.

I stood between the stars and the glowing plankton in that old, dark place. Down at my feet, a still pool on the rock reflected stars above, as green plankton stars below swirled in the unceasing cycle of the waves, each wave the same, each wave different from the last. Both sets of stars came from the same atomic dust. I was at home.

I was on the line connecting chaos and cosmos. A story ran through it all like a river. Earth and heaven, and past and present, flowed together as I stood there, infinitesimally small in the face of it, but grateful to glimpse its arc. Time passed away, unnoticed, until it was time to go in.

GLOSSARY

Abbreviations, Elements, Molecules, and People

Adenosine triphosphate (ATP): A molecule formed from the nucleoside adenosine with three linked phosphates. The reaction that breaks the phosphates apart is spontaneous and can drive other reactions forward. DNA is built from ATP and other triphosphates.

Aluminum (Al): Metal element with atomic number 13, forming the Al^{+3} ion in water.

Amino acid: Molecule that polymerizes into proteins. The 20 natural amino acids are made of CHON and sulfur.

Anoxia: A state of very low or absent oxygen. The Earth was anoxic before the Great Oxidation Event 2.5 billion years ago.

Antioxidant: An electron-rich molecule that gives electrons to counterbalance the disruptive effects of other electron-hungry molecules.

Arsenic (As): Metalloid element with atomic number 33, forming the AsO_4^{-3} (arsenate) ion in water.

ATP synthase: Membrane protein in bacteria, mitochondria, and plasmids that uses protons to make ATP for energy; also called Complex V.

Banded iron formations (BIFs): Oxygen-iron rock layers formed from reduced plus-two iron ions reacting with oxygen, coinciding with the two Great Oxidation Events.

Bejan, Adrian: Engineer who developed "constructal theory" to describe why similar branching flows carry matter and energy in different contexts. Author of *Shape and Structure, from Engineering to Nature* (2000) and coauthor of *Design in Nature* (2013).

Big Six: Robert Hazen's term for the six most prominent elements in the Earth's surface rocks (oxygen, iron, calcium, magnesium, aluminum, and silicon).

Cadmium (Cd): Transition metal element with atomic number 48, forming the Cd^{+2} ion in water.

Caetano-Anollés, Gustavo: Computational biologist who developed a molecular clock to date the invention of different protein activities.

Calcium (Ca): Metal element with atomic number 20, forming the Ca^{+2} ion in water.

Cambrian Explosion: A sudden increase in fossilized life beginning 542 million years ago.

Chaisson, Eric: Astrophysicist who developed energy rate density (ERD) as a measure of the energy flow through a system and therefore its structural complexity.

Chlorine (Cl): Non-metal element with atomic number 17, forming the Cl^- (chloride) ion in water.

CHON: Acronym for carbon, hydrogen, oxygen, and nitrogen, the four most common elements used in life.

Citric acid cycle: A metabolic cycle of eight reactions that breaks acetate molecules down to carbon dioxide, producing electrons in the form of NADH for respiration. Some microbes run it in the reverse direction to make more complex molecules from carbon dioxide. Also called the tricarboxylic acid (TCA) cycle or the Krebs cycle.

Cobalt (Co): Transition metal element with atomic number 27, forming the Co^{+2} ion in water.

Convergence: The evolution of similar traits or functions in different species living in different contexts.

Conway Morris, Simon: Paleontologist who promotes evolutionary convergence and wrote *Life's Solution* (2003).

Copper (Cu): Transition metal element with atomic number 29, forming the Cu^+ or Cu^{+2} ion in water.

Cretaceous extinction: The most recent of the five great extinctions, also known as the Cretaceous-Paleogene (K-Pg) or Cretaceous-Triassic (K-T), which occurred about 60 million years ago and included the extinction of the dinosaurs.

Cytochromes: A class of proteins that uses iron in the form of heme, often found in mitochondria's respiration enzymes.

Deoxyribonucleic acid (DNA): A long chain made of phosphate-linked nucleotides that encodes instructions for making proteins in every cell.

Energy rate density (ERD): The amount of energy processed by a system per unit time per unit mass, as used by Eric Chaisson to quantify complexity over time for diverse systems.

Enzyme: A protein that catalyzes and speeds up a specific chemical reaction.

Eusociality: The highest level of animal sociality, involving care for the offspring and innate social roles, exemplified by the hive.

FeMoCo: The catalytic center of nitrogenase, built from iron (Fe), molybdenum (Mo), and sulfur, illustrated in Figure 8.3 in Chapter 8.

Fermenter: General term for a microbe that breaks down carbon molecules for energy, usually without oxygen.

Genome: The complete sequence of nucleotides and therefore genes in an organism's DNA.

Glycolysis: The metabolic process of breaking sugar in half for energy.

Gold (Au): Metal element with atomic number 79, forming the Au^{+3} ion in water.

Gould, Stephen Jay: Paleontologist and evolutionary theorist who wrote *Wonderful Life: The Burgess Shale and the Nature of History* (1990), which argued that if evolution were rewound and replayed like a tape, it would take different paths and life would look very different.

Great Oxidation Event (GOE): A worldwide change 2.5 billion years ago, when the atmosphere gained steady and detectable levels of oxygen from photosynthesizing microbes. A lesser event occurred about 700 million years ago, sometimes called the second GOE.

Green leaf volatiles: Small, carbon-oxygen molecules released from plants for communication and scent, also called volatile organic compounds (VOCs).

Hazen, Robert: Geologist who wrote *The Story of Earth* (2012), which describes his theory of mineral evolution.

Heme: The iron-bound porphyrin that, as part of the protein hemoglobin, carries oxygen in blood.

Hydrogen (H): Non-metal element with atomic number 1. In water it will lose an electron to form H^+, which is a proton. Acidic materials often form excess H^+.

Hydroxide (OH-): Alkaline materials form excess OH^- in water and are called bases. Bases neutralize protons from acids, making H_2O (water).

Ion: An atom or molecule that has lost or gained electrons so that it carries an electric charge. Ions are stable dissolved in water or crystallized next to other ions.

Iron (Fe): Transition metal element with atomic number 26, forming the Fe^{+2} or Fe^{+3} ion in water, depending on the oxidation of the environment.

Irving-Williams series: The binding strength of transition metals, as shown in Figure 2.3 in Chapter 2, which follows a predictable order across the periodic table.

Magnesium (Mg): Metal element with atomic number 12, forming the Mg^{+2} ion in water.

Manganese (Mn): Transition metal element with atomic number 25, forming the Mn^{+2}, Mn^{+4}, or other ions in water.

Methanogen: Microbe that generates methane as waste.

Methanophage: Microbe that consumes methane as food.

Mitochondrion: Red, iron-rich organelle that oxidizes acetate and fats to carbon dioxide, making a proton gradient that is used to make ATP with ATP synthase. Both the citric acid cycle and the respiration proteins are found here.

Nickel (Ni): Transition metal element with atomic number 28, forming the Ni^{+2} ion in water.

Nicotinamide adenine dinucleotide hydride (NADH): Metabolite that can carry electron pairs to different places in the cell. These pairs can be used for building chemical bonds or "burning" (respiration for energy by combination with oxygen). Most cells also use a second functionally identical form with an added phosphate called NADPH.

Nitrogenase: Metalloenzyme that breaks the stable triple bond of N_2 (nitrogen) to make the more chemically available NH_3 (ammonia) molecule, a process known as "nitrogen fixation."

Nucleotide: Three-part molecule that polymerizes into DNA and RNA. The three parts are sugar, phosphate, and nitrogen-rich base.

Oxidized: Chemical term for a substance with fewer electrons and/or more oxygens.

Permian extinction: The third and largest of the five great extinctions, also known as the Permian-Triassic (P-Tr), which occurred about 250 million years ago.

Phosphorus (P): Non-metal element with atomic number 15, forming the PO_4^{-3} (phosphate) ion in water.

Photosynthesis: The metabolic process that uses light energy to move electrons and build chemical structures such as sugars from carbon dioxide.

Plastid: Green, magnesium- and iron-rich organelle in plants, which harvests light energy, moves electrons, and pumps protons in order to make ATP, NADH, and sugars from water and carbon dioxide, outputting oxygen as waste.

Porphyrin: A flat square made from CHON atoms with four nitrogens around a hole that fits many of the metals in the middle of the periodic table.

Potassium (K): Metal element with atomic number 19, forming the K^+ ion in water.

Reactive oxygen species (ROS): Unstable variants of O_2 (oxygen) with extra electrons (and protons) such as H_2O_2 (hydrogen peroxide) and O_2^- (superoxide).

Redox potential: The voltage at which a redox reaction will be balanced between products and reactants. As the environment oxidizes, reactions with lower redox potentials will oxidize first.

Redox reactions: Short for "reduction-oxidation" reactions, which transfer electrons (frequently with protons or oxygens).

Reduced: Chemical term for a substance with more electrons and/or more hydrogens.

Respiration: The metabolic process of moving electrons to oxygen for energy. This is often the result of oxidation (replacing carbon-hydrogen or carbon-carbon bonds with carbon-oxygen bonds to make carbon dioxide).

Ribonucleic acid (RNA): A long chain made of phosphate-linked nucleotides with one more OH hydroxyl group per nucleotide than DNA. One form called mRNA carries the instructions for making proteins from DNA to ribosomes.

Ribosomes: Large assemblies of RNA and proteins that use information from DNA, carried by mRNA, to make protein chains.

Ribozyme: An RNA chain that catalyzes and speeds up a specific chemical reaction. Ribozymes are less common than enzymes but can perform crucial functions for life.

Sarcoplasmic reticulum: A specialized version of the endoplasmic reticulum found in muscle cells, which releases a wave of calcium upon activation.

Sodium (Na): Metal element with atomic number 11, forming the Na^+ ion in water.

Sterol: Chemical structure of four joined carbon rings that forms the core of cholesterol, steroids, and other important molecules. Our processes for making these structures require oxygen.

Sulfur (S): Non-metal element with atomic number 16, forming the S^{-2} (sulfide) or SO_4^{-2} ion in water.

Vesicles: Spherical chemical structures with a layer of oil enclosing a core of water.

Williams, R. J. P.: Inorganic chemist who cowrote *The Chemistry of Evolution* (2006) and *Evolution's Destiny* (2012), which argue that the chemical rules of abundance and availability in water set a chemical sequence that limited evolution's possible paths.

Wolfe-Simon, Felisa: Arsenic life proponent and lead author of "A Bacterium That Can Grow by Using Arsenic Instead of Phosphorus."

Zinc (Zn): Transition metal element with atomic number 30, forming the Zn^{+2} ion in water.

REFERENCES

Author's Note

"As Darwin said of natural selection, . . ." Charles Darwin, *The Variation of Animals and Plants Under Domestication*. 1905, John Murray, p. 16.

Chapter 1

"A press conference at 11 a.m. had announced . . . " Press release titled "NASA-funded research discovers life built with toxic chemical," http://www.nasa.gov/topics/universe/features/astrobiology_toxic_chemical.html.

". . . raises the entire Sierra Nevada range by about a millimeter a year . . ." W. C. Hammond et al. "Contemporary uplift of the Sierra Nevada, western United States, from GPS and InSAR measurements." 2012. *Geology* 40(7), p. 667. DOI: 10.1130/G32968.1.

". . . a bird sitting on the water too long will calcify." See "#17 Calcifying Lake" at http://sfglobe.com/?id=2163&src=share_fb_new_2163.

"You can make your own by mixing . . . " Adapted from "Mono's Secret Recipe" at http://www.monolake.org/about/geolake.

". . . bacteria from Mono Lake could use arsenic instead of phosphorus." F. Wolfe-Simon et al. "A bacterium that can grow by using arsenic instead of phosphorus." Initially published in *Science* Express, December 2, 2010, corrected in *Science* on June 3, 2011. DOI: 10.1126/science.1197258.

". . . a new superconductor from lanthanum . . ." Eric R. Scerri. *The Periodic Table: A Very Short Introduction*. 2012, Oxford University Press, p.11.

". . . Wolfe-Simon asked the same question . . ." F. Wolfe-Simon et al. "Did nature also choose arsenic?" 2009. *Int J Astrobiology* 8(2), p. 69. DOI: 10.1017/S1473550408004394.

". . . arsenate binds and then drops off in a matter of seconds." H. B. F. Dixon. "The biochemical action of arsonic acids especially as phosphate analogues." 1997. *Adv Inorg Chem*. 44, p. 191.

". . . yeast cells were fed arsenic . . ." I. Litwin et al. "Oxidative stress and replication-independent DNA breakage induced by arsenic in saccharomyces cerevisiae." 2013. *PLOS Genetics* 9(7), p. e1003640. DOI: 10.1371/journal.pgen.1003640.

"Arsenic also pushes proteins out of shape . . ." A. Sapra et al. "Multivalency in the inhibition of oxidative protein folding by arsenic (III) species." 2015. *Biochemistry* 54, p. 612. DOI: 10.1021/bi501360e.

"It published eight separate critiques, . . ." See Comments on "A bacterium that can grow by using arsenic instead of phosphorus." 2011. *Science* 332(6034), p. 1149.

". . . the Mono Lake bacteria grew at "very low" concentrations of phosphate . . ." T. J. Erb et al. "GFAJ-1 is an arsenate-resistant, phosphate-dependent organism." 2012. *Science* 337, p. 467. DOI: 10.1126/science.1218455.

"They also found that the Mono Lake bacteria would not grow ..." M. L. Reaves et al. "Absence of detectable arsenate in DNA from arsenate-grown GFAJ-1 cells." 2012. *Science* 337, p. 470. DOI: 10.1126/science.1219861.

"Researchers from Israel used a complex and slow technique ..." M. Elias et al. "The molecular basis of phosphate discrimination in arsenate-rich environments." 2012. *Nature* 491(7422), p. 134. DOI: 10.1038/nature11517.

"... these bacteria scraped up phosphate by cannibalizing the ribosome ..." G. N. Basturea et al. "Growth of a bacterium that apparently uses arsenic instead of phosphorus is a consequence of massive ribosome breakdown." 2012. *J Biol Chem.* 287(34), p. 28816. DOI: 10.1074/jbc.C112.394403.

"Some fascinating microbes eat chlorate molecules ..." R. Nerenberg. "Breathing perchlorate." 2013. *Science* 340(6128), p. 38. DOI: 10.1126/science.1236336.

"... two chained phosphates will last a thousand days." R. B. Stockbridge and R. Wolfenden. "Enhancement of the rate of pyrophosphate hydrolysis by nonenzymatic catalysts and by inorganic pyrophosphatase." 2011. *J Biol Chem.* 286(21), p. 18538. DOI: 10.1074/jbc. M110.214510.

"... the naturally occurring base structure 'is a stupid design' that he can improve." R. Kwok. "DNA's new alphabet." 2012. *Nature* 491(7425), p. 516. DOI: 10.1038/491516a.

"Other scientists have introduced positive nitrogen-based charges ..." R. Dempcy et al. "Design and synthesis of deoxynucleic guanidine: a polycation analogue of DNA." 1994. *Proc Natl Acad Sci.* 91(17), p. 7864.

"Beryllium allergy is caused when it sticks to an oxygen-rich pocket ..." M. T. Falta et al. "Identification of beryllium-dependent peptides recognized by CD4+ T cells in chronic beryllium disease." 2013. *J Exp Med.* 210(7), p. 1403. DOI: 10.1084/jem.20122426.

"It puts magnesium in a protein toxin ..." A. B. Min et al. "The crystal structure of the Rv0301-Rv0300 VapBC-3 toxin-antitoxin complex from M. tuberculosis reveals a Mg^{2+} ion in the active site and a putative RNA-binding site." 2012. *Protein Sci.* 21(11), p. 1754. DOI: 10.1002/pro.2161.

"Ribosomes in plants without magnesium fall apart, ..." Norman P. A. Huner and William Hopkins. *Introduction to Plant Physiology* (4th edition). John Wiley & Sons.

"... know the value of magnesium." See for example J. L. Fiore et al. "Entropic origin of Mg2+-facilitated RNA folding." 2012. *Proc Natl Acad Sci.* 109(8), p. 2902. DOI: 10.1073/ pnas.1114859109.

"... tested how different metals can tie together the phosphate chain of an RNA ribozyme." S. A. Pabit et al. "Role of ion valence in the submillisecond collapse and folding of a small RNA domain." 2013. *Biochemistry* 52(9), p. 1539. DOI: 10.1021/bi3016636.

"... an essential ingredient when enzymes proofread DNA." B. D. Freudenthal et al. "Observing a DNA polymerase choose right from wrong." 2013. *Cell* 154(1), p. 157. DOI: 10.1016/j.cell.2013.05.048.

"One particular magnesium transporter opens in response to an ATP-magnesium 'key,' ..." Y. Hirata et al. "Mg2+-dependent interactions of ATP with the cystathionine-β-synthase (CBS) domains of a magnesium transporter." 2014. *J Biol Chem.* 289(21), p. 14731. DOI: 10.1074/jbc.M114.551176.

"Another protein patrols the interior of the bacterium ..." K. Rehman and H. Naranmandura. "Arsenic metabolism and thioarsenicals." 2012. *Metallomics* 4, p. 881. DOI: 10.1039/c2mt00181k.

"... *Leishmania* parasites are susceptible to treatments with the element antimony ..." M. R. Perry et al. "Chronic exposure to arsenic in drinking water can lead to resistance to antimonial drugs in a mouse model of visceral leishmaniasis." 2013. *Proc Natl Acad Sci.* 110(49), p. 19932. DOI: 10.1073/pnas.1311535110.

". . . at least five different transporters bring iron inside the cell . . ." K. J. Waldron and N. J. Robinson. "How do bacterial cells ensure that metalloproteins get the correct metal?" 2009. *Nat Rev Microbiol.* 7(1), p. 25. DOI: 10.1038/nrmicro2057.

"Gould's book *Wonderful Life* describes . . ." Stephen Jay Gould. *Wonderful Life: The Burgess Shale and the Nature of History.* 1990. W. W. Norton, pp. 47, 238; "tape of life," p. 14.

"R. J. P. Williams cowrote a book . . ." R. J. P. Williams and R. E. M. Rickaby, *Evolution's Destiny: Co-evolving Chemistry of the Environment and Life.* 2012. RSC Publishing.

Chapter 2

"Scientists have made a molecule that works just like heparin . . ." T. H. Nguyen et al. "A heparin-mimicking polymer conjugate stabilizes basic fibroblast growth factor." 2013. *Nat Chem.* 5(3), p. 221. DOI: 10.1038/nchem.1573.

". . . sulfur-rich bubbles inside yeast cells that store glutathione . . ." B. Morgan et al. "Multiple glutathione disulfide removal pathways mediate cytosolic redox homeostasis." 2013. *Nat Chem Biol.* 9(2), p. 119. DOI: 10.1038/nchembio.1142.

". . . in tadpoles, signaling with electron-imbalanced peroxides leads to growth." H. Sies. "Role of metabolic H2O2 generation: redox signaling and oxidative stress." 2014. *J Biol Chem.* 289(13), p. 8735. DOI: 10.1074/jbc.R113.544635.

". . . one kind of male dragonfly turns from yellow to red . . ." R. Futahashi et al. "Redox alters yellow dragonflies into red." 2012. *Proc Natl Acad Sci.* 109(31), p. 12626. DOI: 10.1073/pnas.1207114109.

". . . one important enzyme (pyruvate kinase) that breaks down sugar . . ." D. Anastasiou et al. "Inhibition of pyruvate kinase M2 by reactive oxygen species contributes to cellular antioxidant responses." 2011. *Science* 334(6060), p. 1278. DOI: 10.1126/science.1211485.

"So does an enzyme that sends proteins to the trash." J. G. Lee et al. "Reversible inactivation of deubiquitinases by reactive oxygen species in vitro and in cells." 2013. *Nat Commun.* 4, p. 1568. DOI: 10.1038/ncomms2532.

". . . one bacterial protein senses bleach with a sulfur . . ." M. J. Gray et al. "NemR is a bleach-sensing transcription factor." 2013. *J Biol Chem.* 288(19), p. 13789. DOI: 10.1074/jbc.M113.454421.

". . . an art installation called 'The Great Work of the Metal Lover.'" N. Drake. "Golden microbial art." 2013. *Proc Natl Acad Sci.* 110(15), p. 5738. DOI: 10.1073/pnas.1304918110.

"Eucalyptus trees have learned a version of this trick." M. Lintern et al. "Natural gold particles in Eucalyptus leaves and their relevance to exploration for buried gold deposits." 2013. *Nat Commun.* 4, p. 2614. DOI: 10.1038/ncomms3614.

". . . you can do it in your kitchen with green tea to make gold nanoparticles." R. K. Sharma et al. "Preparation of gold nanoparticles using tea: a green chemistry experiment." 2012. *J Chem Educ.* 89, p. 1316. DOI: 10.1021/ed2002175.

". . . but also patrols for cadmium." C. D. Klaassen et al. "Metallothionein: An intracellular protein to protect against cadmium toxicity." 1999. *Annu Rev Pharmacol Toxicol.* 39, p. 267.

"In these microbes, cadmium sticks in the place of zinc *and does zinc's chemistry.*" Y. Xu et al. "Structure and metal exchange in the cadmium carbonic anhydrase of marine diatoms." 2008. *Nature* 452(7183), p. 56. DOI: 10.1038/nature06636.

"... mopping up extra lead and mercury ions ..." J. Sobrino-Plata et al. "The role of glutathi-one in mercury tolerance resembles its function under cadmium stress in Arabidopsis." 2014. *Metallomics* 6, p. 356. DOI: 10.1039/C3MT00329A.

"... a red alga named *Galdieria sulphuraria*." G. Schönknecht et al. "Gene transfer from bacteria and archaea facilitated evolution of an extremophilic eukaryote." 2013. *Science* 339(6124), p. 1207. DOI: 10.1126/science.1231707.

"At a certain point, the cells froze up and turned to thick jelly." Personal communication from Rick Ridgway.

"Some bacteria build biofilm structures outside the cell ..." T. Das et al. "Influence of cal-cium in extracellular DNA mediated bacterial aggregation and biofilm formation." 2014. *PLoS ONE* 9(3), p. e91935. DOI: 10.1371/journal.pone.0091935.

"A worm called *C. elegans* that is grown in high levels of +3 aluminum ..." K. E. Page et al. "Aluminium exposure disrupts elemental homeostasis in *Caenorhabditis elegans*." 2012. *Metallomics* 4, p. 512. DOI: 10.1039/c2mt00146b.

"... detoxifies itself from uranium poisoning by collecting phosphates in one place." C. Acharya and S. K. Apte. "Novel surface associated polyphosphate bodies seques-ter uranium in the filamentous, marine cyanobacterium, *Anabaena torulosa*." 2013. *Metallomics* 5(12), p. 1595. DOI: 10.1039/c3mt00139c.

"One microbe keeps copper ions out of the way ..." A. Hong-Hermesdorf et al. "Subcellular metal imaging identifies dynamic sites of Cu accumulation in Chlamydomonas." 2014. *Nat Chem Biol.* 10(12), p. 1034. DOI: 10.1038/nchembio.1662.

"Yeast with high phosphate levels suffer from iron depletion ..." A. Seguin et al. "Co-precipitation of phosphate and iron limits mitochondrial phosphate availability in Saccharomyces cerevisiae lacking the yeast frataxin homologue (YFH1)." 2011. *J Biol Chem.* 286(8), p. 6071. DOI: 10.1074/jbc.M110.163253.

"Recent work suggests that 70% of the metals ..." A. W. Foster et al. "Metal preferences and metallation." 2014. *J Biol Chem.* 289(41), p. 28095. DOI: 10.1074/jbc.R114.588145.

"Williams has published several books with co-authors ..." See especially (1) *The Chemistry of Evolution: The Development of Our Ecosystem*, with J. J. R. Frausto da Silva, 2006, Elsevier; (2) *Evolution's Destiny: Co-evolving Chemistry of the Environment and Life*, with R. E. M. Rickaby, 2012, RSC Publishing; and (3) *Bringing Chemistry to Life: From Matter to Man*, with J. J. R. Frausto da Silva, 1999, Oxford University Press.

"... the fine structure constant in 7-billion-year-old light ..." J. Bagdonaite et al. "A strin-gent limit on a drifting proton-to-electron mass ratio from alcohol in the early universe." 2013. *Science* 339, p. 46. DOI: 10.1126/science.1224898.

"A 2010 study of all co-factors ..." J. D. Fischer et al. "The structures and physicochemi-cal properties of organic cofactors in biocatalysis." 2010. *J Mol Biol.* 403(5), p. 803. DOI: 10.1016/j.jmb.2010.09.018.

"... chemists have made a porous rock-like structure ..." R. Bermejo-Deval et al. "Metalloenzyme-like catalyzed isomerizations of sugars by Lewis acid zeolites." 2012. *Proc Natl Acad Sci.* 109(25), p. 9727. DOI: 10.1073/pnas.1206708109.

"One group of chemists uses dissolved hemoglobin to make plastic polymers ..." T. B. Silva et al. "Hemoglobin and red blood cells catalyze atom transfer radical polymerization." 2013. *Biomacromolecules* 14(8), p. 2703. DOI: 10.1021/bm400556x.

"Your body uses the iron in hemoglobin make nitric oxide for signaling ..." C. J. Roche et al. "Generating S-nitrosothiols from hemoglobin: mechanisms, conformational dependence, and physiological relevance." 2013. *J Biol Chem.* 288(31), p. 22408. DOI: 10.1074/jbc.M113.482679.

"Medicines like Pepto-Bismol will target sulfur-metal bonds like these." S. Cun and H. Sun. "A zinc-binding site by negative selection induces metallodrug susceptibility in an essential chaperonin." 2010. *Proc Natl Acad Sci.* 107(11), p. 4943. DOI: 10.1073/pnas.0913970107.

"... nickel and chromium ... may cause some forms of cancer ..." Y. Chervona et al. "Carcinogenic metals and the epigenome: understanding the effect of nickel, arsenic, and chromium." 2012. *Metallomics* 4(7), p. 619. DOI: 10.1039/c2mt20033c.

"If manganese is put in there instead of zinc, ..." A. Fernández-Gacio et al. "Transforming carbonic anhydrase into epoxide synthase by metal exchange." 2006. *Chembiochem.* 7(7), p. 1013. DOI: 10.1002/cbic.200600127.

"If rhodium is put in, ..." Q. Jing et al. "Stereoselective hydrogenation of olefins using rhodium-substituted carbonic anhydrase—a new reductase." 2009. *Chemistry* 15(6), p. 1370. DOI: 10.1002/chem.200801673.

"One frugal beetle has figured out ..." S. Frick et al. "Metal ions control product specificity of isoprenyl diphosphate synthases in the insect terpenoid pathway." 2013. *Proc Natl Acad Sci.* 110(11), p. 4194. DOI: 10.1073/pnas.1221489110.

"Likewise, a bacterium named *Citrobacter freundii* ..." I. Sánchez-Moreno et al. "From kinase to cyclase: an unusual example of catalytic promiscuity modulated by metal switching." 2009. *Chembiochem.* 10(2), p. 225. DOI: 10.1002/cbic.200800573.

"... some members of the family add oxygen-hydrogen groups ..." R. Chowdhury et al. "Ribosomal oxygenases are structurally conserved from prokaryotes to humans." 2014. *Nature* 510(7505), p. 422. DOI: 10.1038/nature13263.

"One DNA-cutting enzyme cuts DNA in different places with magnesium, ..." K. Vasu et al. "Endonuclease active site plasticity allows DNA cleavage with diverse alkaline Earth and transition metal ions." 2011. *ACS Chem Biol.* 6(9), p. 934. DOI: 10.1021/cb200107y.

"Bacteria that grow in soils with different levels ..." Z. Charlop-Powers et al. "Chemical-biogeographic survey of secondary metabolism in soil." 2014. *Proc Natl Acad Sci.*111(10) p. 3757. DOI: 10.1073/pnas.1318021111.

"We changed our hypothesis and then found success." C. S. Lengyel et al. "Mutations designed to destabilize the receptor-bound conformation increase MICA-NKG2D association rate and affinity." 2007. *J Biol Chem.* 282(42), p. 30658.

"... scientists put sulfurs and nitrogens in a pattern ..." D. Shiga et al. "Creation of a binuclear purple copper site within a de novo coiled-coil protein." 2012. *Biochemistry* 51(40), p. 7901. DOI: 10.1021/bi3007884.

"In 2010 a team of scientists designed a protein so that zinc would bring four copies ..." E. N. Salgado et al. "Metal templated design of protein interfaces." 2010. *Proc Natl Acad Sci.* 107(5), p.1827. DOI: 10.1073/pnas.0906852107.

"Some scientists used this zinc zipper design to string proteins end-on-end ..." J. D. Brodin et al. "Metal-directed, chemically tunable assembly of one-, two- and three-dimensional crystalline protein arrays." 2012. *Nat Chem.* 4(5), p. 375. DOI: 10.1038/nchem.1290.

"No one had set out to create an enzyme, but even so, they had, ..." B. S. Der et al. "Catalysis by a de novo zinc-mediated protein interface: implications for natural enzyme evolution and rational enzyme engineering." 2012. *Biochemistry* 51(18), p. 3933. DOI: 10.1021/bi201881p.

"... to make another type of enzyme called a beta-lactamase ..." W. J. Song and F. A. Teczan. "A designed supramolecular protein assembly with in vivo enzymatic activity." 2014. *Science* 346(6216), p. 1525. DOI: 10.1126/science.1259680.

"... a different laboratory made a part-random protein with a quartet of sulfur atoms ..." B. Seelig and J. W. Szostak. "Selection and evolution of enzymes from a partially randomized non-catalytic scaffold." 2007. *Nature* 448(7155), p. 828.

"One lab made a long, thin protein with a three-nitrogen claw at one end ..." M. L. Zastrow et al. "Hydrolytic catalysis and structural stabilization in a designed metalloprotein." 2011. *Nat Chem.* 4(2), p. 118. DOI: 10.1038/nchem.1201.

"Another lab randomly mutated the shapes in the cores ..." S. C. Patel et al. "Cofactor binding and enzymatic activity in an unevolved superfamily of de novo designed 4-helix bundle proteins." 2009. *Protein Sci.* 18(7), p. 1388. DOI: 10.1002/pro.147.

Chapter 3

"Primo Levi said the periodic table is poetry, ..." Primo Levi. *The Periodic Table.* 1984. Schocken Books, p. 45.

"Eric Scerri writes that this orderly repetition is almost musical: ..." Eric R. Scerri, *The Periodic Table: A Very Short Introduction.* 2012, Oxford University Press, p.19.

"C. P. Snow describes understanding the periodic table, ..." Quoted in Scerri, 2012, p. xviii.

"Anything that is as small as an electron can cohere as a particle ..." See, for example, Peter Atkins and Julio de Paula, *Physical Chemistry for the Life Sciences* (2nd edition). 2011, Chapter 9.

"Once temperature dropped enough, protons and electrons ..." See, for example, M. S. Turner. "Origins of the Universe." 2009. *Sci Amer.* 301(3), p. 36.

"Carbon's rung is lowered by a special, "resonant" energy level ..." This level is called the "Hoyle state." See, for example, U. G. Meißner. "Anthropic considerations in nuclear physics." 2015. *Sci Bull.* 60(1), p. 43. DOI 10.1007/s11434-014-0670-2.

"In 2015, four years of data from the Planck satellite ..." Links to all papers available at http://www.cosmos.esa.int/web/planck/publications under the heading "Planck 2015 Results." A comprehensive summary for the general reader is by C. M. Carlisle, "Planck upholds standard cosmology," published February 10, 2015, at skyandtelescope.com. See also Martin Rees. *Just Six Numbers.* 2000, Basic Books.

"Other calculations suggest that if the light quark mass were ..." U. G. Meißner. "Anthropic considerations in nuclear physics." 2015. *Sci Bull.* 60(1), p. 43. DOI: 10.1007/s11434-014-0670-2.

"... a universe missing one of the four fundamental forces of physics ..." See David Toomey. *Weird Life* 2014, Norton, p. 201.

"In 2015, Peter Behroozi and Molly S. Peeples..." P. Behroozi and M. S. Peeples, "On the history and future of cosmic planet formation." 2015. *Mon Not Royal Astron Soc.* 454(2), p. 1811. DOI: 10.1093/mnras/stv1817.

"One theory is two dead neutron stars ..." E. Berger et al. "An r-process kilonova associated with the short-hard GRB." 2013. *Astrophys J Lett.* 774, p. L23. DOI: 10.1088/2041-8205/774/2/L23.

"The geologist Robert Hazen calls these the 'Big Six' elements of geology." Robert M. Hazen. *The Story of Earth: The First 4.5 Billion Years, from Stardust to Living Planet.* 2012, Viking.

"Three of these elements . . . will increase in groundwater . . ." A. Skelton et al. "Changes in groundwater chemistry before two consecutive earthquakes in Iceland." 2014. *Nature Geo.* 7, p. 752. DOI: 10.1038/ngeo2250.

"Robert Hazen has assembled a geological theory called mineral evolution, . . ." R. M. Hazen et al. "Mineral evolution." 2008. *Amer Mineral.* 93(11–12), p. 1693. DOI: 10.2138/am.2008.2955.

"Inside were purple crystals of halite . . ." Zolensky et al. "Asteroidal water within fluid inclusion-bearing halite in an H5 chondrite, Monahans (1998)." 1999. *Science* 285(5432), p. 1377. DOI: 10.1126/science.285.5432.1377.

". . . a planetoid the size of Mars came out of nowhere and slammed into the proto-Earth, . . ." See R. Canup. "Lunar conspiracies." 2013. *Nature* 504, p. 27. DOI: 10.1038/504027a and Forum section in the same issue.

". . . making the first rock in Earth's foundations out of four of the Big Six elements . . ." See Hazen, *The Story of Earth.*

"This steam-driven motion added about 250 more minerals to the early earth, . . ." See Hazen, "Mineral evolution."

Chapter 4

". . . the red planet was first blue with water." See overview by J. P. Grotzinger. "Analysis of surface materials by the Curiosity Mars Rover." 2013. *Science* 341, p. 1475ff. DOI: 10.1126/science.1244258.

". . . an ocean hundreds of feet deep." G. L. Villanueva et al. "Strong water isotopic anomalies in the martian atmosphere: Probing current and ancient reservoirs." 2015. *Science* 348(6231) p. 218. DOI: 10.1126/science.aaa3630.

"One recent theory casts Venus as Icarus, . . ." K. Hamano et al. "Emergence of two types of terrestrial planet on solidification of magma ocean." 2013. *Nature* 497, p. 607. DOI: 10.1038/nature12163.

"Solar radiation shatters its bonds, . . ." D. C. Catling and K. J. Zahnle. "The planetary air leak." 2009. *Sci Amer.* 300(5), p. 36.

"The color difference reveals a chemical difference . . ." J. P. Grotzinger et al. "A habitable fluvio-lacustrine environment at Yellowknife Bay, Gale Crater, Mars." 2014. *Science* 343(6169), p. 1242777-1. DOI: 10.1126/science.1242777.

"Scientists are struggling to explain how Mars stayed wet . . ." A. Witze. "Mars slow to yield its secrets." 2014. *Nature* 511, p. 396. DOI: 10.1038/511396a.

"An online video titled 'Huygens: Titan Descent Movie' shows what Huygens saw." Available on YouTube at https://www.youtube.com/watch?v=HtYDPj6eFLc.

"The watching scientists described it this way in *Scientific American*: . . ." R. Lorenz and C. Sotin. "The moon that would be a planet." 2010. *Sci Amer.* 302(3), p. 36.

". . . simulations have come up empty." See in particular the simulation of D. F. Strobel. "Molecular hydrogen in Titan's atmosphere: Implications of the measured tropospheric and thermospheric mole fractions." 2010. *Icarus* 208(2), p. 878. DOI: 10.1016/j.icarus.2010.03.003. These results are contextualized by C. McKay, "Have we discovered evidence for life on Titan?" 2011. http://www.ciclops.org/news/making_sense.php?id=6431.

". . . cyanide storms scour the surface . . ." R. J. de Kok et al. "HCN ice in Titan's high-altitude southern polar cloud." 2014. *Nature* 514, p. 65. DOI: 10.1038/nature13789.

"Europa and Ganymede both have liquid oceans . . . Likewise for Saturn's moon Enceladus, . . ." K. Chang. "Suddenly, it seems, water is everywhere in solar system." March 12, 2015. *New York Times.*

"A group of Russian scientists carefully drilled into it . . ." Y. M. Shtarkman et al. "Subglacial Lake Vostok (Antarctica) accretion ice contains a diverse set of sequences from aquatic, marine and sediment-inhabiting bacteria and eukarya." 2013. *PLoS One* 8(7), p. e67221. DOI: 10.1371/journal.pone.0067221.

". . . if a planet like Titan orbits a cooler, bigger type of star called a red dwarf." Y. Bhattacharjee. "Almost-Earth tantalizes astronomers with promise of worlds to come." 2014. *Science* 344(6181), p. 249. DOI: 10.1126/science.344.6181.249.

". . . videos from the vantage point of the International Space Station." Available on YouTube at https://www.youtube.com/watch?v=IpvtTlZMt7Y.

". . . some scientists have calculated that only 10% of galaxies . . ." T. Piran and R. Jimenez. "Possible role of gamma ray bursts on life extinction in the universe." *Phys Rev Lett.* 113, p. 231102. DOI: 10.1103/PhysRevLett.113.231102.

"Venus has a sulfuric acid cycle . . ." Summarized in http://www.skepticalscience.com/Venus-runaway-greenhouse-effect.htm.

". . . Mars has a peroxide cycle . . ." R. T. Clancy et al. "A measurement of the 362 GHz absorption line of Mars atmospheric $H2O2$." 2004. *Icarus* 168(1), p. 116. DOI: 10.1016/j.icarus.2003.12.003.

"But the oceans didn't acidify that much, and the $CO2$ blanket was transformed, . . ." N. H. Sleep et al. "Initiation of clement surface conditions on the earliest Earth." 2001. *Proc Natl Acad Sci.* 98(7), p. 3666. DOI: 1073/pnas.071045698.

". . . nitrogen was released from the mantle over time . . ." S. Mikhail and D. A. Sverjensky. "Nitrogen speciation in upper mantle fluids and the origin of Earth's nitrogen-rich atmosphere." 2014. *Nature Geo.* 7, p. 816. DOI: 10.1038/ngeo2271.

". . . in thermodynamic terms as a state of increased *entropy*." See, for example, Peter Atkins and Julio de Paula, *Physical Chemistry for the Life Sciences* (2nd edition). 2011, Chapter 2.

". . . special antennas could capture this infrared heat energy . . ." S. J. Byrnes et al. "Harvesting renewable energy from Earth's mid-infrared emissions." *Proc Natl Acad Sci.* 111(11), p. 3927. DOI: 10.1073/pnas.1402036111.

"Complex cycles can form . . ." A provocative theory connecting thermodynamics to life cycles can be found in J. L. England. "Statistical physics of self-replication." 2013. *J Chem Phys.* 139, p. 121923. DOI: 10.1063/1.4818538.

"The engineer Adrian Bejan has developed a theory . . ." For the general reader: Adrian Bejan and J. Peder Zane. *Design in Nature: How the Constructal Law Governs Evolution in Biology, Physics, Technology, and Social Organizations.* 2013, Knopf Doubleday. For the reader with a scientific background: Adrian Bejan. *Shape and Structure, from Engineering to Nature.* 2000, Cambridge University Press. Specific examples in this section taken from the latter title.

"Chaisson's hypothesis is that an increased ERD goes along with increased structural complexity to process that energy." See E. J. Chaisson. "Energy rate density as a complexity metric and evolutionary driver." 2011. *Complexity* 16(3), p. 27. DOI: 10.1002/cplx.20323; "Energy rate density. II. Probing further a new complexity metric." 2011. *Complexity* 17(1), p. 44. DOI: 10.1002/cplx.20373; and "Using complexity science to search for unity in the natural sciences," in *Complexity and the Arrow of Time*, ed. by Lineweaver, Davies, and Ruse. 2013, Cambridge University Press. DOI: 10.1017/CBO9781139225700.006.

"A newborn star gives out 0.0001 watts/kg, ..." Data from Figure 3 in E. J. Chaisson. "Energy rate density as a complexity metric and evolutionary driver." 2011. *Complexity* 16(3), p. 27. DOI: 10.1002/cplx.20323.

"The same pattern of increasing ERD over time is seen in biological processes." Data from Table 1 in. J. Chaisson. "Energy rate density as a complexity metric and evolutionary driver." 2011. *Complexity* 16(3), p. 27. DOI: 10.1002/cplx.20323.

"Three and a half billion years ago the average temperature at the ocean surface was about 35°C." M. T. Hren et al. "Oxygen and hydrogen isotope evidence for a temperate climate 3.42 billion years ago." 2009. *Nature* 462, p. 205. DOI: 10.1038/nature08518.

"Closer normal experience, you can see fossilized raindrops, ..." S. M. Som et al. "Air density 2.7 billion years ago limited to less than twice modern levels by fossil raindrop imprints." 2012. *Nature* 484(7394), p. 359. DOI: 10.1038/nature10890.

"Carl Sagan and colleagues pointed out that ..." C. Sagan and G. Mullen. "Earth and Mars: Evolution of atmospheres and surface temperatures." 1972. *Science* 177(4043), p. 52. DOI: 10.1126/science.177.4043.52.

"There are several promising leads involving many gases ..." See R. Wordsworth and R. Pierrehumbert. "Hydrogen-nitrogen greenhouse warming in Earth's early atmosphere." 2013. *Science* 339, p. 64. DOI: 10.1126/science.1225759.

"... the ocean surface stayed below 40°C." Y. Sun et al. "Lethally hot temperatures during the Early Triassic greenhouse." 2012. *Science* 338(6105), p. 366. DOI: 10.1126/science.1224126.

"... as shown by study of a certain class of plants in Australia." J. M. Sniderman et al. "Fossil evidence for a hyperdiverse sclerophyll flora under a non-Mediterranean-type climate." 2013. *Proc Natl Acad Sci.* 110(9), p. 3423. DOI: 10.1073/pnas.1216747110.

"... even a small temperature shift would change the soil microbial ecosystem ..." F. Garcia-Pichel et al. "Temperature drives the continental-scale distribution of key microbes in topsoil communities." 2013. *Science* 340(6140), p. 1574. DOI: 10.1126/science.1236404.

"Some have proposed that this concept be expanded to add a geothermal habitable zone, ..." R. Barnes et al. "Tidal limits to planetary habitability." 2009. *Astrophys J Lett.* 700(1), p. L30. DOI: 10.1088/0004-637X/700/1/L30.

"About 10% of sun-like stars have an Earth-like planet ... " E. A. Petiguraa et al. "Prevalence of Earth-size planets orbiting Sun-like stars." 2013. *Proc Natl Acad Sci.* 110(48), p. 19273. DOI: 10.1073/pnas.1319909110.

"... planets only up to 1.6 times the mass of Earth can turn out rocky ..." D. Clery. "How to make a planet just like Earth." 2013. *Science* 468, p. 342. DOI: 10.1126/science.aaa6296.

"... bearing "scant resemblance" to other systems, ..." K. Batygin and G. Laughlin. "Jupiter's decisive role in the inner Solar System's early evolution." 2015. *Proc Natl Acad Sci.* 112(14), p. 4214. DOI: 10.1073/pnas.1423252112.

"Two possibilities exist ..." As quoted by Michio Kaku, *Visions: How Science Will Revolutionize the Twenty-First Century.* 1998, Anchor, p. 295.

"Caleb Scharf calls this equipoise ..." C. Scharf. "Cosmic (in)significance." 2014. *Sci Amer..* 311(2), p. 74.

"... the Smithsonian website *This Dynamic Planet.*" http://nhb-arcims.si.edu/ThisDynamicPlanet/index.html.

"... a large asteroid impact got the continents moving." N. H. Sleep and D. R. Lowe. "Physics of crustal fracturing and chert dike formation triggered by asteroid impact, ~3.26 Ga,

Barberton greenstone belt, South Africa." 2014. *Geochem Geophys Geosys.* 15(4), p. 1054. DOI: 10.1002/2014GC005229.

". . . because it was released from the mantle via the motion of plate tectonics." S. Mikhail and D. A. Sverjensky. "Nitrogen speciation in upper mantle fluids and the origin of Earth's nitrogen-rich atmosphere." 2014. *Nature Geo.* 7, p. 816. DOI: 10.1038/ngeo2271.

"Water also forms a complex melted substance with rock . . ." B. Schmandt et al. "Dehydration melting at the top of the lower mantle." 2014. *Science* 344(6189), p. 1265. DOI: 10.1126/science.1253358.

". . . the kinds of rock that formed later in Earth's history are less dense and more buoyant . . ." Robert M. Hazen. *The Story of Earth: The First 4.5 Billion Years, from Stardust to Living Planet.* 2012, Viking, pp. 103–105.

"Using water chemistry as a guide, geologists trace ancient paths of water . . ." R. M. Hazen et al. "Mineral evolution." 2008. *Amer Mineral.* 93(11–12), p. 1693. DOI: 10.2138/am.2008.2955.

"The ocean needed negative charges from the right side of the table to cancel out the positives." R. J. P. Williams and J. J. R. Frausto da Silva. *The Natural Selection of the Chemical Elements.* 1997, Clarendon Press (Oxford), pp. 305–306.

"On the left side, manganese often pairs with oxygen in rocks . . ." See Figure 1.8 in R. J. P. Williams and J. J. R. Frausto da Silva. *The Chemistry of Evolution: The Development of Our Ecosystem.* 2006, Elsevier, p. 20.

"Darwin himself employed its cousin 'the scale of nature' at times." In *Origin of Species* this phrase is found in Chapters 3, 5, 8, 13, and 15.

Chapter 5

". . . various lines of evidence . . . can only be explained by life 3.5, 3.6, or even 3.8 billion years ago." Robert M. Hazen. *The Story of Earth: The First 4.5 Billion Years, from Stardust to Living Planet.* 2012, Viking, p. 149.

"A study of phosphorus in rocks 3.5 to 3.2 billion years old . . ." R. E. Blake et al. "Phosphate oxygen isotopic evidence for a temperate and biologically active Archaean ocean." 2010. *Nature* 464(7291), p. 1029. DOI: 10.1038/nature08952.

". . .projecting that the first proteins were built 3.8 billion years ago." M. Wang et al. "A universal molecular clock of protein folds and its power in tracing the early history of aerobic metabolism and planet oxygenation." 2011. *Mol Biol Evol.* 28(1), p. 567. DOI: 10.1093/molbev/msq232.

". . . looked at an essential protein called Elongation Factor . . ." E. A. Gaucher et al. "Palaeotemperature trend for Precambrian life inferred from resurrected proteins." 2008. *Nature* 451(7179), p. 704. DOI: 10.1038/nature06510.

"He resurrected two types of ancient enzymes called thioredoxins and beta-lactamases . . ." A. Ingles-Prieto et al. "Conservation of protein structure over four billion years." 2013. *Structure* 21(9), p. 1690. DOI: 10.1016/j.str.2013.06.020; and V. A. Risso et al. "Hyperstability and substrate promiscuity in laboratory resurrections of Precambrian β-lactamases." 2013. *J Am Chem Soc.* 135(8), p. 2899. DOI: 10.1021/ja311630a.

"Much of this water predated even the sun." A. R. Sarafian et al. "Early solar system: Early accretion of water in the inner solar system from a carbonaceous chondrite-like source." 2014. *Science* 346(6209), p. 623. DOI: 10.1126/science.1256717.

"The carbon circles can be found intermingled with nitrogen, . . ." L. J. Allamandola. "Chemical evolution in the interstellar medium: Feedstock of solar systems," in

Chemical Evolution across Space and Time: From the Big Bang to Prebiotic Chemistry, ed. by Zaikowski and Friedrich. 2008, American Chemical Society. Volume 981, Chapter 5, p. 80. DOI: 10.1021/bk-2008-0981.ch005.

"Meterorites have a mineral called schreibersite, ..." M. A. Pasek et al. "Evidence for reactive reduced phosphorus species in the early Archean ocean." 2013. *Proc Natl Acad Sci.* 110(25), p. 10089. DOI: 10.1073/pnas.1303904110.

"... the sheer abundance of amino acids in meteorites is remarkable." O. Botta. "Extraterrestrial organic chemistry as recorded in carbonaceous chondrites," in *Chemical Evolution across Space and Time: From the Big Bang to Prebiotic Chemistry*, ed. by Zaikowski and Friedrich. 2008, American Chemical Society. Volume 981, Chapter 13, p. 246. DOI: 10.1021/bk-2008-0981.ch013.

"(One study found that a subset of 12 amino acids was *better* than our set of 20 for building proteins.)" J. Tanaka et al. "Comparative characterization of random-sequence proteins consisting of 5, 12, and 20 kinds of amino acids." 2010. *Protein Sci.* 19(4), p. 786. DOI: 10.1002/pro.358.

"Miller and Urey found a mixture of amino acids in their flasks, ..." S. L. Miller et al. "Origin of organic compounds on the primitive earth and in meteorites." 1976. *J Mol Evol.* 9(1), p. 59.

"When an old batch of vials from 1958 experiments ..." E. T. Parker et al. "Primordial synthesis of amines and amino acids in a 1958 Miller H2S-rich spark discharge experiment." 2011. *Proc Natl Acad Sci.* 108(14), p. 5526. DOI: 10.1073/pnas.1019191108.

"These researchers added carbonate ..." H. J. Cleaves et al. "Prebiotic organic synthesis in neutral planetary atmospheres," in *Chemical Evolution across Space and Time: From the Big Bang to Prebiotic Chemistry*, ed. by Zaikowski and Friedrich. 2008, American Chemical Society. Volume 981, Chapter 15, p. 282. DOI: 10.1021/bk-2008-0981.ch015.

"... they blow bubbles as they dissolve." J. Dworkin et al. "Self-assembling amphiphilic molecules: Synthesis in simulated interstellar/precometary ices." 2001. *Proc Natl Acad Sci.* 98(3), p. 815.

"Another meteorite, found near Sutter's Mill, ..." S. Pizzarello et al. "Processing of meteoritic organic materials as a possible analog of early molecular evolution in planetary environments." 2013. *Proc Natl Acad Sci.* 110(39), p. 15614. DOI: 10.1073/pnas.1309113110.

"Ocean bacteria communicate and transfer material ..." S. J. Biller et al. "Bacterial vesicles in marine ecosystems." 2014. *Science* 343(6167), p. 183. DOI: 10.1126/science.1243457.

"... about 80 components in all, and mixed them together with vesicles." P. Stano et al. "A remarkable self-organization process as the origin of primitive functional cells." 2013. *Angew Chem.* 52(50), p. 13397. DOI: 10.1002/anie.201306613.

"... many parasitic RNA strands interfered with the desired reaction, sabotaging the results." Y. Bansho et al. "Importance of parasite RNA species repression for prolonged translation-coupled RNA self-replication." 2012. *Chem Biol.* 19(4), p. 478. DOI: 10.1016/j.chembiol.2012.01.019.

"For example, shearing forces in the liquid tore vesicles ..." I. Budin and J. W. Szostak. "Physical effects underlying the transition from primitive to modern cell membranes." 2011. *Proc Natl Acad Sci.* 108(13), p. 5249. DOI: 10.1073/pnas.1100498108.

"A gentler chemical process based on sulfur's electron-grabbing properties ..." T. F. Zhu et al. "Photochemically driven redox chemistry induces protocell membrane pearling and division." 2012. *Proc Natl Acad Sci.* 109(25), p. 9828. DOI: 10.1073/pnas.1203212109.

"Some bacteria reproduce themselves through a similar mechanical process called 'bleb-bing.'" R. Mercier et al. "Excess membrane synthesis drives a primitive mode of cell proliferation." 2013. *Cell* 152(5), p. 997. DOI: 10.1016/j.cell.2013.01.043.

"One type of clay mineral helps form Szostak's bubbles and also makes long RNA chains." J. P. Ferris. "Montmorillonite-catalysed formation of RNA oligomers: the possible role of catalysis in the origins of life." 2006. *Philos Trans R Soc Lond B*. 361(1474): p. 1777. DOI: 10.1098/rstb.2006.1903.

"In the right conditions, heat flowing across an open pore …" M. Kreysing et al. "Heat flux across an open pore enables the continuous replication and selection of oligonucleotides towards increasing length." 2015. *Nat Chem.* 7(3), p. 203. DOI: 10.1038/nchem.2155.

"Some scientists think we can run serpentinization in the lab …" M. Andreani et al. "Aluminum speeds up the hydrothermal alteration of olivine." 2013. *Amer Mineral.* 98(10), p. 1738. DOI: 10.2138/am.2013.4469.

"… together to explain how this clue may work best at a particular place deep underwater." W. F. Martin et al. "Energy at life's origin." 2014. *Science* 344(6188), p. 1092. DOI: 10.1126/science.1251653.

"… patterns like the "magic garden" sets sold in science toy stores." F. Haudin et al. "Spiral precipitation patterns in confined chemical gardens." 2014. *Proc Natl Acad Sci.* 111(49), p. 17363. DOI: 10.1073/pnas.1409552111.

"Lane's story even works with suboptimal, leaky membranes that tighten up over time." N. Lane and W. F. Martin. "The origin of membrane bioenergetics." 2012. *Cell* 151(7), p. 1406. DOI: 10.1016/j.cell.2012.11.050; and V. Sojo et al. "A bioenergetic basis for membrane divergence in archaea and bacteria." 2014. *PLoS Biol.* 12(8), p. e1001926. DOI: 10.1371/journal.pbio.1001926.

"… iron sulfide rocks at 100°C, along with CO carbon monoxide …" C. Huber and G. Wächtershäuser. "Peptides by activation of amino acids with CO on (Ni,Fe)S surfaces: implications for the origin of life." 1998. *Science* 281(5377), p. 670.

"One group put two iron-sulfur centers at either end …" A. Roy et al. "De novo design of an artificial bis[4Fe-4S] binding protein." 2013. *Biochemistry* 52(43), p. 7586. DOI: 10.1021/bi401199s.

"One group put polarized metal electrodes into compost heaps …" O. Nercessian et al. "Harvesting electricity with Geobacter bremensis isolated from compost." 2012. *PLoS One* 7(3), p. e34216. DOI: 10.1371/journal.pone.0034216.

"You can see iron's two forms in the colors on a Grecian urn." L. Lühl et al. "Confocal XANES and the Attic black glaze: The three-stage firing process through modern repro-duction." 2014. *Anal Chem.* 86(14), p. 6924. DOI: 10.1021/ac500990k.

"One group of scientists at Amgen noticed …" J. F. Valliere-Douglass et al. "Photochemical degradation of citrate buffers leads to covalent acetonation of recombinant protein ther-apeutics." 2010. *Protein Sci.* 19(11), p. 2152. DOI: 10.1002/pro.495.

"Even sugars floating in the ocean have inherent iron-binding activity …" C. S. Hassler et al. "Saccharides enhance iron bioavailability to Southern Ocean phytoplankton." 2011. *Proc Natl Acad Sci.* 108(3), p. 1076. DOI: 10.1073/pnas.1010963108.

"Iron-acquisition genes are located on DNA in places that evolve more slowly, …" S. Paul et al. "Accelerated gene evolution through replication-transcription conflicts." 2013. *Nature* 495(7442), p. 512. DOI: 10.1038/nature11989.

"The bacterium *Staphylococcus aureus* has such a hunger for iron …" G. Pishchany et al. "Specificity for human hemoglobin enhances Staphylococcus aureus infection." 2010. *Cell Host Microbe* 8(6), p. 544. DOI: 10.1016/j.chom.2010.11.002.

"A group of scientists attached a protein that absorbs iron to a porous silica material." F. Carmona et al. "A bioinspired hybrid silica-protein material with antimicrobial activity by iron uptake." 2013. *Metallomics* 5(3), p. 193. DOI: 10.1039/c3mt20211a.

"In 2012, a California businessman ran a large-scale experiment . . ." A. C. Revkin. "A fresh look at iron, plankton, carbon, salmon and ocean engineering." July 18, 2014. *New York Times* (Opinion).

"Nanoparticles of pyrite . . . float around in abundance." M. Yücel et al. "Hydrothermal vents as a kinetically stable source of iron-sulphide-bearing nanoparticles to the ocean." 2011. *Nat Geosci.* 4, p. 367. DOI: 10.1038/ngeo1148.

"These flakes are clumps of fat and happy bacteria, . . ." See YouTube video "Deep-Sea Snowblower Vents," www.youtube.com/watch?v=Ki0boGH4-Kc.

"One study concluded that only about 125 reactions . . ." A. Barve et al. "Superessential reactions in metabolic networks." 2012. *Proc Natl Acad Sci.* 109(18), p. E1121. DOI: 10.1073/pnas.1113065109.

"A team led by Shelley Copley perfected a gas-inlet compressor valve system . . ." Y. Novikov and S. D. Copley. "Reactivity landscape of pyruvate under simulated hydrothermal vent conditions." 2013. *Proc Natl Acad Sci.* 110(33), p. 13283. DOI: 10.1073/pnas.1304923110.

"More molecules have been made by mixing sugar into chemicals that simulate the ancient ocean, . . ." M.A. Keller et al. "Non-enzymatic glycolysis and pentose phosphate pathway-like reactions in a plausible Archean ocean." 2014. *Mol Syst Biol.* 10, p. 725. DOI: 10.1002/msb.20145228.

". . . with nine peptides, for example, and with RNA." B. Rubinov et al. "Self-replicating amphiphilic β-sheet peptides." 2009. *Angew Chem.* 48(36), p. 6683. DOI: 10.1002/anie.200902790; and T. A. Lincoln and G. F. Joyce. "Self-sustained replication of an RNA enzyme." 2009. *Science* 323(5918), p. 1229. DOI: 10.1126/science.1167856.

"Stuart Kauffman described how this type of situation . . ." S. A. Kauffman. "Systems chemistry sketches," in *Chemical Evolution across Space and Time: From the Big Bang to Prebiotic Chemistry*, ed. by Zaikowski and Friedrich. 2008, American Chemical Society. Volume 981, Chapter 17, p. 319. DOI: 10.1021/bk-2008-0981.ch017.

"The ice caves are captured in photographs by Denis Bud'ko . . ." See http://www.huffingtonpost.com/2013/08/27/russian-ice-cave-mutnovsky-volcano_n_3822632.html.

"However, in 2012 a team led by Eugene Koonin published a paper . . ." A. Y. Mulkidjanian et al. "Origin of first cells at terrestrial, anoxic geothermal fields." 2012. *Proc Natl Acad Sci.* 109(14), p. E821. DOI: 10.1073/pnas.1117774109.

"Each balance has a chemical reason behind it . . ." See Figure 4.8 in R. J. P. Williams and J. J. R. Frausto da Silva. *The Chemistry of Evolution: The Development of Our Ecosystem.* 2006, Elsevier, p. 155ff.

"What works is not running three separate steps, but throwing the chemicals together . . ." M. W. Powner et al. "Synthesis of activated pyrimidine ribonucleotides in prebiotically plausible conditions." 2009. *Nature* 459, p. 239. DOI: 10.1038/nature08013.

"Recently, Sutherland and colleagues published an update . . ." B. H. Patel et al. "Common origins of RNA, protein and lipid precursors in a cyanosulfidic protometabolism." 2015. *Nat Chem.* 7, p. 301. DOI: 10.1038/nchem.2202.

"The proteins that work with RNA show a pattern of evolution . . ." L. Li et al. "Aminoacylating urzymes challenge the RNA world hypothesis." 2013. *J Biol Chem.* 288(37), p. 26856. DOI: 10.1074/jbc.M113.496125.

"... into a sticky tar-like mass (the 'asphalt problem')." S. A. Benner et al. "Asphalt, water, and the prebiotic synthesis of ribose, ribonucleosides, and RNA." *Acc Chem Res.* 45(12), p. 2025. DOI: 10.1021/ar200332w.

"Several labs have made nucleotide chains that compete and even evolve, ..." M. P. Robertson and G. F. Joyce. "Highly efficient self-replicating RNA enzymes." 2014. *Chem Biol.* 21(2), p. 238. DOI: 10.1016/j.chembiol.2013.12.004.

"In his excellent book *The Secrets of Alchemy,* ..." Lawrence M. Principe. *The Secrets of Alchemy.* 2012, University of Chicago Press.

"Carbon chains tipped with a particular common 'head group' chemical ..." L. Toppozini et al. "Adenosine monophosphate forms ordered arrays in multilamellar lipid matrices: insights into assembly of nucleic acid for primitive life." 2013. *PLoS One* 8(5), p. e62810. DOI: 10.1371/journal.pone.0062810.

"the form of iron with two charges ... will sit in the same places." S. S. Athavale et al. "RNA folding and catalysis mediated by iron (II)." 2012. *PLoS One* 7(5), p. e38024. DOI: 10.1371/journal.pone.0038024.

"This problem went away when citrate was added ..." K. Adamala and J. W. Szostak. "Nonenzymatic template-directed RNA synthesis inside model protocells." 2013. *Science* 342(6162), p. 1098. DOI: 10.1126/science.1241888.

"Sarah Keller and a team mixed oily oxygen-tagged carbon chains ..." R. A. Black et al. "Nucleobases bind to and stabilize aggregates of a prebiotic amphiphile, providing a viable mechanism for the emergence of protocells." 2013. *Proc Natl Acad Sci.* 110(33), p. 13272. DOI: 10.1073/pnas.1300963110.

"... R. J. P. Williams writes that the chemical reactions that create life ..." *Evolution's Destiny: Co-evolving Chemistry of the Environment and Life*, with R. E. M. Rickaby. 2012, RSC Publishing; p. 260.

Chapter 6

"A pocket of liquid water hides behind Blood Falls, ..." J. A. Mikucki et al. "A contemporary microbially maintained subglacial ferrous 'ocean.'" 2009. *Science* 324(5925), p. 397. DOI: 10.1126/science.1167350; and J. A. Mikucki et al. "Deep groundwater and potential subsurface habitats beneath an Antarctic dry valley." 2015. *Nature Comm.* 6, p. 6831. DOI: 10.1038/ncomms7831.

"Some old cannonballs spontaneously explode when exposed to surface air." G. Muyzer and A. J. M. Stams. "The ecology and biotechnology of sulphate-reducing bacteria." 2008. *Nat Rev Micro.* 6, p. 441. DOI: 10.1038/nrmicro1892; and http://superbeefy.com/why-do-old-cast-iron-cannonballs-brought-up-from-the-sea-sometimes-explode-and-what-causes-the-oxidization/.

"Multiple species coorperate to live in fluctuating oxygen and salt levels." D. Ionescu et al. "Microbial and chemical characterization of underwater fresh water springs in the Dead Sea." 2012. *PLoS One* 7(6), p. e38319. DOI: 10.1371/journal.pone.0038319.

"Black lichens grow at the high-tide line, ..." Tristan Gooley. *The Walker's Guide to Outdoor Clues and Signs.* 2014, Sceptre Press, pp. 307–308.

"To make a Winogradsky column, ..." You can also purchase one from the Hiram Genomics Store: http://anthony-marchi.squarespace.com/customized-winogradsky-columns-microcosms/.

"The chemical loops that happen in a bottle also happen in the muck at the bottom of the ocean." See Figure 1 in A. P. Teske. "Tracking microbial habitats in subseafloor sediments." 2012. *Proc Natl Acad Sci.* 109(42), p. 16756. DOI: 10.1073/pnas.1215867109.

"Life is a loop, not a helix." J. Davies. "A closed loop." September 26, 2014. *Aeon Magazine*, http://aeon.co/magazine/science/why-the-symbol-of-life-is-a-loop-not-a-helix/.

"In the early twentieth century, the Union Oil Company ..." Tristan Gooley. *The Walker's Guide to Outdoor Clues and Signs.* 2014, Sceptre Press, p. 242.

"If on your walks, you sniff a pong, ..." Tristan Gooley. *The Walker's Guide to Outdoor Clues and Signs.* 2014, Sceptre Press, p. 159.

"So scientists added a waste-eating bacterium ..." K. L. Hillesland and D. A. Stahl. "Rapid evolution of stability and productivity at the origin of a microbial mutualism." 2010. *Proc Natl Acad Sci.* 107(5), p. 2124. DOI: 10.1073/pnas.0908456107.

"Even different kingdoms of life can cooperate, ..." E. F. Y. Horn and A. W. Murray. "Niche engineering demonstrates a latent capacity for fungal-algal mutualism." 2014. *Science* 345(6192), p. 94. DOI: 10.1126/science.1253320.

"Some have worked out a thermodynamic theory for how making these complex cycles ..." D. H. Wolpert, "Information width: A way for the Second Law to increase complexity." Chapter 11, p. 251 in *Complexity and the Arrow of Time*, ed. by Lineweaver, Davies, and Ruse. 2013, Cambridge University Press. DOI: 10.1017/CBO9781139225700.006.

"One man got a yeast infection in his gut." B. Cordell and J. McCarthy. "A case study of gut fermentation syndrome (auto-brewery) with *Saccharomyces cerevisiae* as the causative organism." 2013. *Int J Clin Med.* 4(7), p. 309. DOI: 10.4236/ijcm.2013.47054.

"... their colleagues who stayed home found even weirder shapes ..." Peter Cimermancic, conference presentation in Chemical Biology & Enzymology, July 28, 2014. 28th Annual Symposium of the Protein Society, San Diego, CA.

"It is even possible that non-caloric sweeteners change your gut bacteria so much ..." J. Suez et al. "Artificial sweeteners induce glucose intolerance by altering the gut microbiota." 2014. *Nature* 514(7521), p. 181. DOI: 10.1038/nature13793.

"Even the lowly cockroach has a complex gut ..." C. M. Agapakis et al. "Natural strategies for the spatial optimization of metabolism in synthetic biology." 2012. *Nat Chem Biol.* 8(6), p. 527. DOI: 10.1038/nchembio.975.

"Some stromatolites look, from the side, like a stack of plates ..." J. W. Schopf. "Fossil evidence of Archaean life." 2006. *Philos Trans R Soc B.* 361(1470), pp. 869–885. DOI: 10.1098/rstb.2006.1834.

"Very old stromatolites have been found with sulfur isotope patterns ..." T. R. Bontognali et al. "Sulfur isotopes of organic matter preserved in 3.45-billion-year-old stromatolites reveal microbial metabolism." 2012. *Proc Natl Acad Sci.* 109(38), p. 15146. DOI: 10.1073/pnas.1207491109.

"The preferred technique is to dip a gummi bear in molten potassium chlorate in a test tube." See www.youtube.com/watch?v=7Xu2YZzufTM.

"As Annie Dillard wrote in *Pilgrim at Tinker Creek*, ..." Annie Dillard, *Pilgrim at Tinker Creek.* 1974, Bantam Books, p. 10.

"... as shown in another chemical YouTube mainstay, the Volta experiment." See "Volta Experiment at Rutgers," www.youtube.com/watch?v=e2Gz-h35HCE.

"The scientist Linda Kah describes this period ..." Quoted in Robert M. Hazen. *The Story of Earth: The First 4.5 Billion Years, from Stardust to Living Planet.* 2012, Viking, p. 199.

"Life's hunger for electrons led to reactions that oxidized the environment." R. J. P. Williams and R. E. M. Rickaby. *Evolution's Destiny: Co-evolving Chemistry of the*

Environment and Life, RSC Publishing, p. 160: "As this cytoplasmic (in cell) organic chemistry was reductive of necessity, because the ingredients were made of the oxides of carbon, it was inevitable that oxidizing compounds, which became oxygen, would be released. There followed in the environment an unavoidable and predictable sequence of the oxidation of minerals and non-metal elements in solution, limited by diffusion but generally following sequentially equilibrium constants, redox potentials...."

"This process is called the Wood-Ljungdahl pathway, ..." S. W. Ragsdale and E. Pierce. "Acetogenesis and the Wood-Ljungdahl pathway of CO(2) fixation." 2008. *Biochim Biophys Acta.* 1784(12), p. 1873. DOI: 10.1016/j.bbapap.2008.08.012.

"The Wood-Ljungdahl pathway may have been the first metabolism ..." R. Braakman and E. Smith. "The emergence and early evolution of biological carbon-fixation." *PLoS Comput Biol.* 8(4), p. e1002455. DOI: 10.1371/journal.pcbi.1002455.

"There is a story that the first monk ..." Mary Ruefle. *Madness, Rack, and Honey.* 2012, Wave Books, p. 101.

"In glycolysis, each of the ten steps follows from the next with relentless chemical logic." A. Bar-Even et al. "Rethinking glycolysis: on the biochemical logic of metabolic pathways." 2012. *Nat Chem Biol.* 8(6), p. 509. DOI: 10.1038/nchembio.971.

"Glycolysis is surprisingly *evolvable.*" A. Barve and A. Wagner. "A latent capacity for evolutionary innovation through exaptation in metabolic systems." 2013. *Nature* 500(7461), p. 203. DOI: 10.1038/nature12301.

"A cell could use this reaction to release energy from the earth, ..." A. Mulkidjanian and M. Y. Galperin in *Structural bioinformatics of membrane proteins,* ed. by D. Frishman. 2009, Springer, pp. 108–109: "... the main initial [energy] source [before oxygen] was probably the mineral (pyrite-forming) reaction H2S + FeS → FeS2 (solid) + H2 ..."

"Geological evidence suggests certain rocks dissolved ..." S. A. Crowe et al. "Sulfate was a trace constituent of Archean seawater." 2014. *Science* 346(6210), p. 735. DOI: 10.1126/science.1258966.

"... they digested a surprising amount of the oil spilled ..." J. D. Kessler et al. "A persistent oxygen anomaly reveals the fate of spilled methane in the deep Gulf of Mexico." 2011. *Science* 331(6015), p. 312. DOI: 10.1126/science.1199697.

"Even this part of life's circle has been fossilized, ..." These were also sulfate-reducing bacteria. See D. Wacey et al. "Nanoscale analysis of pyritized microfossils reveals differential heterotrophic consumption in the ~1.9-Ga Gunflint chert." 2013. *Proc Natl Acad Sci.* 110(20), p. 8020. DOI: 10.1073/pnas.1221965110.

"To quote Annie Dillard again, ..." Annie Dillard, *Pilgrim at Tinker Creek.* 1974, Bantam Books, p. 232.

"The crucial chemical role is filled by tungsten ..." S. W. Ragsdale and E. Pierce. "Acetogenesis and the Wood-Ljungdahl pathway of CO(2) fixation." 2008. *Biochim Biophys Acta.* 1784(12), p. 1873. DOI: 10.1016/j.bbapap.2008.08.012.

"Organic chemists made a small CHON framework ..." S. Ogo et al. "A functional [NiFe] hydrogenase mimic that catalyzes electron and hydride transfer from H2." 2013. *Science* 339(6120), p. 682. DOI: 10.1126/science.1231345.

"Another hydrogen-electron reaction catalyst was built around a single nickel atom ..." A. Dutta et al. "Amino acid modified Ni catalyst exhibits reversible H2 oxidation/production over a broad pH range at elevated temperatures." 2014. *Proc Natl Acad Sci.* 111(46), pp. 16286–16291. DOI: 10.1073/pnas.1416381111.

"... iron, nickel, and cobalt can catalyze hydrogen production for less cost ..." R. M. Bullock. "Abundant metals give precious hydrogenation performance." 2013. *Science* 342(6162), p. 1054. DOI: 10.1126/science.1247240.

"Primo Levi tells a story in his memoir ..." Primo Levi. *The Periodic Table*. 1984. Schocken Books, pp. 82–83.

"... the salt licks that farmers and zookeepers ..." R. J. P. Williams and R. E. M. Rickaby. *Evolution's Destiny: Co-evolving Chemistry of the Environment and Life*. 2012, RSC Publishing, p. 241.

"Scientists couldn't grow human gut methanogens until ..." K. C. Wrighton et al. "Fermentation, hydrogen, and sulfur metabolism in multiple uncultivated bacterial phyla." 2012. *Science* 337(6102), pp. 1661–1665. DOI: 10.1126/science.1224041.

"... nickel alone can spur methanogen growth." E. M. Hausrath et al. "The effect of methanogen growth on mineral substrates: will Ni markers of methanogen-based communities be detectable in the rock record?" 2007. *Geobiology* 5(1), p. 49. DOI: 10.1111/j.1472-4669.2007.00095.x.

"... nickel binds to and activates immune proteins, ..." B. Roediger and W. Weninger. "How nickel turns on innate immune cells." 2011. *Immun Cell Bio*. 89, p. 1. DOI: 10.1038/icb.2010.114.

"R. J. P.. Williams writes that when a porphyrin ring ..." R. J. P. Williams and R. E. M. Rickaby. *Evolution's Destiny: Co-evolving Chemistry of the Environment and Life*. 2012, RSC Publishing, p. 219.

"Some labs made porphyrins in water from simple combinations ..." G. W. Hodgson and C. Ponnamperuma. "Prebiotic porphyrin genesis: porphyrins from electric discharge in methane, ammonia, and water vapor." 1968. *Proc Natl Acad Sci*. 59(1), p. 22.

"Zinc porphyrin in your blood is a red flag ..." R. F. Labbé et al. "Zinc protoporphyrin: A metabolite with a mission." 1999. *Clin Chem*. 45(12), p. 2060.

"R. J. P. Williams explains this strange pattern by looking at zinc's chemistry." R. J. P. Williams and R. E. M. Rickaby. *Evolution's Destiny: Co-evolving Chemistry of the Environment and Life*. 2012, RSC Publishing, p. 219: "We note that neither Cu nor Zn ions have porphyrin complexes and these ions became very important only later in evolution."

"The geologist Kurt Konhauser and colleagues found zinc " C. Scott et al. "Bioavailability of zinc in marine systems through time." 2013. *Nature Geo*. 6, p. 125. DOI: 10.1038/ngeo1679.

"Its binding to sulfide is the tightest of all the first-row elements, ..." R. J. P. Williams and R. E. M. Rickaby. *Evolution's Destiny: Co-evolving Chemistry of the Environment and Life*. 2012, RSC Publishing, p. 222.

"Christopher Dupont and colleagues published two papers, ..." C. L. Dupont et al. "Modern proteomes contain putative imprints of ancient shifts in trace metal geochemistry." 2006. *Proc Natl Acad Sci*. 103(47), p. 17822. DOI: 10.1073/pnas.0605798103; and C. L. Dupont et al. "History of biological metal utilization inferred through phylogenomic analysis of protein structures." 2010. *Proc Natl Acad Sci*. 107(23), p. 10567. DOI: 10.1073/pnas.0912491107.

"... published by Lawrence David and Eric Alm in 2011." L. A. David and E. J. Alm. "Rapid evolutionary innovation during an Archaean genetic expansion." 2011. *Nature* 469(7328), p. 93. DOI: 10.1038/nature09649.

"This trend is a theory of Kurt Konhauser ..." K. O. Konhauser et al. "Oceanic nickel depletion and a methanogen famine before the Great Oxidation Event." 2009. *Nature* 458(7239), p. 750. DOI: 10.1038/nature07858.

"... the accompanying news article ..." M. A. Saito. "Less nickel for more oxygen." 2009. *Nature* 458(7239), p. 714. DOI: 10.1038/458714a.

"(... sulfate metabolism before oxygen was a limited affair, but today, it is one of the major metabolic activities in the ocean.)" S. A. Crowe et al. "Sulfate was a trace constituent of Archean seawater." 2014. *Science* 346(6210), p. 735. DOI: 10.1126/science.1258966; and M. W. Bowles et al. "Global rates of marine sulfate reduction and implications for sub-sea-floor metabolic activities." 2014. *Science* 344(6186), p. 889. DOI: 10.1126/science.1249213.

Chapter 7

"... covered with an orange haze ..." A. L. Zerkle et al. "A bistable organic-rich atmosphere on the Neoarchaean Earth." 2012. *Nature Geo.* 5, p. 359. DOI: 10.1038/ngeo1425.

"... evidence for wispy amounts of free oxygen ..." S. A. Crowe et al. "Atmospheric oxygenation three billion years ago." 2013. *Nature* 501, p. 535. DOI: 10.1038/nature12426.

"Canfield found surprisingly high amounts of oxygen ..." D. E. Canfield et al. "Oxygen dynamics in the aftermath of the Great Oxidation of Earth's atmosphere." 2013. *Proc Natl Acad Sci.* 110(42), p. 16736. DOI: 10.1073/pnas.1315570110.

"The lake is surrounded by green forest and sits near the blue ocean, ..." H. Collis. "The bizarre PINK lakes from around the world ..." June 19, 2013. *Mail Online.* http://www.dailymail.co.uk/news/article-2344399/The-bizarre-PINK-lakes-world-look-like-milkshakes-freak-nature.html.

"You may have seen some in San Francisco Bay ..." M. Sharlach. "New genes = new Archaea?" October 15, 2014. *The Scientist Online.* http://www.the-scientist.com/?articles.view/articleNo/41230/title/New-Genes---New-Archaea-/

"At Yellowstone hot springs, scientists saw pink gel ..." David Toomey. *Weird Life.* 2014, Norton, p. 12.

"On the other side of the world is a bright magenta river ..." N. Bonaccorso. "Beautiful photos of the River of Five Colors." March 25, 2014. http://www.weather.com/travel/news/river-five-colors-20140324.

"... it is perfectly natural to find the occasional lobster that is an unsettlingly bright blue." S. Begum et al. "On the origin and variation of colors in lobster carapace." March 23, 2015. *Phys Chem Chem Phys.* DOI: 10.1039/C4CP06124A.

"Paul Falkowski describes photosynthesis ..." Quoted in M. Leslie. "On the origin of photosynthesis." 2009. *Science* 323(5919), p. 1286. DOI: 10.1126/science.323.5919.1286.

"First, actually, it may have turned purple." E. Sanromá et al. "Characterizing the purple Earth: Modelling the globally-integrated spectral variability of the Archean Earth." 2014. *Astrophys J.* 780(1), p. 52. DOI: 10.1088/0004-637X/780/1/52.

"... match the shape of the sunlight falling upon the system." This is seen in other kingdoms of life as well: N. C. Rockwell et al. "Eukaryotic algal phytochromes span the visible spectrum." 2014. *Proc Natl Acad Sci.* 111(10), p. 3871. DOI: 10.1073/pnas.1401871111.

"This halo is an invisible pigment ..." S. Moser et al. "Blue luminescence of ripening bananas." 2008. *Angew Chem.* 47(46), p. 8954. DOI: 10.1002/anie.200803189.

"Rhodopsin is found in the membrane of Lake Hiller's pinkish-purple salt-loving microbes, ..." From genomic analysis of similar microbes in E. A. Becker et al. "Phylogenetically driven sequencing of extremely halophilic archaea reveals strategies for static and dynamic osmo-response." 2014. *PLoS Genet.* 10(11), p. e1004784. DOI: 10.1371/journal.pgen.1004784.

"Through proteins like this, day and night cycles are detected by the smallest organisms …" E. A. Ottesen et al. "Multispecies diel transcriptional oscillations in open ocean heterotrophic bacterial assemblages." 2014. *Science* 345(6193), p. 207. DOI: 10.1126/science.1252476.

"These pigments are arranged so that energy harvesting is maximized …" R. Croce and H. van Amerongen. "Natural strategies for photosynthetic light harvesting." 2014. *Nat Chem Bio.* 10, p. 492. DOI: 10.1038/nchembio.1555.

"Recently scientists found that these pigments flex in sync …" R. Hildner et al. "Quantum coherent energy transfer over varying pathways in single light-harvesting complexes." 2013. *Science* 340(6139), p. 1448. DOI: 10.1126/science.1235820.

"Green photosystems use more iron than their purple counterparts, …" I. Yruela. "Transition metals in plant photosynthesis." 2013. *Metallomics* 5(9), p. 1090. DOI: 10.1039/c3mt00086a: "… PSI has a high requirement for Fe."

"One set of experiments shows that ferredoxin …" M. Kunugi et al. "Evolutionary changes in chlorophyllide a oxygenase (CAO) structure contribute to the acquisition of a new light-harvesting complex in micromonas." 2013. *J Biol Chem.* 288(27), p. 19330. DOI: 10.1074/jbc.M113.462663.

"About a tenth of the energy in the oceans is provided by these 'alternate' photosynthesizers." O. Béjà et al. "Unsuspected diversity among marine aerobic anoxygenic phototrophs." 2002. *Nature* 415(6872), p. 630.

"An amazing example of using light to make powerful electrons is found in aphids." J. C. Valmalette. "Light- induced electron transfer and ATP synthesis in a carotene synthesizing insect." 2012. *Sci Rep.* 2, p. 579. DOI: 10.1038/srep00579.

"We don't know yet exactly how the purple and green combined, …" See M. F. Hohmann-Marriott and R. E. Blankenship. "Evolution of photosynthesis." 2011. *Annu Rev Plant Biol.* 62, p. 515. DOI: 10.1146/annurev-arplant-042110-103811.

"Oxygen's reactivity was shown, or smelled, at the Apollo moon landings." E. K. Wilson. "Fireworks fashion, extraterrestrial smells." 2014. *Chem Eng News.* 92(44), p. 56.

"The cell throws out the protein and builds a new one every half hour." D. C. I. Yao et al. "Photosystem II component lifetimes in the cyanobacterium Synechocystis sp. strain PCC 6803." 2012. *J Biol Chem.* 287(1), p. 682. DOI: 10.1074/jbc.M111.320994.

"Gustavo Caetano-Anollés analyzed the protein folds …" M. Wang et al. "A universal molecular clock of protein folds and its power in tracing the early history of aerobic metabolism and planet oxygenation." 2011. *Mol Biol Evol.* 28(1), p. 567. DOI: 10.1093/molbev/msq232.

"… living microbes produce at least as much superoxide as the sun." J. M. Diaz et al. "Widespread production of extracellular superoxide by heterotrophic bacteria." 2013. *Science* 340(6137), p. 1223. DOI: 10.1126/science.1237331.

"Near popular beaches, sunscreens can generate enough ROS …" D. Sánchez-Quiles and A. Tovar-Sánchez. "Sunscreens as a source of hydrogen peroxide production in coastal waters." 2014. *Environ Sci Technol.* 48(16), p. 9037. DOI: 10.1021/es5020696.

"Scientists are working on sunscreen pigments …" E. M. M. Tan et al. "Excited-state dynamics of isolated and microsolvated cinnamate-based UV-B sunscreens." 2014. *J Phys Chem Lett.* 5(14), p. 2464. DOI: 10.1021/jz501140b.

"Iron reacts best with peroxides to make oxygen radicals …" This set of reactions is called Fenton reactions; they create several kinds of ROS. When iron is protected by heme, its Fenton chemistry is diminished.

"Chemistry professors impress their classes . . ." J. J. Dolhun. "Observations on manganese dioxide as a catalyst in the decomposition of hydrogen peroxide: A safer demonstration." 2014. *J Chem Educ.* 91(5), p. 760. DOI: 10.1021/ed4006826.

"It made a versatile chemical called PLP that comes from vitamin B6." K. M. Kim et al. "Protein domain structure uncovers the origin of aerobic metabolism and the rise of planetary oxygen." 2012. *Structure* 20(1), p. 67. DOI: 10.1016/j.str.2011.11.003.

"Some go so far as to call potassium permanganate the 'survival chemical.'" See www.intherabbithole.com/potassium-permanganate-the-most-useful-survival-chemical/ (but be advised that its fire-starting abilities are often overstated).

"Two thousand years ago, Romans used a black powder called *sapo vitri* . . ." John Emsley. *Nature's Building Blocks: An A–Z Guide to the Elements.* 2001, Oxford University Press, p. 251.

". . . to make a blanket of black manganese-oxide crust around the microbe colony." C. N. Butterfield et al. "Mn(II,III) oxidation and MnO2 mineralization by an expressed bacterial multicopper oxidase." 2013. *Proc Natl Acad Sci.* 110(29), p. 11731. DOI: 10.1073/pnas.1303677110; and W. Yang et al. "Population structure of manganese-oxidizing bacteria in stratified soils and properties of manganese oxide aggregates under manganese-complex medium enrichment." 2013. *PLoS One* 8(9), p. e73778. DOI: 10.1371/journal.pone.0073778.

"These 'eggs' were found around 1875 . . ." John Emsley. *Nature's Building Blocks: An A–Z Guide to the Elements.* 2001, Oxford University Press, p. 253.

"This second set of proteins uses manganese . . ." E. M. Bruch et al. "Variations in Mn(II) speciation among organisms: What makes D. radiodurans different." 2015. *Metallomics* 7(1), pp. 136–144. DOI: 10.1039/c4mt00265b.

"Small molecules stuck to manganese and free manganese help protect *D. radiodurans.*" Sharma et al. "Responses of Mn2+ speciation in Deinococcus radiodurans and Escherichia coli to γ-radiation by advanced paramagnetic resonance methods." 2013. *Proc Natl Acad Sci.* 110(15), p. 5945. DOI: 10.1073/pnas.1303376110.

"One study injected diabetic rats with manganese ions, . . ." E. Burlet and S. K. Jain. "Manganese supplementation reduces high glucose-induced monocyte adhesion to endothelial cells and endothelial dysfunction in Zucker diabetic fatty rats." 2013. *J Biol Chem.* 288(9), p. 6409. DOI: 10.1074/jbc.M112.447805.

"The microbe that causes Lyme disease, . . ." J. D. Aguirre et al. "A manganese-rich environment supports superoxide dismutase activity in a Lyme disease pathogen, Borrelia burgdorferi." 2013. *J Biol Chem.* 288(12), p. 8468. DOI: 10.1074/jbc.M112.433540.

". . . a manganese-binding protein that is a sort of manganese sponge." S. M. Damo et al. "Molecular basis for manganese sequestration by calprotectin and roles in the innate immune response to invading bacterial pathogens." 2013. *Proc Natl Acad Sci.* 110(10), p. 3841. DOI: 10.1073/pnas.1220341110.

"Researchers showed that cells can tolerate high levels of sticky, toxic cadmium . . ." S. L. Begg et al. "Dysregulation of transition metal ion homeostasis is the molecular basis for cadmium toxicity in Streptococcus pneumoniae." 2015. *Nat Commun.* 6, p. 6418. DOI: 10.1038/ncomms7418.

"The oldest catalases were probably built around manganese's special chemistry, . . ." M. Zámocký et al. "Molecular evolution of hydrogen peroxide degrading enzymes." 2012. *Arch Biochem Biophys.* 525(2), p. 131. DOI: 10.1016/j.abb.2012.01.017.

"Basically, it acted as a catalase on itself." S. S. Kamat et al. "The catalase activity of diiron adenine deaminase." 2011. *Protein Sci.* 20(12), p. 2080. DOI: 10.1002/pro.748.

"... antibodies that hold onto heme rings also make peroxide, ..." V. Muñoz Robles et al. "Crystal structure of two anti-porphyrin antibodies with peroxidase activity." 2012. *PLoS One* 7(12), p. e51128. DOI: 10.1371/journal.pone.0051128.

"For example, the iron in plant hemoglobins can remove an oxygen ..." R. Sturms et al. "Hydroxylamine reduction to ammonium by plant and cyanobacterial hemoglobins." 2011. *Biochemistry* 50(50), p. 10829. DOI: 10.1021/bi201425f.

"When a lot of respiration happens, ROS increase enough ..." M. Kathiresan et al. "Respiration triggers heme transfer from cytochrome c peroxidase to catalase in yeast mitochondria." 2014. *Proc Natl Acad Sci.* 111(49), p. 17468. DOI: 10.1073/pnas.1409692111.

"A fungal protein uses the same carbon pigment structure ..." J. Lokhandwala et al. "Structural biochemistry of a fungal LOV domain photoreceptor reveals an evolutionarily conserved pathway integrating light and oxidative stress." 2015. *Structure* 23(1), p. 116. DOI: 10.1016/j.str.2014.10.020.

"These connections are deep enough to influence the path of evolution." O. Méhi et al. "Perturbation of iron homeostasis promotes the evolution of antibiotic resistance." 2014. *Mol Biol Evol.* 31(10), p. 2793. DOI: 10.1093/molbev/msu223.

"Just four changes next to the heme in hemoglobin ..." N. Yeung et al. "Rational design of a structural and functional nitric oxide reductase." 2009. *Nature* 462(7276), p. 1079. DOI: 10.1038/nature08620.

"One group took a purple photosystem and added a manganese-binding site ..." J. P. Allen et al. "Light-driven oxygen production from superoxide by Mn-binding bacterial reaction centers." 2012. *Proc Natl Acad Sci.* 109(7), p. 2314. DOI: 10.1073/pnas.1115364109.

"Manganese when attached to oxygen has a special electron orbital ..." J. Suntivich et al. "Design principles for oxygen-reduction activity on perovskite oxide catalysts for fuel cells and metal-air batteries." 2011. *Nat Chem.* 3(7), p. 546. DOI: 10.1038/nchem.1069.

"Some researchers made the manganese work better by mixing calcium into the nanoparticles, ..." N. Birkner et al. "Energetic basis of catalytic activity of layered nanophase calcium manganese oxides for water oxidation." 2013. *Proc Natl Acad Sci.* 110(22), p. 8801. DOI: 10.1073/pnas.1306623110.

"They built a carbon-based lattice of small holes ..." B. Nepal and S. Das. "Sustained water oxidation by a catalyst cage-isolated in a metal–organic framework." 2013. *Ang Chem.* 52(28), p. 7224. DOI: 10.1002/anie.201301327.

"Daniel Nocera heads a group of researchers that study metal catalysis of water splitting." J. Marshall. "Springtime for the artificial leaf." 2014. *Nature* 510(7503), p. 22. DOI: 10.1038/510022a; but also see J. Hitt. "The artificial leaf is here. Again." March 29, 2014, *New York Times.*

"This metal leaf may eventually turn sunlight and water into ..." The most recent advance in this area at press time combines this with microbes: J. P. Torella et al. "Efficient solar-to-fuels production from a hybrid microbial-water-splitting catalyst system." 2015. *Proc Natl Acad Sci U S A.* 112(8), p. 2337. DOI: 10.1073/pnas.1424872112.

"By changing the shape of the binding pocket, ..." A. J. Reig et al. "Alteration of the oxygen-dependent reactivity of de novo Due Ferri proteins." 2012. *Nat Chem.* 4(11), p. 900. DOI: 10.1038/nchem.1454.

"Eastman Chemical's dye library at North Carolina State University ..." C. Drahl. "North Carolina State University receives dye library donation from Eastman Chemical." 2014. *Chem Eng News.* 92(22), p. 26.

"The iron protein ferredoxin adds oxygen to one version of chlorophyll to change its color." A. H. Saunders et al. "Characterization of BciB: A ferredoxin-dependent 8-vinyl-protochlorophyllide reductase from the green sulfur bacterium Chloroherpeton thalassium." 2013. *Biochemistry* 52(47), p. 8442. DOI: 10.1021/bi401172b.

"Also, the process of making many different pigments is dependent on manganese." R. E. Wilkinson and K. Ohki. "Influence of manganese deficiency and toxicity on isoprenoid syntheses." 1988. *Plant Physiol.* 87, p. 841.

"It does so with two iron ions arranged ..." R. Banerjee et al. "Structure of the key species in the enzymatic oxidation of methane to methanol." 2015. *Nature* 518(7539), p. 431. DOI: 10.1038/nature14160.

"For example, when yeast evolve to withstand higher temperatures ..." L. Caspeta et al. "Altered sterol composition renders yeast thermotolerant." 2014. *Science* 346(6205), p. 75. DOI: 10.1126/science.1258137.

"Caetano-Anollés found that molecules like this were made soon after oxygen became available." M. Wang et al. "A universal molecular clock of protein folds and its power in tracing the early history of aerobic metabolism and planet oxygenation." 2011. *Mol Biol Evol.* 28(1), p. 567. DOI: 10.1093/molbev/msq232.

"... release a wave of iron ions and die from the ROS that the iron automatically makes." S. J. Dixon and B. R. Stockwell. "The role of iron and reactive oxygen species in cell death." 2014. *Nat Chem Biol.* 10(1), p. 9. DOI: 10.1038/nchembio.1416.

"The simplest kind of compartment is an oil drop." A. R. Thiam et al. "The biophysics and cell biology of lipid droplets." 2013. *Nat Rev Mol Cell Biol.* 14(12), p. 775. DOI: 10.1038/nrm3699.

"These 'chlorosomes' have oily pigments that are stacked together ..." M. F. Hohmann-Marriott and R. E. Blankenship. "Hypothesis on chlorosome biogenesis in green photosynthetic bacteria." *FEBS Lett.* 581(5), p. 800.

"Both compartments also involve oxygen and appeared at this stage of the story, ..." E. Pennisi. "Modern symbionts inside cells mimic organelle evolution." 2014. *Science* 346(6209), p. 532. DOI: 10.1126/science.346.6209.532.

"In leaf cells, plastids contain *eighty percent* of the iron." I. Yruela. "Transition metals in plant photosynthesis." *Metallomics* 5(9), p. 1090. DOI: 10.1039/c3mt00086a.

"Some plastid-containing plants and algae have been found in various states ..." D. Smith. "Steal my sunshine." 2013. *The Scientist* 27(1), p. 22.

"They can even be picked up by animals ..." F. Jabr. "Leafy green 'solar-powered' sea slugs begin to reveal their true colors." December 16, 2013, *Scientific American Brainwaves.* http://blogs.scientificamerican.com/brainwaves/leafy-green-e28098solar-powerede28099-sea-slugs-begin-to-reveal-their-true-colors/.

"But then a different kind of plastid in a microbe called *Paulinella* upended this assumption." D. Smith. "Steal my sunshine." 2013. *The Scientist* 27(1), p. 22.

"Researchers have evolved pea aphids and *Buchnera* bacteria ..." N. A. Moran and Y. Yun. "Experimental replacement of an obligate insect symbiont." 2015. *Proc Natl Acad Sci.* 112(7), p. 2093. DOI: 10.1073/pnas.1420037112.

"Another group evolved a cyanobacterium ..." K. Hosoda et al. "Adaptation of a cyanobacterium to a biochemically rich environment in experimental evolution as an initial step toward a chloroplast-like state." 2014. *PLoS One* 9(5), p. e98337. DOI: 10.1371/journal.pone.0098337.

"Mitochondria take up a quarter of a cell's volume but more than half of its iron." N. D. Jhurry. "Biophysical investigation of the ironome of human jurkat cells and mitochondria." 2012. *Biochemistry* 51(26), p. 5276.

"Also, inside the mitochondria an electron box . . ." U. Krengel and S. Törnroth-Horsefield. "Coping with oxidative stress." 2015. *Science* 347(6218), p. 125. DOI: 10.1126/science. aaa3602.

". . . another interesting chemical theory explaining why mitochondria are so useful." N. Lane and W. Martin. "The energetics of genome complexity." 2010. *Nature* 467(7318), p. 929. DOI: 10.1038/nature09486.

"If Canfield's geological data holds up, . . ." D. E. Canfield et al. "Oxygen dynamics in the aftermath of the Great Oxidation of Earth's atmosphere." 2013. *Proc Natl Acad Sci.* 110(42), p. 16736. DOI: 10.1073/pnas.1315570110.

Chapter 8

". . . rained down as red jasper and other iron-oxide rocks, . . ." Robert M. Hazen. *The Story of Earth: The First 4.5 Billion Years, from Stardust to Living Planet.* 2012, Viking, p. 179.

"These BIFs are found . . . in the Pilbara region . . ." R. M. Hazen et al. "Mineral evolution." 2008. *Amer Mineral.* 93(11–12), p. 1693. DOI: 10.2138/am.2008.2955.

"In the lab, a mixture of typical microbes from this era . . ." E. Chi Fru et al. "Fossilized iron bacteria reveal a pathway to the biological origin of banded iron formation." 2013. *Nat Commun.* 4, p. 2050. DOI: 10.1038/ncomms3050.

". . . during what is called the 'boring billion.'" Hazen often uses this term, as does H. D. Holland. "The oxygenation of the atmosphere and oceans." 2006. *Phil Trans R Soc B.* 361, p. 903. DOI: 10.1098/rstb.2006.1838.

"Right after the first burst of BIF formation, the first few fossils of complex life appear." A. El Albani et al. "The 2.1 Ga old Francevillian biota: biogenicity, taphonomy and biodiversity." 2014. *PLoS One* 9(6), p. e99438. DOI: 10.1371/journal.pone.0099438.

". . . break and loosen parts distant to the active site and still deactivate the protein." C. E. Bobst et al. "Impact of oxidation on protein therapeutics: conformational dynamics of intact and oxidized acid-β-glucocerebrosidase at near-physiological pH." 2010. *Protein Sci.* 19(12), p. 2366. DOI: 10.1002/pro.517.

"When blood cells burst, the leaked heme iron . . ." C. B. Andersen et al. "Structure of the haptoglobin-haemoglobin complex." 2012. *Nature* 489(7416), p. 456. DOI: 10.1038/ nature11369.

"The tropical parasite *Leishmania* can only cause Leishmaniasis . . ." B. Mittra et al. "Iron uptake controls the generation of Leishmania infective forms through regulation of ROS levels." 2013. *J Exp Med.* 210(2), p. 401. DOI: 10.1084/jem.20121368.

". . . the mitochondrial proteins that carry electrons and pump protons are broken." Y. Zhuo et al. "Leber hereditary optic neuropathy and oxidative stress." 2012. *Proc Natl Acad Sci.* 109(49), p. 19882. DOI: 10.1073/pnas.1218953109.

"This happens in age-related macular degeneration, . . ." D. Weismann et al. "Complement factor H binds malondialdehyde epitopes and protects from oxidative stress." 2011. *Nature* 478(7367), p. 76. DOI: 10.1038/nature10449.

"Proteins called perilipins act like sheepdogs . . ." K. Kuramoto et al. "Perilipin 5, a lipid droplet-binding protein, protects heart from oxidative burden by sequestering fatty acid from excessive oxidation." 2012. *J Biol Chem.* 287(28), p. 23852. DOI: 10.1074/jbc.M111.328708.

"Some microbes build a sticky sugar shield called alginate . . ." J. C. Setubal et al. "Genome sequence of azotobacter vinelandii, an obligate aerobe specialized to support diverse anaerobic metabolic processes." 2009. *J Bacter.* 191(14), p. 4534. DOI: 10.1128/JB.00504-09.

"Others decorate sensitive proteins with sacrificial atoms, . . ." T. Goris et al. "A unique iron-sulfur cluster is crucial for oxygen tolerance of a [NiFe]-hydrogenase." 2011. *Nat Chem Biol.* 7(5), p. 310. DOI: 10.1038/nchembio.555.

"One prominent theory suggests that such bioluminescence developed to deliberately waste oxygen . . ." G. S. Timmins et al. "The evolution of bioluminescent oxygen consumption as an ancient oxygen detoxification mechanism." 2001. *J Mol Evol.* 52(4), p. 321.

"Oxygen in the air activates the sensor . . ." H. M. Girvan and A. W. Munro. "Heme sensor proteins." 2013. *J Biol Chem.* 288(19), p. 13194. DOI: 10.1074/jbc.R112.422642.

"As the oxygen filled the planet from the top down, the oceans became patchy." C. T. Reinhard et al. "Proterozoic ocean redox and biogeochemical stasis." 2013. *Proc Natl Acad Sci.* 110(14), p. 5357. DOI: 10.1073/pnas.1208622110.

"The step up in oxygen was followed by a swift step down in temperature, . . ." Robert M. Hazen. *The Story of Earth: The First 4.5 Billion Years, from Stardust to Living Planet.* 2012, Viking, p. 224.

". . . their manganese-binding site has been papered over . . ." I. Krastanova et al. "Structural and functional insights into the DNA replication factor Cdc45 reveal an evolutionary relationship to the DHH family of phosphoesterases." 2012. *J Biol Chem.* 287(6), p. 4121. DOI: 10.1074/jbc.M111.285395.

". . . incorporated extra sulfur-moving proteins to mitigate oxygen's threat." E. S. Boyd et al. "Interplay between oxygen and Fe-S cluster biogenesis: Insights from the Suf pathway." 2014. *Biochemistry* 53(37), p. 5834. DOI: 10.1021/bi500488r.

"One study of iron-poor parts of the Indian Ocean . . ." D. B. Rusch et al. "Characterization of *Prochlorococcus* clades from iron-depleted oceanic regions." 2010. *Proc Natl Acad Sci.* 107(37), p. 16184. DOI: 10.1073/pnas.1009513107.

"One study saw this by watching bacteria respond to a deadly antibiotic." H. A. Lindsey et al. "Evolutionary rescue from extinction is contingent on a lower rate of environmental change." 2013. *Nature* 494(7438), p. 463. DOI: 10.1038/nature11879.

"A moderate amount of oxygen gives the bacteria energy . . ." P. Kallio et al. "An engineered pathway for the biosynthesis of renewable propane." 2014. *Nat Commun.* 5, p. 4731. DOI: 10.1038/ncomms5731.

"David and Alm's genomic analysis found these exact patterns, . . ." L. A. David and E. J. Alm. "Rapid evolutionary innovation during an Archaean genetic expansion." 2011. *Nature* 469(7328), p. 93. DOI: 10.1038/nature09649.

"To this day, oxygen-using microbes have to use more energy . . ." T. M. McCollom and J. P. Amend. "A thermodynamic assessment of energy requirements for biomass synthesis by chemolithoautotrophic micro-organisms in oxic and anoxic environments." 2005. *Geobiology* 3(2), p. 135. DOI: 10.1111/j.1472-4669.2005.00045.x.

". . . some microbes "breathe" with nitrogen-oxygen compounds . . ." S. E. Winter et al. "Host-derived nitrate boosts growth of E. coli in the inflamed gut." 2013. *Science* 339(6120), p. 708. DOI: 10.1126/science.1232467.

"A scientist named Fritz Haber heated and pressurized nitrogen and hydrogen . . ." S. Everts. "Who was Fritz Haber?" 2015. *Chem Eng News.* 93(8), p. 18.

". . . half of the people alive today are here because of the food provided by Haber-process ammonia." J. W. Erisman et al. "How a century of ammonia synthesis changed the world." 2008. *Nature Geosci.* 1, p. 636. DOI: 10.1038/ngeo325.

". . . the center of nitrogen catalysis was a metal, . . ." X. Zhang et al. "Nitrogen isotope fractionation by alternative nitrogenases and past ocean anoxia." 2014. *Proc Natl Acad Sci.* 111(13), p. 4782. DOI: 10.1073/pnas.1402976111.

". . . a study in Panama found that trees with nitrogenases grew much better, . . ." S. A. Batterman et al. "Key role of symbiotic dinitrogen fixation in tropical forest secondary succession." 2013. *Nature* 502(7470), p. 224. DOI: 10.1038/nature12525.

". . . the genes for it are arranged in a well-ordered genetic cartridge . . ." K. Temme et al. "Refactoring the nitrogen fixation gene cluster from *Klebsiella oxytoca*." 2012. *Proc Natl Acad Sci.* 109(18), p. 7085. DOI: 10.1073/pnas.1120788109.

"If FeMoCo is removed from its enzyme, it keeps splitting nitrogen by itself, . . ." V. K. Shah and W. J. Brill. "Isolation of an iron-molybdenum cofactor from nitrogenase." 1977. *Proc Natl Acad Sci.* 74(8), p. 3249–3253.

". . . a 'chalcogel' that can turn nitrogen to ammonia at room temperature." A. Banerjee et al. "Photochemical nitrogen conversion to ammonia in ambient conditions with FeMoS-chalcogels." 2015. *J Am Chem Soc.* 137(5), p. 2030. DOI: 10.1021/ja512491v.

". . . molybdenum-based nitrogen processing as early as 3.2 billion years ago . . ." E. E. Stüeken et al. "Isotopic evidence for biological nitrogen fixation by molybdenum-nitrogenase from 3.2 Gyr." 2015. *Nature* 520(7549), p. 666. DOI: 10.1038/nature14180.

"In 2011, a group of scientists made a hole in the nitrogenase protein . . ." Z. Y. Yang et al. "Molybdenum nitrogenase catalyzes the reduction and coupling of CO to form hydrocarbons." *J Biol Chem.* 286(22), p. 19417. DOI: 10.1074/jbc.M111.229344.

"Twisting the pterin framework shifts the electron-pushing power of molybdenum, . . ." R. A. Rothery et al. "Pyranopterin conformation defines the function of molybdenum and tungsten enzymes." *Proc Natl Acad Sci.* 109(37), p. 14773. DOI: 10.1073/pnas.1200671109.

"As the planet oxygenated, it followed a chemical sequence moving from left to right along this arrow." R. J. P. Williams and J. J. R. Frausto da Silva, *The Chemistry of Evolution: The Development of Our Ecosystem*, 2006, Elseiver.

"Research by the lab of John Peters agrees with that sequence . . ." E. S. Boyd et al. "An alternative path for the evolution of biological nitrogen fixation." 2011. *Front Microbiol.* 2, p. 205. DOI: 10.3389/fmicb.2011.00205.

". . . it took a billion years for the tiny microbes to add enough oxygen to change the world." R. J. P. Williams and R. E. M. Rickaby. *Evolution's Destiny: Co-evolving Chemistry of the Environment and Life.* 2012, RSC Publishing; pp. 49–51.

"The Earth changed with life, in a process geologist Robert Hazen describes as co-evolution." Robert M. Hazen. *The Story of Earth: The First 4.5 Billion Years, from Stardust to Living Planet.* 2012, Viking, following quotes from p. 177.

". . . some sulfate-reducing bacteria appear not to have evolved at all . . ." J. W. Schopf et al. "Sulfur-cycling fossil bacteria from the 1.8-Ga Duck Creek Formation provide promising evidence of evolution's null hypothesis." 2015. *Proc Natl Acad Sci.* 112(7), p. 2087. DOI: 10.1073/pnas.1419241112.

". . . then they were destroyed by other more diverse species." J. M. Bernhard et al. "Insights into foraminiferal influences on microfabrics of microbialites at Highborne Cay, Bahamas." 2013. *Proc Natl Acad Sci.* 110(24), p. 9830. DOI: 10.1073/pnas.1221721110.

"The Earth turned red as blood, . . ." Robert M. Hazen. *The Story of Earth: The First 4.5 Billion Years, from Stardust to Living Planet.* 2012, Viking, Chapter 7, "Red Earth."

"Mercury makes ROS that can damage DNA, . . ." I. Pieper et al. "Mechanisms of Hg species induced toxicity in cultured human astrocytes: genotoxicity and DNA-damage response." 2014. *Metallomics* 6(3), p. 662. DOI: 10.1039/c3mt00337j.

"Bacteria can unstick it by reversing what oxygen did to release it . . ." T. Barkay et al. "A thermophilic bacterial origin and subsequent constraints by redox, light and salinity on the evolution of the microbial mercuric reductase." 2010. *Environ Microbiol.* 12(11), p. 2904. DOI: 10.1111/j.1462-2920.2010.02260.x.

"One chromium-cleansing enzyme is proactive, . . ." K. J. Robins et al. "Escherichia coli NemA is an efficient chromate reductase that can be biologically immobilized to provide a cell free system for remediation of hexavalent chromium." 2013. *PLoS One* 8(3), p. e59200. DOI: 10.1371/journal.pone.0059200.

". . . they eject phosphate and make a cerium phosphate crust around the cell . . ." M. Jiang et al. "Biological nano-mineralization of Ce phosphate by *Saccharomyces cerevisiae.*" 2010. *Chem Geo.* 277(1–2), p. 61.

"One of the first pesticides works because copper is sticky." John Emsley. *Nature's Building Blocks: An A–Z Guide to the Elements.* 2001, Oxford University Press, p. 124.

"Today, plant roots in urban areas have surfaces like flypaper for copper, . . ." A. Manceau et al. "Thlaspi arvense binds Cu(II) as a bis-(L-histidinato) complex on root cell walls in an urban ecosystem." 2013. *Metallomics* 5(12), p. 1674. DOI: 10.1039/c3mt00215b.

"Tolkien called this kind of happy turnaround a 'eucatastrophe,' . . ." Tolkien, J. R. R. "On Fairy-Stories." *Essays Presented to Charles Williams.* 1947, Oxford University Press.

"Just encountering a new metal can evolve new protein shapes." H. F. Ji et al. "Evolutionary formation of new protein folds is linked to metallic cofactor recruitment." 2009. *Bioessays* 31(9), p. 975. DOI: 10.1002/bies.200800201.

"Also, old protein shapes can find new uses by picking up a new active metal center, . . ." M. Leslie. "'Dead' enzymes show signs of life." 2013. *Science* 340(6128), p. 25. DOI: 10.1126/science.340.6128.25.

"More often, they *repurpose* old enzymes to serve new functions." A. D. Moore and E. Bornberg-Bauer. "The dynamics and evolutionary potential of domain loss and emergence." 2012. *Mol Biol Evol.* 29(2), p. 787. DOI: 10.1093/molbev/msr250.

"Acid stresses sea urchins, and they double their rate of evolution in response, . . ." M. H. Pespeni et al. "Evolutionary change during experimental ocean acidification." 2013. *Proc Natl Acad Sci.* 110(17), p. 6937. DOI: 10.1073/pnas.1220673110.

"Lindquist calls heat shock proteins 'evolutionary capacitors' . . ." B. P. Trivedi. "Prions and chaperones: Outside the fold." 2012. *Nature* 482(7385), p. 294. DOI: 10.1038/482294a.

"Heat shock proteins gave fish swimming underground . . ." N. Rohner et al. "Cryptic variation in morphological evolution: HSP90 as a capacitor for loss of eyes in cavefish." 2013. *Science* 342(6164), p. 1372. DOI: 10.1126/science.1240276.

"Worms, bacteria, and plants detox toxins by bonding them to sugar molecules." G. S. Stupp et al. "Chemical detoxification of small molecules by Caenorhabditis elegans." 2013. *ACS Chem Biol.* 8(2), p. 309. DOI: 10.1021/cb300520u.

". . . the hungry bacteria in your gut eat the sugar and release the toxin." A. Dall'Erta et al. "Masked mycotoxins are efficiently hydrolyzed by human colonic microbiota releasing their aglycones." 2013. *Chem Res Toxicol.* 26(3), p. 305. DOI: 10.1021/tx300438c.

"The binding protein became a ring-making enzyme." M. N. Ngaki et al. "Evolution of the chalcone-isomerase fold from fatty-acid binding to stereospecific catalysis." 2012. *Nature* 485(7399), p. 530. DOI: 10.1038/nature11009.

"For example, sulfur carrier proteins were transformed to become proteins that built sulfurous sugars." E. Sasaki et al. "Co-opting sulphur-carrier proteins from primary metabolic pathways for 2-thiosugar biosynthesis." 2014. *Nature* 510(7505), p. 427. DOI: 10.1038/nature13256.

"A home experiment shows how copper has a special relationship with oxygen." E. Inglis-Arkell. "Glowing pennies prove that the '80s were the last great decade." 2013, i09.com.

"... a surprising number that break down ammonia by using copper to combine it with oxygen." A. J. Probst et al. "Archaea on human skin." 2013. *PLoS One* 8(6), p. e65388. DOI: 10.1371/journal.pone.0065388.

"Older organisms clean up ROS with iron and manganese, but animals do the same chemistry more safely with copper and zinc." R. J. P. Williams and R. E. M. Rickaby. *Evolution's Destiny: Co-evolving Chemistry of the Environment and Life.* 2012, RSC Publishing, pp. 138–139.

"The general reactivity of laccases makes them highly designable." D. Maté et al. "Laboratory evolution of high-redox potential laccases." 2010. *Chem Biol.* 17(9), p. 1030. DOI: 10.1016/j.chembiol.2010.07.010.

"... Camille Paglia's enjoyable book about poetry, *Break Blow Burn*: ... " Camille Paglia. *Break, Blow, Burn: Camille Paglia Reads Forty-three of the World's Best Poems.* 2006. Vintage.

"Multiple coppers in a laccase can slice up the lignin net ..." A. Levasseur et al. "Exploring laccase-like multicopper oxidase genes from the ascomycete Trichoderma reesei: A functional, phylogenetic and evolutionary study." 2010. *BMC Biochem.* 11, p. 32. DOI: 10.1186/1471-2091-11-32.

"A copper enzyme can also break down chitin, ..." F. L. Aachmann et al. "NMR structure of a lytic polysaccharide monooxygenase provides insight into copper binding, protein dynamics, and substrate interactions." 2012. *Proc Natl Acad Sci.* 109(46), p. 18779. DOI: 10.1073/pnas.1208822109.

"... a protein that pulls electrons off iron ..." J. M. Gutteridge and J. Stocks. "Caeruloplasmin: Physiological and pathological perspectives." 1981. *Crit Rev Clin Lab Sci.* 14(4), p. 257.

"... adding oxygen onto vitamin C as part of a cycle that expands the cell walls." M. C. De Tullio et al. "Ascorbate oxidase is the potential conductor of a symphony of signaling pathways." 2013. *Plant Signal Behav.* 8(3), p. e23213. DOI: 10.4161/psb.23213.

"... copper makes tumors grow because it helps mitochondria burn sugars ..." S. Ishida et al. "Bioavailable copper modulates oxidative phosphorylation and growth of tumors." 2013. *Proc Natl Acad Sci.* 110(48), p. 19507. DOI: 10.1073/pnas.1318431110.

"Also, copper's "breaking" power is active in cancer metastasis, ..." G. MacDonald et al. "Memo is a copper-dependent redox protein with an essential role in migration and metastasis." 2014. *Sci Signal.* 7(329), p. ra56. DOI: 10.1126/scisignal.2004870.

"... therapies that remove copper may starve the cancer of the copper they crave." D. C. Brady et al. "Copper is required for oncogenic BRAF signalling and tumorigenesis." 2014. *Nature* 509(7501), p. 492. DOI: 10.1038/nature13180.

"Plants and fungi have many polyphenol oxidases for different purposes . . ." A. M. Mayer. "Polyphenol oxidases in plants and fungi: going places? A review." 2006. *Phytochemistry* 67(21), p. 2318.

". . . similar molecules called chalkophores do the same for copper." G. E. Kenney and A. C. Rosenzweig. "Chemistry and biology of the copper chelator methanobactin." 2012. *ACS Chem Biol.* 7(2), p. 260. DOI: 10.1021/cb2003913.

"Primitive immune systems in insects and crabs . . ." L. Cerenius and K. Söderhäll. "The prophenoloxidase-activating system in invertebrates." 2004. *Immunol Rev.* 198, p. 116.

"Humans have immune cells called macrophages . . ." V. Hodgkinson and M. J. Petris. "Copper homeostasis at the host-pathogen interface." 2012. *J Biol Chem.* 287(17), p. 13549. DOI: 10.1074/jbc.R111.316406.

". . . copper links the protein together with a strong, visceral interaction." F. Hane et al. "Cu(2+) affects amyloid-β (1-42) aggregation by increasing peptide-peptide binding forces." 2013. *PLoS One* 8(3), p. e59005. DOI: 10.1371/journal.pone.0059005.

"Zinc helps blood clot by bridging protein backbones." J. C. Fredenburgh et al. "Zn2+ mediates high affinity binding of heparin to the αC domain of fibrinogen." 2013. *J Biol Chem.* 288(41), p. 29394. DOI: 10.1074/jbc.M113.469916.

"Copper and zinc's partnership can be seen in history as well." M. A. Saito et al. "The bioinorganic chemistry of the ancient ocean: the co-evolution of cyanobacterial metal requirements and biogeochemical cycles at the Archean/Proterozoic boundary?" 2003. *Inorg Chim Acta.* 356, p. 308.

"The green version prefer the "old" elements iron, cobalt, and manganese, while the red use more copper and zinc." A. Quigg et al. "Evolutionary inheritance of elemental stoichiometry in phytoplankton." 2011. *Proc Biol Sci.* 278(1705), p. 526. DOI: 10.1098/rspb.2010.1356; see also "The evolutionary inheritance of elemental stoichiometry in marine phytoplankton." 2003. *Nature* 425(6955), p. 291.

"Organisms that live in oxygen have copper genes, and organisms that live without oxygen do not." P. G. Ridge et al. "Comparative genomic analyses of copper transporters and cuproproteomes reveal evolutionary dynamics of copper utilization and its link to oxygen." 2008. *PLoS One* 3(1), p. e1378. DOI: 10.1371/journal.pone.0001378; L. Decaria et al. "Copper proteomes, phylogenetics and evolution." 2011. *Metallomics* 3(1), p. 56. DOI: 10.1039/c0mt00045k.

"Copper is outside the cells in your nose, where it sticks to molecules . . ." X. Duan et al. "Crucial role of copper in detection of metal-coordinating odorants." 2012. *Proc Natl Acad Sci.* 109(9), p. 3492. DOI: 10.1073/pnas.1111297109.

"But zinc can kill nose cells, possibly by sticking too much." J. H. Lim et al. "Zicam-induced damage to mouse and human nasal tissue." 2009. *PLoS One* 4(10), p. e7647. DOI: 10.1371/journal.pone.0007647.

"Zinc deficiency in humans can delay puberty, . . ." R. J. P. Williams and R. E. M. Rickaby. *Evolution's Destiny: Co-evolving Chemistry of the Environment and Life.* 2012, RSC Publishing; p. 148: "In humans it is known that the metamorphosis known as going through puberty, as well as features parallel with it, can be induced by zinc in food."

"Likewise, plants watered with a zinc sulfate solution better survive drought." H. Upadhyaya et al. "Zinc modulates drought-induced biochemical damages in tea." 2013. *J Agric Food Chem.* 61(27), p. 6660. DOI: 10.1021/jf304254z.

"Mice without enough zinc don't make enough antibodies." M. Maiguma et al. "Dietary zinc is a key environmental modifier in the progression of IgA nephropathy." 2014. *PLoS One* 9(2), p. e90558. DOI: 10.1371/journal.pone.0090558.

"Finally, when a mammal's egg is fertilized, ..." E. L. Que et al. "Quantitative mapping of zinc fluxes in the mammalian egg reveals the origin of fertilization-induced zinc sparks." 2014. *Nat Chem.* 7(2), p. 130. DOI: 10.1038/nchem.2133.

"... two levels of zinc biochemistry, simple and advanced." R. J. P. Williams and R. E. M. Rickaby. *Evolution's Destiny: Co-evolving Chemistry of the Environment and Life.* 2012, RSC Publishing, p. 136.

"duplicated with abandon in higher organisms." R. J. P. Williams and R. E. M. Rickaby. *Evolution's Destiny: Co-evolving Chemistry of the Environment and Life.* 2012, RSC Publishing, p. 239, Figure 6.13.

"... even a set of random proteins with added zinc contained several proteins that bound DNA." A. D. Keefe and J. W. Szostak. "Functional proteins from a random-sequence library." 2001. *Nature* 410(6829), p. 715.

"... the result is a precise and efficient set of scissors for DNA." C. A. Gersbach. "Genome engineering: The next genomic revolution." 2014. *Nat Methods.* 11(10): 1009. DOI: 10.1038/nmeth.3113.

"According to Hazen, ..." Robert M. Hazen. *The Story of Earth: The First 4.5 Billion Years, from Stardust to Living Planet.* 2012, Viking, p. 224.

"... the supercontinent Rodinia split in half, ..." Hazen, p. 210, and I. W. D. Dalziel. "Cambrian transgression and radiation linked to an Iapetus-Pacific oceanic connection?" 2014. *Geology* 42(11), p. 979. DOI: 10.1130/G35886.1.

"... as much as five percent of the liquid water on Earth solidified." Hazen, p. 273.

"... the earth froze 720, 650, and 580 million years ago." Hazen, pp. 220–222.

"... (increasing sunlight and a cooling planet decreasing seafloor spreading)." B. Mills et al. "Proterozoic oxygen rise linked to shifting balance between seafloor and terrestrial weathering." 2014. *Proc Natl Acad Sci.* 111(25), p. 9073. DOI: 10.1073/pnas.1321679111.

"This rise in oxygen was significant enough that it could be called the Second Great Oxidation Event." For example, from R. M. Hazen et al. "Mineral evolution." 2008. *Amer Mineral.* 93(11–12), p. 1693. DOI: 10.2138/am.2008.2955: "An increase in atmospheric oxygen, from <2% to ~15% of modern values, appears to have occurred immediately following glacial melting, ..."

Chapter 9

"The first of these are unlike anything we've seen, ..." See The Burgess Shale Geoscience Foundation: www.burgess-shale.bc.ca.

"The tan layers on top are 525-million-year-old Cambrian Tapeats sandstone ... Dissolved rocks brought more elements into the ocean, ..." S. E. Peters and R. R. Gaines. "Formation of the 'Great Unconformity' as a trigger for the Cambrian explosion." 2012. *Nature* 484(7394), p. 363. DOI: 10.1038/nature10969.

"In particular, a class of rocks called skeletal minerals ..." R. M. Hazen et al. "Mineral evolution." 2008. *Amer Mineral.* 93(11–12), p. 1693. DOI: 10.2138/am.2008.2955.

"On your skin, each physical niche holds a different community of bacteria." J. Oh et al. "Biogeography and individuality shape function in the human skin metagenome." 2014. *Nature* 514(7520), p. 59. DOI: 10.1038/nature13786.

"... a review of Stephen Meyer's book ..." C. R. Marshall. "When prior belief trumps scholarship." 2013. *Science* 341(6152), p. 1344. DOI: 10.1126/science.1244515.

"This article, by Paul Smith and David Harper, ..." M. P. Smith and D. A. T. Harper.
"Causes of the Cambrian Explosion." 2013. *Science* 341(6152), p. 1355. DOI: 10.1126/
science.1239450.

"... as far back as 650 million years ago, but no further." Robert M. Hazen. *The Story of
Earth: The First 4.5 Billion Years, from Stardust to Living Planet.* 2012, Viking, p. 228.

"While surface ocean oxygen increased 10 times, deep ocean oxygen increased a *million
times*." C. Scott et al. "Tracing the stepwise oxygenation of the Proterozoic ocean." 2008.
Nature 452(7186), p. 456. DOI: 10.1038/nature06811.

"The proliferation of fossils seen in the rocks did not invent new genes, but repurposed
old genes." This results in early genetic complexity of even simple animals like sponges,
as in A. Riesgo et al. "The analysis of eight transcriptomes from all poriferan classes
reveals surprising genetic complexity in sponges." 2014. *Mol Biol Evol.* 31(5), p. 1102.
DOI: 10.1093/molbev/msu057.

"Comparing the genome of a multicelled alga ..." S. E. Prochnik et al. "Genomic analysis
of organismal complexity in the multicellular green alga Volvox carteri." 2010. *Science*
329(5988), p. 223. DOI: 10.1126/science.1188800.

"Such genomic similarity was also found ..." N. King et al. "The genome of the choanoflagellate Monosiga brevicollis and the origin of metazoans." 2008. *Nature* 451(7180),
p. 783. DOI: 10.1038/nature06617.

"Cyanobacteria learned to stick together in strings as far back as the first Great Oxidation
Event, ..." B. E. Schirrmeister et al. "Evolution of multicellularity coincided with
increased diversification of cyanobacteria and the Great Oxidation Event." 2013. *Proc
Natl Acad Sci.* 110(5), p. 1791. DOI: 10.1073/pnas.1209927110.

"Microbes with mitochondria would get the energy to run large genomes from oxygen, ..."
N. Lane and W. Martin. "The energetics of genome complexity." 2010. *Nature* 467(7318),
p. 929. DOI: 10.1038/nature09486.

"... cells keep their insides low in oxygen." R. J. P. Williams and R. E. M. Rickaby. *Evolution's
Destiny: Co-evolving Chemistry of the Environment and Life.* 2012, RSC Publishing,
p. 160.

"Not just a few, but hundreds of these proteins." M. Srivastava et al. "The *Amphimedon
queenslandica* genome and the evolution of animal complexity." 2010. *Nature* 466(7307),
p. 720. DOI: 10.1038/nature09201.

"Animals have only 17 major cell signaling pathways, ..." M. Lamothe et al. *The Where, the
Why, and the How: 75 Artists Illustrate Wondrous Mysteries of Science.* 2012, Chronicle
Books, Question 74 (Cancer).

"... eerily similar to that of an embryo." G. Johnson. "A tumor, the embryo's evil twin."
March 17, 2014. *New York Times.*

"... even the sponge's cousin *Hydra* gets tumors." T. Domazet-Lošo et al. "Naturally occurring tumours in the basal metazoan Hydra." 2014. *Nat Commun.* 5, p. 4222. DOI: 10.1038/
ncomms5222.

"One of most important tumor suppressor genes, p53, ..." A. C. Joerger et al. "Tracing
the evolution of the p53 tetramerization domain." 2014. *Structure* 22(9), p. 1301.
DOI: 10.1016/j.str.2014.07.010.

"In humans, the enzyme that puts oxygen onto collagen fibers ..." C. Loenarz et al.
"The hypoxia-inducible transcription factor pathway regulates oxygen sensing in the

simplest animal, *Trichoplax adhaerens*." 2011. *EMBO Rep.* 12(1), p. 63. DOI: 10.1038/embor.2010.170.

"Sponges can live in the same low levels of oxygen." D. B. Mills et al. "Oxygen requirements of the earliest animals." 2014. *Proc Natl Acad Sci.* 111(11), p. 4168. DOI: 10.1073/pnas.1400547111.

"Despite being sharp as a knife, chitin has no metal." A. Miserez et al. "Cross-linking chemistry of squid beak." 2010. *J Biol Chem.* 285(49), p. 38115. DOI: 10.1074/jbc.M110.161174.

"... linked and waterproofed by the oxidizing activity of peroxidases." F. A. Dias et al. "Ovarian dual oxidase (Duox) activity is essential for insect eggshell hardening and waterproofing." 2013. *J Biol Chem.* 288(49), p. 35058. DOI: 10.1074/jbc.M113.522201.

"The chemists called this an 'aerogel' ..." X. L. Wu et al. "Biomass-derived sponge-like carbonaceous hydrogels and aerogels for supercapacitors." 2013. ACS *Nano.* 7(4), p. 3589. DOI: 10.1021/nn400566d.

"Accordingly, sponge fossils have been found in rocks ..." C. K. Brain et al. "The first animals: ca. 760-million-year-old sponge-like fossils from Namibia." 2012. *S Afr J Sci.* 108(1/2), p. 658. DOI: /10.4102/sajs.v108i1/2.658.

"Collagen strands are built by copper enzymes, ..." E. D. Harris et al. "Copper and the synthesis of elastin and collagen." 1980. *Ciba Found Symp.* 79, p. 163.

"... they are cut by zinc enzymes, ..." D. H. Pashley et al. "Collagen degradation by host-derived enzymes during aging." 2004. *J Dent Res.* 83(3), p. 216.

"In animals, these enzymes play a role in the unwanted motion of cells ..." Zinc is also involved in pathogenic movement, as in J. Li and K. Q. Zhang. "Independent expansion of zincin metalloproteinases in Onygenales fungi may be associated with their pathogenicity." 2014. *PLoS One* 9(2), p. e90225. DOI: 10.1371/journal.pone.0090225.

"... don't think burrowing can pay off ..." P. Ward and J. Kirschvink. *A New History of Life: The Radical New Discoveries about the Origins and Evolution of Life on Earth.* 2015, Bloomsbury, p. 118.

"... the bacterium *S. marcescens* breaks insect shell chitin ..." F. L. Aachmann et al. "NMR structure of a lytic polysaccharide monooxygenase provides insight into copper binding, protein dynamics, and substrate interactions." 2012. *Proc Natl Acad Sci.* 109(46), p. 18779. DOI: 10.1073/pnas.1208822109.

"... making and breaking in plant-eating fungi and plant cell walls." P. C. Brunner et al. "Coevolution and life cycle specialization of plant cell wall degrading enzymes in a hemibiotrophic pathogen." 2013. *Mol Biol Evol.* 30(6), p. 1337. DOI: 10.1093/molbev/mst041.

"Minerals made from oxidized iron are fragile and erode easily." Robert M. Hazen. *The Story of Earth: The First 4.5 Billion Years, from Stardust to Living Planet.* 2012, Viking, p. 179.

"... aluminum-silicon clays like smectite and kaolinite are found increasingly ..." R. M. Hazen et al. "Mineral evolution." 2008. *Amer Mineral.* 93(11–12), p. 1693. DOI: 10.2138/am.2008.2955.

"Life used calcium in new ways after the explosion." A. H. Knoll. "Biomineralization and evolutionary history." 2003. *Rev Mineral Geochem.* 54(1), p. 329. DOI: 10.2113/0540329.

"... several different types of mollusks developed calcium-based shells at the same time, ..." R. J. P. Williams and R. E. M. Rickaby. *Evolution's Destiny: Co-evolving Chemistry of the Environment and Life.* 2012, RSC Publishing, p. 151.

"Overall, calcification evolved at least 28 times in separate species." A. H. Knoll. "Biomineralization and evolutionary history." 2003. *Rev Mineral Geochem.* 54(1), p. 329. DOI: 10.2113/0540329.

"Robert Hazen relates how "massive" limestone deposits, ..." R. M. Hazen et al. "Mineral evolution." 2008. *Amer Mineral.* 93(11–12), p. 1693. DOI: 10.2138/am.2008.2955.

"... adventurous, science-minded chefs cook with 'spherification.'" Jeff Potter. *Cooking for Geeks: Real Science, Great Hacks, and Good Food.* 2010, O'Reilly Media.

"One kind of sponge doesn't make the charged material itself ..." D. J. Jackson et al. "An evolutionary fast-track to biocalcification." 2010. *Geobiology* 8(3), p. 191. DOI: 10.1111/j.1472-4669.2010.00236.x.

"Now corals have a "toolkit" of 36 such proteins ..." J. L. Drake et al. "Proteomic analysis of skeletal organic matrix from the stony coral *Stylophora pistillata*." 2013. *Proc Natl Acad Sci.* 110(10), p. 3788. DOI: 10.1073/pnas.1301419110.

"Two shells that look very different were secreted by the same machinery, ..." R. Chirat et al. "Mechanical basis of morphogenesis and convergent evolution of spiny seashells." 2013. *Proc Natl Acad Sci.* 110(15), p. 6015. DOI: 10.1073/pnas.1220443110.

"... airways may form from the same type of simple, tunable process." A. E. Shyer et al. "Villification: how the gut gets its villi." 2013. *Science* 342(6155), p. 212. DOI: 10.1126/science.1238842.

"When calcium is the building material, the branches have rectangular, sharp edges." Adrian Bejan. *Shape and Structure, from Engineering to Nature.* 2000, Cambridge University Press.

"Similar chemical processes form these by arranging negative charges and collecting calcium on a surface." D. J. Murdock et al. "The origin of conodonts and of vertebrate mineralized skeletons." 2013. *Nature* 502(7472), p. 546. DOI: 10.1038/nature12645.

"... an early animal called Cloudina built calcium reefs ..." A. M. Penny et al. "Ediacaran metazoan reefs from the Nama Group, Namibia." 2014. *Science* 344(6191), p. 1504. DOI: 10.1126/science.1253393.

"Phosphorus is also concentrated at the reefs, in some cases by microbes ..." F. Zhang et al. "Phosphorus sequestration in the form of polyphosphate by microbial symbionts in marine sponges." 2015. *Proc Natl Acad Sci.* 112(14), p. 4381. DOI: 10.1073/pnas.1423768112.

"... bone evolved by repurposing old proteins ..." C. S. Viegas et al. "Sturgeon osteocalcin shares structural features with matrix Gla protein: Evolutionary relationship and functional implications." 2013. *J Biol Chem.* 288(39), p. 27801. DOI: 10.1074/jbc.M113.450213; and B. Venkatesh et al. "Elephant shark genome provides unique insights into gnathostome evolution." 2014. *Nature* 505(7482), p. 174. DOI: 10.1038/nature12826.

"... by mixing calcium sulfate with ground oyster shells." Y. Shen et al. "Engineering scaffolds integrated with calcium sulfate and oyster shell for enhanced bone tissue regeneration." 2014. *ACS Appl Mater Inter.* 6(15), p. 12177. DOI: 10.1021/am501448t.

"These moving charges create an electromagnetic field." R. J. P. Williams and R. E. M. Rickaby. *Evolution's Destiny: Co-evolving Chemistry of the Environment and Life.* 2012, RSC Publishing, p. 195.

"... listed in a single review on biomimetic materials, ..." M. A. Meyers et al. "Structural biological materials: Critical mechanics-materials connections." 2013. *Science* 339(6121), p. 773. DOI: 10.1126/science.1220854.

"When a predator bites into it, internal holding cells are ripped open and spill out dopamine." K. L. Van Alstyne et al. "Dopamine release by *Ulvaria obscura* (phylum chlorophyta): environmental triggers and impacts on the photosynthesis, growth, and survival of the releaser." 2013. *J Phyc.* 49, p. 719. DOI: 10.1111/jpy.12081.

". . . neurons maintain a protective sulfur enzyme as an antidote." N. P. Sidharthan et al. "Cytosolic sulfotransferase 1A3 is induced by dopamine and protects neuronal cells from dopamine toxicity: Role of D1 receptor-N-methyl-D-aspartate receptor coupling." 2013. *J Biol Chem.* 288(48), p. 34364. DOI: 10.1074/jbc.M113.493239.

"Mussels oxidize dopamine to glue themselves in place." H. Lee et al. "Mussel-inspired surface chemistry for multifunctional coatings." 2007. *Science* 318(5849), p. 426.

"Algae use another oxidative network . . ." M. Welling et al. "A desulfatation-oxidation cascade activates coumarin-based cross-linkers in the wound reaction of the giant unicellular alga *Dasycladus vermicularis.*" 2011. *Angew Chem* 50(33), p. 7691. DOI: 10.1002/anie.201100908.

"Wendell Lim has shown how a rudimentary secrete-and-detect cycle . . ." H. Youk and W. A. Lim. "Secreting and sensing the same molecule allows cells to achieve versatile social behaviors." 2014. *Science* 343(6171), p. 628. DOI: 10.1126/science.1242782.

"Animals cooperate with bacteria against fungi . . ." M. McFall-Ngai et al. "Animals in a bacterial world, a new imperative for the life sciences." 2013. *Proc Natl Acad Sci.* 110(9):3229. DOI: 10.1073/pnas.1218525110.

"In another example, fungi 'farm' bacteria . . ." M. Pion et al. "Bacterial farming by the fungus *Morchella crassipes.*" 2013. *Proc Biol Sci.* 280(1773), p. 2242. DOI: 10.1098/rspb.2013.2242.

". . . the smell of cut grass is six carbons with an oxygen on one end, . . ." See "What causes the smell of fresh-cut grass?" on Compound Interest, April 25, 2014. www.compoundchem.com.

"Tomato plants release a similar carbon-chain molecule when chewed on by insects." K. Sugimoto et al. "Intake and transformation to a glycoside of (Z)-3-hexenol from infested neighbors reveals a mode of plant odor reception and defense." 2014. *Proc Natl Acad Sci.* 111(19), p. 7144. DOI: 10.1073/pnas.1320660111.

"Mosquitoes home in on human blood . . ." C. S. McBride et al. "Evolution of mosquito preference for humans linked to an odorant receptor." 2014. *Nature* 515(7526), p. 222. DOI: 10.1038/nature13964.

"The diversity of smells from a spice rack . . ." See "Chemical compounds in herbs & spices" on Compound Interest, March 13, 2014. www.compoundchem.com.

"It travels to the darker side of the plant and signals for growth there." Tristan Gooley. *The Walker's Guide to Outdoor Clues and Signs.* 2014, Sceptre Press, p. 55.

". . . coordinate the multiple arms and functions of the immune system." N. J. Spann and C. K. Glass. "Sterols and oxysterols in immune cell function." 2013. *Nat Immunol.* 14(9), p. 893. DOI: 10.1038/ni.2681.

"Some tumors survive by making oxysterol hormones . . ." A. G. York and S. J. Bensinger. "Subverting sterols: Rerouting an oxysterol-signaling pathway to promote tumor growth." 2013. *J Exp Med.* 210(9), p. 1653. DOI: 10.1084/jem.20131335.

"Now we know the connection runs through an oxygen-sterol." E. R. Nelson et al. "27-Hydroxycholesterol links hypercholesterolemia and breast cancer pathophysiology." 2013. *Science* 342(6162), p. 1094. DOI: 10.1126/science.1241908.

". . . reconstructed a timeline of when new oxidized hormones appeared." Y. Y. Jiang et al. "The impact of oxygen on metabolic evolution: a chemoinformatic investigation." 2012. *PLoS Comput Biol.* 8(3), p. e1002426. DOI: 10.1371/journal.pcbi.1002426.

"... 97.5% of the molecules that send signals to the nucleus were made possible by oxygen ..." Y. Y. Jiang et al. "How does oxygen rise drive evolution? Clues from oxygen-dependent biosynthesis of nuclear receptor ligands." 2010. *Biochem Biophys Commun.* 391(2), p. 1158. DOI: 10.1016/j.bbrc.2009.11.041.

"... went extinct at this stage because it wasn't efficient enough." J. F. Hoyal Cuthill and S. Conway Morris. "Fractal branching organizations of Ediacaran rangeomorph fronds reveal a lost Proterozoic body plan." 2014. *Proc Natl Acad Sci.* 111(36), p. 13122. DOI: 10.1073/pnas.1408542111.

"This shape is particularly good at *slowing down* evolution ..." L. Vermeulen et al. "Defining stem cell dynamics in models of intestinal tumor initiation." *Science* 342(6161), p. 995. DOI: 10.1126/science.1243148.

"Long-lived plants protect themselves with a similar structural damper ..." D. K. Aanen. "How a long-lived fungus keeps mutations in check." 2014. *Science* 346(6212), p. 922. DOI: 10.1126/science.1261401.

"As a result, the nuclei of plants like *P. japonica* ..." M. Scudellari. "Genomes gone wild." January 1, 2014. *The Scientist*, www.the-scientist.com.

"Whole genome duplications happened when plants learned to produce flowers ..." K. Adams. "Genomic clues to the ancestral flowering plant." 2013. *Science* 342(6165), p. 1456. DOI: 10.1126/science.1248709.

"Such damaged plants can duplicate their genome in response, ..." D. R. Scholes and K. N. Paige. "Plasticity in ploidy underlies plant fitness compensation to herbivore damage." 2014. *Mol Ecol.* 23(19), p. 4862. DOI: 10.1111/mec.12894.

"Fossils of ferns found in Sweden from the early Jurassic period, ..." B. Bomfleur et al. "Fossilized nuclei and chromosomes reveal 180 million years of genomic stasis in royal ferns." 2014. *Science* 343(6177), p. 1376. DOI: 10.1126/science.1249884.

"Even the simple act of grafting can create a genetically distinct species." I. Fuentes et al. "Horizontal genome transfer as an asexual path to the formation of new species." 2014. *Nature* 511(7508), p. 232. DOI: 10.1038/nature13291.

"... the two organisms converged in their choices, picking the same targets." M. S. Mukhtar et al. "Independently evolved virulence effectors converge onto hubs in a plant immune system network." 2011. *Science* 333(6042), p. 596. DOI: 10.1126/science.1203659.

"This new cell type and new C4 process evolved independently more than *sixty* times ..." R. F. Sage et al. "The C4 plant lineages of planet Earth." 2011. *J Exp Bot.* 62, p. 3155. DOI: 10.1093/jxb/err048.

"... maize and sugarcane, with ERDs of about two watts per kilogram or more." See Figure 4 in E. J. Chaisson. "Energy rate density as a complexity metric and evolutionary driver." 2011. *Complexity* 16(3), p. 27. DOI: 10.1002/cplx.20323.

"The pattern is the same as when multiple genes change together in discrete 'modules': ..." D. Heckmann et al. "Predicting C4 photosynthesis evolution: Modular, individually adaptive steps on a Mount Fuji fitness landscape." 2013. *Cell* 153(7), p. 1579. DOI: 10.1016/j.cell.2013.04.058.

"Even the *microbes* were big in the Carboniferous." J. L. Payne et al. "Late paleozoic fusulinoidean gigantism driven by atmospheric hyperoxia." 2012. *Evolution* 66(9), p. 2929. DOI: 10.1111/j.1558-5646.2012.01626.x.

"... direct measurements of the oxygen content of bubbles in amber, ..." Robert M. Hazen. *The Story of Earth: The First 4.5 Billion Years, from Stardust to Living Planet.* 2012, Viking, p. 248.

"Large predatory fish appear in the fossil record ..." T. W. Dahl et al. "Devonian rise in atmospheric oxygen correlated to the radiations of terrestrial plants and large predatory fish." 2010. *Proc Natl Acad Sci.* 107(42), p. 17911. DOI: 10.1073/pnas.1011287107.

"... some enterprising plant invented the carbon-oxygen armor lignin ..." R. J. P. Williams and R. E. M. Rickaby. *Evolution's Destiny: Co-evolving Chemistry of the Environment and Life.* 2012, RSC Publishing, p. 150.

"... both are laid down by similar-looking rings of copper-containing oxidative enzymes." Y. Lee et al. "A mechanism for localized lignin deposition in the endodermis." 2013. *Cell* 153(2), p. 402. DOI: 10.1016/j.cell.2013.02.045.

"More than 10% of the calories listed for high-fiber foods ..." R. Dunn. "Everything you know about calories is wrong." 2013. *Sci Amer.* 309(3), p. 56.

"... when moss first grew tough enough to grow on land ..." T. M. Lenton et al. "First plants cooled the Ordovician." 2012. *Nature Geo.* 5, p. 86. DOI: 10.1038/ngeo1390.

"It may have recurred 30 million years ago ..." P. G. Falkowski et al. "The evolution of modern eukaryotic phytoplankton." 2004. *Science* 305(5682), p. 354.

"Finally, a protein evolved that was up to the task of lignin decomposition ..." D. Floudas et al. "The Paleozoic origin of enzymatic lignin decomposition reconstructed from 31 fungal genomes." 2012. *Science* 336(6089), p. 1715. DOI: 10.1126/science.1221748.

"Waves of anoxia swept through the oceans and species went extinct at alarming rates." B. C. Gill et al. "Geochemical evidence for widespread euxinia in the later Cambrian ocean." 2011. *Nature* 469(7328), p. 80. DOI: 10.1038/nature09700.

"Butterfield notes how we see transitions from anoxic to oxic environments today ..." N. J. Butterfield. "Oxygen, animals and oceanic ventilation: An alternative view." 2009. *Geobiology* 7(1), p. 1. DOI: 10.1111/j.1472-4669.2009.00188.x; and "Animals and the invention of the Phanerozoic Earth system." 2011. *Trends Ecol Evol.* 26(2), p. 81. DOI: 10.1016/j.tree.2010.11.012.

"... that advanced organisms caused the increase in oxygen, rather than vice versa." T. M. Lenton et al. "Co-evolution of eukaryotes and ocean oxygenation in the Neoproterozoic era." 2014. *Nature Geo.* 7, p. 257. DOI: 10.1038/ngeo2108.

"Butterfield may be right to be skeptical about the Carboniferous explosion, ..." N. J. Butterfield. "Was the Devonian radiation of large predatory fish a consequence of rising atmospheric oxygen concentration?" 2011. *Proc Natl Acad Sci.* 108(9), p. E28. DOI: 10.1073/pnas.1018072108.

"... allowed us to trace pre-Cambrian oxygen levels with unprecedented accuracy." N. J. Planavsky et al. "Low mid-Proterozoic atmospheric oxygen levels and the delayed rise of animals." 2014. *Science* 346(6209), p. 635. DOI: 10.1126/science.1258410.

"Moreover, in another study, the burrowing activity described ..." R. A. Boyle et al. "Stabilization of the coupled oxygen and phosphorus cycles by the evolution of bioturbation." 2014. *Nature Geo.* 7, p. 671. DOI: 10.1038/ngeo2213.

"I agree with Nick Lane that mitochondria allowed the energy ..." N. Lane and W. Martin. "The energetics of genome complexity." 2010. *Nature* 467(7318), p. 929. DOI: 10.1038/nature09486.

"Most important is the coincidence of all these events." R. J. P. Williams and R. E. M. Rickaby. *Evolution's Destiny: Co-evolving Chemistry of the Environment and Life.* 2012, RSC Publishing, p. 97.

Chapter 10

". . . some calculate that more oxygen in the late Cambrian made more predators evolve." E. A. Sperling et al. "Oxygen, ecology, and the Cambrian radiation of animals." 2013. *Proc Natl Acad Sci.* 110(33), p. 13446. DOI: 10.1073/pnas.1312778110.

"Plants and insects colonized dry land at the same time." B. Misof et al. "Phylogenomics resolves the timing and pattern of insect evolution." 2014. *Science* 346(6210), p. 763. DOI: 10.1126/science.1257570.

". . . land insects breathing that air to enjoy a 10-fold metabolic increase . . ." E. Pennisi. "All in the (bigger) family." 2015. *Science* 347(6219), p. 220. DOI: 10.1126/science.347.6219.220.

"Nick Lane explains in *Life Ascending*: . . ." Nick Lane. *Life Ascending: The Ten Great Inventions of Evolution.* 2009, Norton, p. 62.

"The old Cambrian predator–prey network was modeled in a computer, . . ." J. A. Dunne et al. "Compilation and network analyses of Cambrian food webs." 2008. *PLoS Biol.* 6(4), p. e102. DOI: 10.1371/journal.pbio.0060102.

"This process of building an ecosystem can be seen as a different kind of selection . . ." Conor Cunningham. *Darwin's Pious Idea: Why the Ultra-Darwinists and Creationists Both Get It Wrong.* 2010, Eerdmans, p. 168: Quoting Wandschneider, "On the Problem of Direction . . . in Darwinian Evolution": "two types of selection, horizontal and vertical. Horizontal selection involves the occupation of existing ecological niches; . . . diversification. By contrast, vertical selection encourages the development of new biospheres."

"For example, calcium glues a gel together . . ." T. Phan-Xuan et al. "Tuning the structure of protein particles and gels with calcium or sodium ions." 2013. *Biomacromolecules* 14(6), p. 1980. DOI: 10.1021/bm400347d.

"Calcium is also found tangled up and stuck tight in arterial plaques." J. Weaver et al. "Insights into how calcium forms plaques in arteries pave the way for new treatments for heart disease." 2013. *PLoS Biol.* 11(4), p. e1001533. DOI: 10.1371/journal.pbio.1001533.

". . . reshaping the protein to open a canyon . . ." R. A. Hanna et al. "Calcium-bound structure of calpain and its mechanism of inhibition by calpastatin." 2008. *Nature* 456(7220), p. 409. DOI: 10.1038/nature07451.

"Brains solve this problem by shutting down everything during a period of sleep." A. Varshavsky. "Augmented generation of protein fragments during wakefulness as the molecular cause of sleep: A hypothesis." 2012. *Protein Sci.* 21(11), p. 1634. DOI: 10.1002/pro.2148; and L. Xie et al. "Sleep drives metabolite clearance from the adult brain." 2013. *Science* 342(6156), p. 373. DOI: 10.1126/science.1241224.

"Some receptors on the outside of cells use calcium . . ." W. Xia and T. A. Springer. "Metal ion and ligand binding of integrin α5β1." 2014. *Proc Natl Acad Sci.* 111(50), p. 17863. DOI: 10.1073/pnas.1420645111.

"Some antibodies grab onto bacterial molecules using sticky calcium ions as a bridge." J. M. Wojciak et al. "The crystal structure of sphingosine-1-phosphate in complex with a Fab fragment reveals metal bridging of an antibody and its antigen." 2009. *Proc Natl Acad Sci.* 106(42), p. 17717. DOI: 10.1073/pnas.0906153106.

"Calcium moves the muscle protein myosin most efficiently, . . ." Y. V. Tkachev et al. "Metal cation controls myosin and actomyosin kinetics." 2013. *Protein Sci.* 22(12), p. 1766–1774. DOI: 10.1002/pro.2376.

". . . interferes with this pump and flips it 'on' . . ." Y. A. Mahmmoud. "Capsaicin stimulates uncoupled ATP hydrolysis by the sarcoplasmic reticulum calcium pump." 2008. *J Biol Chem.* 283(31), p. 21418. DOI: 10.1074/jbc.M803654200.

". . . in a rhythmic "calcium clock" pattern." E. G. Lakatta and V. A. Maltsev. "Reprogramming paces the heart." 2013. *Nat Biotech.* 31(1), p. 31. DOI: 10.1038/nbt.2480.

". . . by sending a signal to close the calcium channels that bring calcium in." C. H. Zhang et al. "The cellular and molecular basis of bitter tastant-induced bronchodilation." 2013. *PLoS Biol.* 11(3), p. e1001501. DOI: 10.1371/journal.pbio.1001501.

"When a wound punctures a fly's cells, calcium pours into the wounded cells." W. Razzell et al. "Calcium flashes orchestrate the wound inflammatory response through DUOX activation and hydrogen peroxide release." 2013. *Curr Biol.* 23(5), p. 424. DOI: 10.1016/j.cub.2013.01.058.

"Even a plant may communicate with long-distance calcium waves." S. A. Mousavi et al. "Glutamate receptor-like genes mediate leaf-to-leaf wound signalling." 2013. *Nature* 500(7463), p. 422. DOI: 10.1038/nature12478.

"Salt stress also sends rapid calcium-based signals from the root to the shoot of a plant." W. G. Choi et al. "Salt stress-induced Ca2+ waves are associated with rapid, long-distance root-to-shoot signaling in plants." 2014. *Proc Natl Acad Sci.* 111(17), p. 6497. DOI: 10.1073/pnas.1319955111.

". . . a ghostly blue called 'death fluorescence.'" C. Coburn et al. "Anthranilate fluorescence marks a calcium-propagated necrotic wave that promotes organismal death in *C. elegans.*" 2013. *PLoS Biol.* 11(7), p. e1001613. DOI: 10.1371/journal.pbio.1001613.

"If calcium signaling is impaired in zebrafish, . . ." W. Dai et al. "Calcium deficiency-induced and TRP channel-regulated IGF1R-PI3K-Akt signaling regulates abnormal epithelial cell proliferation." 2014. *Cell Death Differ.* 21(4), p. 568. DOI: 10.1038/cdd.2013.177.

"A huge study searched the genes of 60,000 people . . ." Cross-Disorder Group of the Psychiatric Genomics Consortium. "Identification of risk loci with shared effects on five major psychiatric disorders: A genome-wide analysis." 2013. *Lancet* 381(9875), p. 1371. DOI: 10.1016/S0140-6736(12)62129-1.

". . . new fluorescent sensors that track where and when neurons activate." B. F. Fosque et al. "Labeling of active neural circuits in vivo with designed calcium integrators." 2015. *Science* 347(6223), p. 755. DOI: 10.1126/science.1260922.

"In one blood-clotting protein, seven sites in a row bind calcium, . . ." K. Vadivel et al. "Structural and functional studies of γ-carboxyglutamic acid domains of factor VIIa and activated Protein C: Role of magnesium at physiological calcium." 2013. *J Mol.Biol.* 425(11), p. 1961. DOI: 10.1016/j.jmb.2013.02.017.

"Immune system cells called T cells can pump in a wave of magnesium to send a signal." F. Y. Li et al. "Second messenger role for Mg2+ revealed by human T-cell immunodeficiency." 2011. *Nature* 475(7357), p. 471. DOI: 10.1038/nature10246.

"Nose nerves may use waves of zinc to communicate, too." L. J. Blakemore et al. "Zinc released from olfactory bulb glomeruli by patterned electrical stimulation of the olfactory nerve." 2013. *Metallomics* 5(3), p. 208. DOI: 10.1039/c3mt20158a.

"In a world without as much calcium, phosphate could have caused suboptimal protein movements . . ." J. L. Woodhead et al. "Structural basis of the relaxed state of a Ca2+-regulated myosin filament and its evolutionary implications." 2013. *Proc Natl Acad Sci.* 110(21), p. 8561. DOI: 10.1073/pnas.1218462110.

"Animals have many calcium channels and fungi have fewer, . . ." X. Cai and D. E. Clapham. "Ancestral Ca2+ signaling machinery in early animal and fungal evolution." 2012. *Mol Biol Evol.* 29(1), p. 91. DOI: 10.1093/molbev/msr149.

"Mitochondria use calcium as an 'on-switch' ..." B. Glancy and R. S. Balaban. "Role of mitochondrial Ca2+ in the regulation of cellular energetics." 2012. *Biochemistry* 51(14), p. 2959. DOI: 10.1021/bi2018909.

"Parkinson's Disease is associated with overloaded mitochondria in neurons, ..." D. J. Surmeier and P. T. Schumacker. "Calcium, bioenergetics, and neuronal vulnerability in Parkinson's disease." 2013. *J Biol Chem.* 288(15), p. 10736. DOI: 10.1074/jbc.R112.410530.

"These shells both protect the crayfish and store ..." A. Sato et al. "Glycolytic intermediates induce amorphous calcium carbonate formation in crustaceans." 2011. *Nat Chem Biol.* 7(4), p. 197. DOI: 10.1038/nchembio.532.

"Calcium reorganizes and stabilizes the proteins, ..." For another similar calcium-driven system, see Figure 5 in C. Ma et al. "Reconstitution of the vital functions of Munc18 and Munc13 in neurotransmitter release." 2013. *Science* 339(6118), p. 421. DOI: 10.1126/science.1230473.

"A calcium pulse moves vesicles containing neurotransmitters ..." N. P. Vyleta and P. Jonas. "Loose coupling between Ca2+ channels and release sensors at a plastic hippocampal synapse." 2014. *Science* 343(6171), p. 665. DOI: 10.1126/science.1244811.

"Medically, mutations in this pump ..." M. S. Toustrup-Jensen et al. "Relationship between intracellular Na+ concentration and reduced Na+ affinity in Na+,K+-ATPase mutants causing neurological disease." 2014. *J Biol Chem.* 289(6), p. 3186. DOI: 10.1074/jbc. M113.543272; and W. Kopec et al. "Molecular mechanism of Na(+),K(+)-ATPase malfunction in mutations characteristic of adrenal hypertension." 2014. *Biochemistry* 53(4), p. 746. DOI: 10.1021/bi401425g.

"Sodium/potassium pumps use a full 25% of your energy when you're resting." Albert L. Lehninger and David L. Nelson, *Lehninger Principles of Biochemistry*, 4th edition. 2004, W. H. Freeman, p. 399.

"Then the cloud of positive charge moves down the neuron in a single direction." Albert L. Lehninger and David L. Nelson, *Lehninger Principles of Biochemistry*, 4th edition. 2004, W. H. Freeman, p. 428.

"For example, a broken sodium channel causes erythromelalgia, ..." M. Eberhardt et al. "Inherited pain: sodium channel Nav1.7 A1632T mutation causes erythromelalgia due to a shift of fast inactivation." *J Biol Chem.* 289(4), p. 1971. DOI: 10.1074/jbc.M113.502211.

"Blocking pain-causing channels with a specific drug may stop the pain." B. Halford. "Changing the channel." 2014. *Chem Eng News.* 92(12), p. 10.

"... so much sodium into the ecosystems that it alters butterfly development, ..." E. C. Snell-Rood et al. "Anthropogenic changes in sodium affect neural and muscle development in butterflies." 2014. *Proc Natl Acad Sci.* 111(28), p. 10221. DOI: 10.1073/pnas.1323607111.

"... a component in crude oil plugs the potassium channels in animal hearts." F. Brette et al. "Crude oil impairs cardiac excitation-contraction coupling in fish." 2014. *Science* 343(6172), p. 772. DOI: 10.1126/science.1242747.

"... chloride channels are implicated in how tumors cause brain seizures, ..." J. Pallud et al. "Cortical GABAergic excitation contributes to epileptic activities around human glioma." 2014. *Sci Transl Med.* 6(244), p. 244ra89. DOI: 10.1126/scitranslmed.3008065.

"In this model, mice treated with drugs that target chloride channels ..." R. Tyzio et al. "Oxytocin-mediated GABA inhibition during delivery attenuates autism pathogenesis in rodent offspring." 2014. *Science* 343(6171), p. 675. DOI: 10.1126/science.1247190.

"This complexity makes a human brain the most complex known object in the universe." For example, Christof Koch, Chief Scientific Officer, Allen Institute for Brain Science, as quoted in "Decoding 'the Most Complex Object in the Universe,'" June 14, 2013, National Public Radio, www.npr.com. Also see the author Marilynne Robinson and physicist Eric Chaisson later in this chapter.

"Even before birth, patterned waves fire from the retinas ..." J. B. Ackman et al. "Retinal waves coordinate patterned activity throughout the developing visual system." 2012. *Nature* 490(7419), p. 219. DOI: 10.1038/nature11529.

"After birth, when a neuron fires repeatedly, that connection grows stronger." With a process involving calcium spikes, see J. Cichon and W. B. Gan. "Branch-specific dendritic Ca(2+) spikes cause persistent synaptic plasticity." 2015. *Nature* 520(7546), p. 180. DOI: 10.1038/nature14251; another mechanism for this may be found in P. Long and G. Corfas. "To learn is to myelinate." 2014. *Science* 346(6207), p. 298. DOI: 10.1126/science.1261127.

"Their brains compensate by remodeling neuron connections ..." D. Kätzel and G. Miesenböck. "Experience-dependent rewiring of specific inhibitory connections in adult neocortex." 2014. *PLoS Biol.* 12(2), p. e1001798. DOI: 10.1371/journal.pbio.1001798.

"... have more taste receptors for bitter molecules ..." D. Li and J. Zhang. "Diet shapes the evolution of the vertebrate bitter taste receptor gene repertoire." 2014. *Mol Biol Evol.* 31(2), p. 303. DOI: 10.1093/molbev/mst219.

"Specifically, they evolved their neurons." A. Wada-Katsumata et al. "Changes in taste neurons support the emergence of an adaptive behavior in cockroaches." 2013. *Science* 340(6135), p. 972. DOI: 10.1126/science.1234854.

"The cells that start the process of building these food-roads are oxygen-sensing *neurons*." G. A. Linneweber et al. "Neuronal control of metabolism through nutrient-dependent modulation of tracheal branching." 2014. *Cell* 156(1–2), p. 69. DOI: 10.1016/j.cell.2013.12.008.

"Immune cells in the brain called microglia don't attack pathogens, ..." Y. Zhan et al. "Deficient neuron-microglia signaling results in impaired functional brain connectivity and social behavior." 2014. *Nat Neurosci.* 17(3), p. 400. DOI: 10.1038/nn.3641.

"Annie Dillard wrote that ..." Annie Dillard. *Pilgrim at Tinker Creek*. 1974, Bantam Books, p. 20.

"Marilynne Robinson writes, ..." Marilynne Robinson. *When I Was a Child I Read Books*. 2012, Picador, p. 8.

"With Eric Chaisson's Energy Rate Density, ..." E. J. Chaisson. "Energy rate density. II. Probing further a new complexity metric." 2011. *Complexity* 17(1), p. 44. DOI: 10.1002/cplx.20373.

"Recent imaging techniques show the mitochondria in mouse neurons moving around ..." D. Safiulina and A. Kaasik. "Energetic and dynamic: How mitochondria meet neuronal energy demands." 2013. *PLoS Biol.* 11(12), p. e1001755. DOI: 10.1371/journal.pbio.1001755.

"Most submarine officers have stories of the amazingly illogical things that can happen." R. Evans and C. Langdale. "6 things movies don't show you about life on a submarine." January 31, 2014. *Cracked.* Cracked.com.

"... a hierarchy in ERD that parallels evolutionary history, ..." See E. J. Chaisson. "Energy rate density as a complexity metric and evolutionary driver." 2011. *Complexity* 16(3), p. 27. DOI: 10.1002/cplx.20323; "Energy rate density. II. Probing further a new complexity metric." 2011. *Complexity* 17(1), p. 44. DOI: 10.1002/cplx.20373; and "Using complexity science to search for unity in the natural sciences," in *Complexity and the Arrow of Time*, ed. by Lineweaver, Davies, and Ruse. 2013, Cambridge University Press, pp. 68–79.

"... all designs of powered flight find similar structures that maximize the efficiency of flight, ..." Adrian Bejan. *Shape and Structure, from Engineering to Nature*. 2000, Cambridge University Press, pp. 237–239.

"Chaisson writes that 'fishing leisurely, ...'" E. J. Chaisson. "Energy rate density. II. Probing further a new complexity metric." 2011. *Complexity* 17(1), p. 44. DOI: 10.1002/cplx.20373.

"Newborns have deposits of brown fat over their heart and kidneys ..." Albert L. Lehninger and David L. Nelson, *Lehninger Principles of Biochemistry*, 4th edition. 2004, W. H. Freeman, pp. 717–718.

"... the precise heat regulation in your body requires tens of thousands of molecules, ..." I. Lestas et al. "Fundamental limits on the suppression of molecular fluctuations." 2010. *Nature* 467(7312), p. 174. DOI: 10.1038/nature09333.

"Observation matches Bejan's theory, ..." Adrian Bejan. *Shape and Structure, from Engineering to Nature*. 2000, Cambridge University Press, p. 265. (Technically, there is a power-law relationship with an exponent 3/4 for birds and most mammals; small mammals have a different type of heat flow that is conductive, so they obey an exponent of 1/3; and cold-blooded lizards and amphibians are warmed by outside convective flow and have a 2/3 exponent. All of these exponents are predicted by Bejan's theory of branching heat flows. Bejan goes one step further, and predicts that super-large mammals with arteries large enough for turbulent flow would need to have mass related to metabolism by an exponent of 7/8. Unfortunately, such mammals are currently hard to find.)

"The best spacing was the one fish already used." K. Park et al. "Optimal lamellar arrangement in fish gills." 2014. *Proc Natl Acad Sci.* 111(22), p. 8067. DOI: 10.1073/pnas.1403621111.

"David Krakauer writes that when using DNA, ..." D. Krakauer. "Chapter 10: The inferential evolution of biological complexity: forgetting nature by learning to nurture," in *Complexity and the Arrow of Time*, ed. by Lineweaver, Davies, and Ruse. 2013, Cambridge University Press, p. 232. DOI: 10.1017/CBO9781139225700.013.

"Neuron size may be optimized." D. Fox. "The limits of intelligence." 2011. *Sci Am.* 305(1), p. 36.

"One hypothesis is that 'tether cells' tied primate neurons ..." R. L. Buckner and F. M. Krienen. "The evolution of distributed association networks in the human brain." 2013. *Trends Cog Sci.* 17(12), p. 846. DOI: 10.1016/j.tics.2013.09.017.

"... our brain can distinguish more than 1 trillion smells, ..." C. Bushdid et al. "Humans can discriminate more than 1 trillion olfactory stimuli." 2014. *Science* 343(6177), p. 1370. DOI: 10.1126/science.1249168.

"Sacks describes how Hull lost even his visual thinking over time." Oliver Sacks. *The Mind's Eye*. 2010, Knopf.

"... bacteria can create faint sodium-ion waves." J. M. Kralj et al. "Electrical spiking in Escherichia coli probed with a fluorescent voltage-indicating protein." 2011. *Science* 333(6040), p. 345. DOI: 10.1126/science.1204763.

"In 2012, a 500-million-year-old fossil of an animal from the crab and lobster family ..." X. Ma et al. "Complex brain and optic lobes in an early Cambrian arthropod." 2012. *Nature* 490(7419), p. 258. DOI: 10.1038/nature11495.

"This early complexity is also supported by a study of zebrafish brains." M. S. Šestak and T. Domazet-Lošo. "Phylostratigraphic profiles in zebrafish uncover chordate origins of the vertebrate brain." 2015. *Mol Biol Evol.* 32(2), p. 299. DOI: 10.1093/molbev/msu319.

"... brain proteins have been consistently evolving ..." A. J. Sander et al. "The evolution of human cells in terms of protein innovation." 2014. *Mol Biol Evol.* 31(6), p. 1364. DOI: 10.1093/molbev/mst139.

"This may be seen in genes related to potassium." A. Rivas-Ubach et al. "Strong relationship between elemental stoichiometry and metabolome in plants." 2012. *Proc Natl Acad Sci.* 109(11), p. 4181. DOI: 10.1073/pnas.1116092109.

"... alligators evolve slowly while mice evolve quickly, ..." R. E. Green et al. "Three crocodilian genomes reveal ancestral patterns of evolution among archosaurs." 2014. *Science* 346(6215), p. 1254449. DOI: 10.1126/science.1254449.

"... bacteria evolve more quickly when they first infect a new organism ..." B. Linz et al. "A mutation burst during the acute phase of Helicobacter pylori infection in humans and rhesus macaques." 2014. *Nat Commun.* 5, p. 4165. DOI: 10.1038/ncomms5165.

"... just how recently polar bears evolved." S. Liu et al. "Population genomics reveal recent speciation and rapid evolutionary adaptation in polar bears." 2014. *Cell* 157(4), p. 785. DOI: 10.1016/j.cell.2014.03.054.

"Evolution speeds up in symbiosis, ..." P. Remigi et al. "Transient hypermutagenesis accelerates the evolution of legume endosymbionts following horizontal gene transfer." 2014. *PLoS Biol.* 12(9), p. e1001942. DOI: 10.1371/journal.pbio.1001942.

"... in more complex environments." D. Zuppinger-Dingley et al. "Selection for niche differentiation in plant communities increases biodiversity effects." 2014. *Nature* 515(7525), p. 108. DOI: 10.1038/nature13869.

"It slows down when environments are filled ..." T. D. Price et al. "Niche filling slows the diversification of Himalayan songbirds." 2014. *Nature* 509(7499), p. 222. DOI: 10.1038/nature13272.

"... during the battle for iron." S. Paul et al. "Accelerated gene evolution through replication-transcription conflicts." 2013. *Nature* 495(7442), p. 512. DOI: 10.1038/nature11989.

"The transcription factors that turn on DNA organize into a robust network that is itself more evolvable." J. L. Payne and A. Wagner. "The robustness and evolvability of transcription factor binding sites." 2014. *Science* 343(6173), p. 875. DOI: 10.1126/science.1249046.

"Scientists conclude from this that sodium channels are the youngest child, ..." H. H. Zakon. "Adaptive evolution of voltage-gated sodium channels: the first 800 million years." 2012. *Proc Natl Acad Sci.* 109(Suppl. 1), p. 10619. DOI: 10.1073/pnas.1201884109.

"... to turn a particular sodium-potassium channel into a chloride channel." A. Berndt et al. "Structure-guided transformation of channelrhodopsin into a light-activated chloride channel." 2014. *Science* 344(6182), p. 420. DOI: 10.1126/science.1252367.

"Likewise, a calcium channel can be changed into a sodium channel with a few tweaks, ..." A. Senatore et al. "T-type channels become highly permeable to sodium ions using an alternative extracellular turret region (S5-P) outside the selectivity filter." 2014. *J Biol Chem.* 289(17), p. 11952. DOI: 10.1074/jbc.M114.551473.

"For one class of protein pumps, evolution can be traced backward ..." K. Wang et al. "Structure and mechanism of Zn2+-transporting P-type ATPases." 2014. *Nature* 514(7523), p. 518. DOI: 10.1038/nature13618.

"Some one-celled organisms have a sodium channel that conducts both sodium *and* calcium." B. J. Liebeskind et al. "Evolution of sodium channels predates the origin of nervous systems in animals." 2011. *Proc Natl Acad Sci.* 108(22), p. 9154. DOI: 10.1073/pnas.1106363108.

"A few bacterial sodium channels (including one from Mono Lake) ..." D. Shava et al. "Structure of a prokaryotic sodium channel pore reveals essential gating elements and an outer ion binding site common to eukaryotic channels." 2014. *J Mol Biol.* 426(2), p. 467. DOI: 10.1016/j.jmb.2013.10.010.

"A sea anemone sodium channel looks suspiciously like an older bacterial calcium channel, ..." M. Gur Barzilai et al. "Convergent evolution of sodium ion selectivity in metazoan neuronal signaling." 2012. *Cell Rep.* 2(2), p. 242. DOI: 10.1016/j.celrep.2012.06.016.

"Sodium channels rapidly duplicated to make electric fish, twice." M. E. Arnegard et al. "Old gene duplication facilitates origin and diversification of an innovative communication system—twice." 2010. *Proc Natl Acad Sci.* 107(51), p. 22172. DOI: 10.1073/pnas.1011803107; J. R. Gallant et al. "Genomic basis for the convergent evolution of electric organs." 2014. *Science* 344(6191), p. 1522. DOI: 10.1126/science.1254432.

"Vampire bats started the process with a normal heat-sensitive calcium channel." E. O. Gracheva et al. "Ganglion-specific splicing of TRPV1 underlies infrared sensation in vampire bats." 2011. *Nature* 476(7358), p. 88. DOI: 10.1038/nature10245.

"... scientists found the genetic footprint were the umami-tasting gene used to be in pandas." H. Zhao et al. "Pseudogenization of the umami taste receptor gene Tas1r1 in the giant panda coincided with its dietary switch to bamboo." 2010. *Mol Biol Evol.* 27(12), p. 2669. DOI: 10.1093/molbev/msq153.

"Hummingbirds have also lost their umami taste." M. W. Baldwin et al. "Evolution of sweet taste perception in hummingbirds by transformation of the ancestral umami receptor." 2014. *Science* 345(6199), p. 929. DOI: 10.1126/science.1255097.

"Penguins, not to be outdone, have lost three distinct tastes." H. Zhao et al. "Molecular evidence for the loss of three basic tastes in penguins." 2015. *Curr Biol.* 25(4), p. R141. DOI: 10.1016/j.cub.2015.01.026.

"... a few general bitter taste receptors can do the job adequately, ..." M. Behrens et al. "Tuning properties of avian and frog bitter taste receptors dynamically fit gene repertoire sizes." 2014. *Mol Biol Evol.* 31(12), p. 3216. DOI: 10.1093/molbev/msu254.

"Opsins are found all over the animal kingdom, ..." E. Pennisi. "Opsins: not just for eyes." 2013. *Science* 339(6121), p. 754. DOI: 10.1126/science.339.6121.754.

"This precise arrangement is seen binding zinc in very different proteins, ..." J. W. Torrance et al. "Evolution of binding sites for zinc and calcium ions playing structural roles." 2008. *Proteins* 71(2), p. 813.

"... a distinctive pattern of amino acids that binds calcium ..." R. O. Morgan et al. "Evolutionary perspective on annexin calcium-binding domains." 2004. *Biochim Biophys Acta.* 1742(1–3), p.133.

"This is why the scorpion genome shows a huge expansion in genes ..." Z. Cao et al. "The genome of Mesobuthus martensii reveals a unique adaptation model of arthropods." 2013. *Nat Commun.* 4, p. 2602. DOI: 10.1038/ncomms3602.

"Scorpion toxin and target channel are tied together in evolutionary flow." A. O. Chugunov et al. "Modular organization of α-toxins from scorpion venom mirrors domain structure of their targets, sodium channels." 2013. *J Biol Chem.* 288(26), p. 19014. DOI: 10.1074/jbc.M112.431650.

"Six species of snakes at three places around the world all reshaped ..." C. R. Feldman et al. "Constraint shapes convergence in tetrodotoxin-resistant sodium channels of snakes." 2012. *Proc Natl Acad Sci.* 109(12), p. 4556. DOI: 10.1073/pnas.1113468109.

"... were all found to make the same shape changes in the same places." Y. Zhen et al. "Parallel molecular evolution in an herbivore community." 2012. *Science* 337(6102), p. 1634. DOI: 10.1126/science.1226630; and S. Dobler et al. "Community-wide convergent evolution in insect adaptation to toxic cardenolides by substitutions in the Na,K-ATPase." 2012. *Proc Natl Acad Sci.* 109(32), p. 13040. DOI: 10.1073/pnas.1202111109.

"... because both evolved the same shape *independently* as a chemical defense." F. Denoeud et al. "The coffee genome provides insight into the convergent evolution of caffeine biosynthesis." 2014. *Science* 345(6201), p. 1181. DOI: 10.1126/science.1255274.

"The genes show that the nerves gathered together in a head twice, independently." K. M. Kocot et al. "Phylogenomics reveals deep molluscan relationships." 2011. *Nature* 477(7365), p. 452. DOI: 10.1038/nature10382.

"In his book *Life's Solution*, Conway Morris lists ..." Simon Conway Morris. *Life's Solution: Inevitable Humans in a Lonely Universe.* 2003, Cambridge University Press, p. 307.

"But katydid leg-ears hear with the same *shapes* as human ears: ..." F. Montealegre-Z and D. Robert. "Biomechanics of hearing in katydids." 2015. *J Comp Physiol A.* 201(1), p. 5. DOI: 10.1007/s00359-014-0976-1.

"... which must have developed independently at least six times." Simon Conway Morris. *Life's Solution: Inevitable Humans in a Lonely Universe.* 2003, Cambridge University Press, p. xii.

"Octopus eyes look like mammal eyes ..." S. C. Morris. "Consider the octopus." 2011. *EMBO Rep.* 12(3), p. 182. DOI: 10.1038/embor.2011.12.

"Australian mammals and North American mammals also look the same, ..." Peter Atkins. *Galileo's Finger: The Ten Great Ideas of Science.* 2003, Oxford University Press. p. 26.

"A group studying lizard evolution on four Carribean islands published a paper ..." D. L. Mahler et al. "Exceptional convergence on the macroevolutionary landscape in island lizard radiations." 2013. *Science* 341(6143), p. 292. DOI: 10.1126/science.1232392.

"Several groups have taken proteins at the beginning and end of an evolutionary path, ..." M. Lunzer et al. "The biochemical architecture of an ancient adaptive landscape." 2005. *Science* 310(5747), pp. 499–501; and S. Noor et al. "Intramolecular epistasis and the evolution of a new enzymatic function." 2012. *PLoS One* 7(6), p. e39822. DOI: 10.1371/journal.pone.0039822; for the order in which proteins assemble: J. A. Marsh et al. "Protein complexes are under evolutionary selection to assemble via ordered pathways." 2013. *Cell* 153(2), p. 461. DOI: 10.1016/j.cell.2013.02.044; see also F. J. Poelwijk et al. "Empirical fitness landscapes reveal accessible evolutionary paths." 2007. *Nature* 445(7126), p. 383.

"One group analyzed several of these experiments ..." M. Carneiro and D. L. Hartl. "Adaptive landscapes and protein evolution." 2010. *Proc Natl Acad Sci.* 107 Suppl 1, p. 1747. DOI: 10.1073/pnas.0906192106.

"One paper even predicts the predictability of bacterial evolution in the lab!" D. Blank et al. "The predictability of molecular evolution during functional innovation." 2014. *Proc Natl Acad Sci.* 111(8), p. 3044. DOI: 10.1073/pnas.1318797111.

"Many times, even the DNA undergoes the same changes in different populations." For example, see L. Pappas et al. "Rapid development of broadly influenza neutralizing antibodies through redundant mutations." 2014. *Nature* 516(7531), p. 418. DOI: 10.1038/nature13764.

"The figures in these papers show genes ebbing and flowing like flowing rivers through time, ..." See Figure 2 in M. D. Herron and M. Doebeli. "Parallel evolutionary dynamics of adaptive diversification in *Escherichia coli*." 2013. *PLoS Biol.* 11(2), p. e1001490. DOI: 10.1371/journal.pbio.1001490.

"One group is anticipating the path of malaria's evolution . . ." A. K. Lukens et al. "Harnessing evolutionary fitness in Plasmodium falciparum for drug discovery and suppressing resistance." 2014. *Proc Natl Acad Sci.* 111(2), pp. 799–804. DOI: 10.1073/pnas.1320886110.

". . . by stopping production of a gene called PTEN." D. Juric et al. "Convergent loss of PTEN leads to clinical resistance to a PI(3)Kα inhibitor." 2015. *Nature* 518(7538), p. 240. DOI: 10.1038/nature13948.

"An enzyme that adds electrons to mercury . . ." A. Sayed et al. "A novel mercuric reductase from the unique deep brine environment of Atlantis II in the Red Sea." 2014. *J Biol Chem.* 289(3), p. 1675. DOI: 10.1074/jbc.M113.493429.

". . . have converged on similar chemical solutions to survive the environmental saltiness." E. F. Mongodin et al. "The genome of *Salinibacter ruber*: convergence and gene exchange among hyperhalophilic bacteria and archaea." 2005. *Proc Natl Acad Sci.* 102(50), p.18147.

". . . show genetic signs that they can quickly adapt to a hot environment . . ." P. Puigbò et al. "Gaining and losing the thermophilic adaptation in prokaryotes." 2008. *Trends Genet.* 24(1), p. 10.

". . . this DNA sequence moved from the ocean to the human gut." J. H. Hehemann et al. "Transfer of carbohydrate-active enzymes from marine bacteria to Japanese gut microbiota." 2010. *Nature* 464(7290), p. 908. DOI: 10.1038/nature08937.

"If yeast are grown in a constant environment, they will jettison *all* their signaling proteins." D. J. Kvitek and G. Sherlock. "Whole genome, whole population sequencing reveals that loss of signaling networks is the major adaptive strategy in a constant environment." 2013. *PLoS Genet.* 9(11), p. e1003972. DOI: 10.1371/journal.pgen.1003972.

"Also, bacteria at some point threw out at least two protein-moving systems." M. T. Bohnsack and E. Schleiff. "The evolution of protein targeting and translocation systems." 2010. *Biochim Biophys Acta.* 1803(10), p. 1115. DOI: 10.1016/j.bbamcr.2010.06.005.

". . . it's simpler to conclude that these fins evolved multiple times." T. A. Stewart et al. "The origins of adipose fins: An analysis of homoplasy and the serial homology of vertebrate appendages." 2014. *Proc Biol Sci.* 281(1781), p. 20133120. DOI: 10.1098/rspb.2013.3120.

". . . from an internal-fertilizing ancestor to fertilize its eggs externally again." J. A. Long et al. "Copulation in antiarch placoderms and the origin of gnathostome internal fertilization." 2015. *Nature* 517(7533), p. 196. DOI: 10.1038/nature13825.

"But in the lab, it took *less than four days* . . ." T. B. Taylor et al. "Evolutionary resurrection of flagellar motility via rewiring of the nitrogen regulation system." 2015. *Science* 347(6225), p. 1014. DOI: 10.1126/science.1259145.

". . . preceded by a huge chain of volcanic eruptions." R. A. Kerr. "Mega-eruptions drove the mother of mass extinctions." 2013. *Science* 342(6165), p. 1424. DOI: 10.1126/science.342.6165.1424.

"As a result, 70% of the species on the land and 96% of the species in the sea vanished." Robert M. Hazen. *The Story of Earth: The First 4.5 Billion Years, from Stardust to Living Planet.* 2012, Viking, p. 250.

"A "reef gap" appears in the fossil record because the reefs were destroyed." Elizabeth Kolbert. *The Sixth Extinction: An Unnatural History.* 2014, Henry Holt, p. 140.

"Elements from previous chapters were injected into this chapter, including nickel, . . ." D. H. Rothman et al. "Methanogenic burst in the end-Permian carbon cycle." 2014. *Proc Natl Acad Sci.* 111(15), p. 5462. DOI: 10.1073/pnas.1318106111.

". . . complex communities survived better and outnumbered simpler communities three to one, . . ." P. J. Wagner et al. "Abundance distributions imply elevated complexity of post-paleozoic marine ecosystems." 2006. *Science* 314(5803), p. 1289. DOI:10.1126/science.1133795.

"They eventually succeed, creating an ecological void that other species fill." R. Bagchi et al. "Pathogens and insect herbivores drive rainforest plant diversity and composition." 2014. *Nature* 506(7486), p. 85. DOI: 10.1038/nature12911.

". . . more precise measurements showed that this event extended through only 5% of the oceans." J. D. Owens et al. "Sulfur isotopes track the global extent and dynamics of euxinia during Cretaceous Oceanic Anoxic Event 2." 2013. *Proc Natl Acad Sci.* 110(46), p.18407. DOI: 10.1073/pnas.1305304110.

"The Deccan province, now in India, spewed . . ." B. Schoene et al. "U-Pb geochronology of the Deccan Traps and relation to the end-Cretaceous mass extinction." 2015. *Science* 347(6218), p. 182. DOI: 10.1126/science.aaa0118.

"Plants shifted toward a 'fast' growth strategy . . ." B. Blonder et al. "Plant ecological strategies shift across the Cretaceous-Paleogene boundary." 2014. *PLoS Biol.* 12(9), p. e1001949. DOI: 10.1371/journal.pbio.1001949.

"This could have led to 'higher rates of ecosystem functioning' according to one reference, . . ." J. M. Chase. "A plant's guide to surviving the Chicxulub impact." 2014. *PLoS Biol.* 12(9), p. e1001948. DOI: 10.1371/journal.pbio.1001948.

"Although they were already around before the impact, mammals diversified afterward." P. G. Falkowski et al. "The rise of oxygen over the past 205 million years and the evolution of large placental mammals." 2005. *Science* 309(5744), p. 2202.

". . . dinosaurs were already on their way out, . . ." S. Conway Morris. "Predicting what extraterrestrials will be like: And preparing for the worst." 2011. *Philos Trans A.* 369(1936), p. 555. DOI: 10.1098/rsta.2010.0276.

". . . each mass extinction advanced the evolutionary clock by 50 million years." S. Conway Morris. "Is convergence becoming too popular?" July 17, 2014, lecture at "Evolution and Historical Explanation: Contingency, Convergence, and Teleology." St. Anne's College, Oxford, UK.

". . . birds started to reduce in body size as far back as 200 million years ago, . . ." M. J. Benton. "How birds became birds." 2014. *Science* 345(6196), p. 508. DOI: 10.1126/science.1257633.

". . . using old genes that predated even dinosaurs." C. B. Lowe et al. "Feather development genes and associated regulatory innovation predate the origin of Dinosauria." 2015. *Mol Biol Evol.* 32(1), p. 23. DOI: 10.1093/molbev/msu309; and P. Godefroit et al. "A Jurassic ornithischian dinosaur from Siberia with both feathers and scales." 2014. *Science* 345(6195), p. 451. DOI: 10.1126/science.1253351.

"The avian body plan was gradually assembled before the meteor impact, . . ." S. L. Brusatte et al. "Gradual assembly of avian body plan culminated in rapid rates of evolution across the dinosaur-bird transition." 2014. *Curr Biol.* 24(20), p. 2386. DOI: 10.1016/j.cub.2014.08.034.

"Genes specifically from this chapter evolved especially quickly . . . A full-genome analysis of 48 different birds . . ." G. Zhang et al. "Comparative genomics reveals insights into avian genome evolution and adaptation." 2014. *Science* 346(6215), p. 1311. DOI: 10.1126/science.1251385.

"This has been described as 'an evolutionary Big Bang.'" R. Williams. "Bird genomes abound." December 11, 2014, *The Scientist*, www.the-scientist.com.

"(Insects may have had their own 'Big Bang' right after the Permian extinction.)" B. Misof et al. "Phylogenomics resolves the timing and pattern of insect evolution." 2014. *Science* 346(6210), p. 763. DOI: 10.1126/science.1257570: "Finally, our analyses suggest that the major diversity within living cockroaches, mantids, termites, and stick insects evolved after the Permian mass extinction."

"Sometime in this chapter, oxygen-driven chemical evolution ended." R. J. P. Williams and R. E. M. Rickaby. *Evolution's Destiny: Co-evolving Chemistry of the Environment and Life.* 2012, RSC Publishing; p. 163.

"Fossilized tracks from 50-million-year old sea urchins ..." D.W. Sims et al. "Hierarchical random walks in trace fossils and the origin of optimal search behavior." 2014. *Proc Natl Acad Sci.* 111(30), p. 11073. DOI: 10.1073/pnas.1405966111.

Chapter 11

"Just as music must be like a hive of bees, ..." Mark Helprin. *Winter's Tale.* 1983, Houghton Mifflin Harcourt, p. 555.

"Evolution invented the hive." E. O. Wilson and M. A. Nowak. "Natural selection drives the evolution of ant life cycles." *Proc Natl Acad Sci.*111(35), p. 12585. DOI: 10.1073/pnas.1405550111.

"This chemical language can be mimicked with organic chemistry: ..." A. Van Oystaeyen et al. "Conserved class of queen pheromones stops social insect workers from reproducing." 2014. *Science* 343(6168), p. 287. DOI: 10.1126/science.1244899.

"Chemicals form the bridges of communication between the species." I. Schoenian et al. "Chemical basis of the synergism and antagonism in microbial communities in the nests of leaf-cutting ants." 2011. *Proc Natl Acad Sci.* 108(5), p. 1955. DOI: 10.1073/pnas.1008441108.

"... even the fungus genes are kept the same." D. K. Aanen et al. "High symbiont relatedness stabilizes mutualistic cooperation in fungus-growing termites." 2009. *Science* 326(5956), p. 1103. DOI: 10.1126/science.1173462.

"The ant has lost all its genes for making arginine, ..." S. Nygaard et al. "The genome of the leaf-cutting ant *Acromyrmex echinatior* suggests key adaptations to advanced social life and fungus farming." 2011. *Genome Res.* 21(8), p. 1339. DOI: 10.1101/gr.121392.111.

"Different castes have different patterns of receptors ..." S. I. Koch et al. "Caste-specific expression patterns of immune response and chemosensory related genes in the leaf-cutting ant, *Atta vollenweideri.*" 2013. *PLoS One* 8(11), p. e81518. DOI: 10.1371/journal.pone.0081518.

"... by changing the DNA of their symbiotic bacteria." C. Arnold. "The other you." 2015. *New Scientist* 2899, p. 31.

"... caused by the environment as part of an animal's stress response ..." D. Cossins. "Stress fractures." January 1, 2015, *The Scientist*, www.the-scientist.com.

"Termite mounds help buffer the surroundings against wild swings of climate." J. A. Bonachela et al. "Termite mounds can increase the robustness of dryland ecosystems to climatic change." 2015. *Science* 347(6222), p. 651. DOI: 10.1126/science.1261487.

"... coral reefs, too, appear to evolve faster than expected ..." S. R. Palumbi et al. "Mechanisms of reef coral resistance to future climate change." 2014. *Science* 344(6186), p. 895. DOI: 10.1126/science.1251336.

"Eusociality ... evolved at least two dozen times ..." E. O. Wilson. "The riddle of the human species." February 24, 2013. *New York Times.* www.nytimes.com.

"What but the energies of the universe ..." Marilynne Robinson. *Absence of Mind: The Dispelling of Inwardness from the Modern Myth of the Self (The Terry Lectures Series).* 2010, Yale University Press, p. 134.

"... separate regions of the brain must talk to each other." M. Boly et al. "Preserved feedforward but impaired top-down processes in the vegetative state." 2011. *Science* 332(6031), p. 858. DOI: 10.1126/science.1202043.

"... the connections *between* separate areas are lost." L. D. Lewis et al. "Rapid fragmentation of neuronal networks at the onset of propofol-induced unconsciousness." 2012. *Proc Natl Acad Sci.* 109(49), p. E3377. DOI: 10.1073/pnas.1210907109; see also D. Godwin et al. "Breakdown of the brain's functional network modularity with awareness." 2015. *Proc Natl Acad Sci.* 112(12):3799. DOI: 10.1073/pnas.1414466112.

"... see if they notice it in a mirror ..." L. Chang et al. "Mirror-induced self-directed behaviors in rhesus monkeys after visual-somatosensory training." 2015. *Curr Biol.* 25(2), p .212. DOI: 10.1016/j.cub.2014.11.016; and H. Prior et al. "Mirror-induced behavior in the magpie (Pica pica): Evidence of self-recognition." 2008. *PLoS Biol.* 6(8), p. e202. DOI: 10.1371/journal.pbio.0060202.

"As children develop, they shunt metabolic resources ..." C. W. Kuzawa et al. "Metabolic costs and evolutionary implications of human brain development." 2014. *Proc Natl Acad Sci.* 111(36), p. 13010. DOI: 10.1073/pnas.1323099111.

"We are relatively half as strong as our primate cousins ..." K. Bozek et al. "Exceptional evolutionary divergence of human muscle and brain metabolomes parallels human cognitive and physical uniqueness." 2014. *PLoS Biol.* 12(5), p. e1001871. DOI: 10.1371/journal.pbio.1001871.

"... our species' bone density has decreased measurably ..." H. Chirchir et al. "Recent origin of low trabecular bone density in modern humans." 2015. *Proc Natl Acad Sci.* 112(2), p. 366. DOI: 10.1073/pnas.1411696112.

"Our family tree is more a braid or a thicket." C. Finlayson. "Viewpoint: Human evolution, from tree to braid." December 31, 2013. *BBC News (Science & Environment)*, www.bbc.co.uk.

"... we catch their colds and flus." B. Natterson-Horowitz and K. Bowers. "Our animal natures." June 9, 2012. *New York Times*, www.nytimes.com.

"This moment was when Nature was created, ..." This concept comes from Walker Percy's writings, especially *Lost in the Cosmos*, and Owen Barfield's writings, especially *Saving the Appearances.*

"Now, with humans, Lamarck's evolution returns, through a brain built by Darwin's rules." P. Clayton. " On the plurality of complexity-producing mechanisms," In *Complexity and the Arrow of Time*, ed. by Lineweaver, Davies, and Ruse. 2013, Cambridge University Press. DOI: 10.1017/CBO9781139225700.018; p. 345: "Cultural complexity does not arise in the same way biological complexity does, nor can it be studied in the same way. Cultural explanations are fundamentally Lamarckian ... through social learning ... culturally transmitted influence of new ideas and theories, the power dynamics of competing groups, the personalities of charismatic leaders, and the conscious intentions of agents. These are not dynamics that we can model in the same ways that we model the dynamics of Darwinian systems. Of course, biology holds culture on a leash."

"Both tools and pigments made by chemistry ..." C. W. Marean et al. "Early human use of marine resources and pigment in South Africa during the Middle Pleistocene." 2007.

Nature 449(7164), p. 905; and K. S. Brown et al. "Fire as an engineering tool of early modern humans." 2009. *Science* 325(5942), p. 859. DOI: 10.1126/science.1175028.

"The sea provides shellfish protein, but to harvest them, one has to *remember* tides." C. W. Marean. "When the sea saved humanity." 2010. *Sci Am.* 303(2), p. 54.

"For Tolkien, as for Barfield, . . ." Verlyn Flieger. *Splintered Light: Logos and Language in Tolkien's World.* 1983, Kent State University Press, pp. 66–67.

"Tolkien created backward, from word to world, . . ." Verlyn Flieger. *Splintered Light: Logos and Language in Tolkien's World.* 1983. Kent State University Press.

"The speech centers in human brains and the song centers in bird brains have converged . . ." A. R. Pfenning et al. "Convergent transcriptional specializations in the brains of humans and song-learning birds." 2014. *Science* 346(6215), p. 1256846. DOI: 10.1126/science.1256846.

". . . genes in songbirds have triply accelerated rates of evolution, . . ." G. Zhang et al. "Comparative genomics reveals insights into avian genome evolution and adaptation." 2014. *Science* 346(6215), p. 1311. DOI: 10.1126/science.1251385.

". . . the hermit thrush constructs a musical scale that matches ours . . ." E. L. Doolittle et al. "Overtone-based pitch selection in hermit thrush song: Unexpected convergence with scale construction in human music." 2014. *Proc Natl Acad Sci.* 111(46), p. 16616. DOI: 10.1073/pnas.1406023111.

"Our sense of rhythm is rare." A. D. Patel. "The evolutionary biology of musical rhythm: Was Darwin wrong?" 2014. *PLoS Biol.* 12(3), p. e1001821. DOI: 10.1371/journal.pbio.1001821.

"Think of how, when you talk to an infant, . . ." L. S. Morris. "Notes of importance." 2014. *Science* 345(6197), p. 630. DOI: 10.1126/science.1256050.

". . . high-altitude languages have more 'ejective consonants' that click loudly in dry air." C. Everett. "Evidence for direct geographic influences on linguistic sounds: the case of ejectives." 2013. *PLoS One* 8(6), p. e65275. DOI: 10.1371/journal.pone.0065275.

". . . the shapes of the brain waves match the shapes of the sound waves." J. Gross *et al.* "Speech rhythms and multiplexed oscillatory sensory coding in the human brain." 2013. *PLoS Biol.* 11(12), p. e1001752. DOI: 10.1371/journal.pbio.1001752.

". . . match natural shapes in the environment, . . ." M. A. Changizi et al. "The structures of letters and symbols throughout human history are selected to match those found in objects in natural scenes." 2006. *Amer Naturalist.* 167(5), p. E117.

". . . it was a relatively simple conversion to apply that software to language words . . ." J. J. Tehrani "The phylogeny of Little Red Riding Hood." 2013. *PLoS One* 8(11), p. e78871. DOI: 10.1371/journal.pone.0078871.

"The bioinformatics methods date modern language to an ancestor language 15,000 years old." M. Pagel et al. "Ultraconserved words point to deep language ancestry across Eurasia." 2013. *Proc Natl Acad Sci.* 110(21), p. 8471. DOI: 10.1073/pnas.1218726110.

"A massive computer analysis has favored a root in Anatolia." R. Bouckaert et al. "Mapping the origins and expansion of the Indo-European language family." 2012. *Science* 337(6097), p. 957. DOI: 10.1126/science.1219669.

"The story of migration told in the genes of UK residents . . ." P. Ralph and G. Coop. "The geography of recent genetic ancestry across Europe." 2013. *PLoS Biol.* 11(5), p. e1001555. DOI: 10.1371/journal.pbio.1001555.

". . . Ashkenazic and Sephardic Jews are genetically similar despite their dispersal." D. M. Behar et al. "The genome-wide structure of the Jewish people." 2010. *Nature* 466(7303), p. 238. DOI: 10.1038/nature09103.

"... the DNA for dog breeds backs up ..." B. van Asch et al. "Pre-Columbian origins of Native American dog breeds, with only limited replacement by European dogs, confirmed by mtDNA analysis." 2013. *Proc Biol Sci.* 280(1766), p. 20131142. DOI: 10.1098/rspb.2013.1142.

"We can read the same story now for horses, tomatoes, rice, and more." See M. Schubert et al. "Prehistoric genomes reveal the genetic foundation and cost of horse domestication." 2014. *Proc Natl Acad Sci.* 111(52), p. E5661. DOI: 10.1073/pnas.1416991111.

"We are part of that same process ..." A. Gibbons. "How we tamed ourselves—and became modern." 2014. *Science* 346(6208), p. 405. DOI: 10.1126/science.346.6208.405.

"Medieval leprosy genes look like modern ones ..." V. J. Schuenemann et al. "Genome-wide comparison of medieval and modern *Mycobacterium leprae*." 2013. *Science* 341(6142), p. 179. DOI: 10.1126/science.1238286. Epub 2013 Jun 13.

"the Justinian plague in the sixth century was caused by the same ..." M. Harbeck et al. "Yersinia pestis DNA from skeletal remains from the 6(th) century AD reveals insights into Justinianic Plague." 2013. *PLoS Pathog.* 9(5), p. e1003349. DOI: 10.1371/journal.ppat.1003349.

"Different human populations develop similar immune systems to fight off the same plagues, ..." H. Laayouni et al. "Convergent evolution in European and Rroma populations reveals pressure exerted by plague on Toll-like receptors." 2014. *Proc Natl Acad Sci.* 111(7), p. 2668. DOI: 10.1073/pnas.1317723111.

"... the ground zero of innovation ..." Steven Johnson. *Where Good Ideas Come From.* 2010, Riverhead, p. 61.

"Technologies increase in ERD over time." E. J. Chaisson. "A singular universe of many singularities: Cultural evolution in a cosmic context," in *The Singularity Hypothesis: A Scientific and Philosophical Assessment (The Frontiers Collection)*, ed. by Eden, Soraker, Moor, and Steinhart. 2012, Springer, pp. 414–440.

"... the extra time provided by fire at night is used to string words together in stories." P. W. Wiessner. "Embers of society: Firelight talk among the Ju/'hoansi Bushmen." 2014. *Proc Natl Acad Sci.* 111(39), p. 14027. DOI: 10.1073/pnas.1404212111.

"Chaisson estimates that foragers use 2000 calories a day ..." E. J. Chaisson. "Energy rate density as a complexity metric and evolutionary driver." 2011. *Complexity* 16(3), p. 27. DOI: 10.1002/cplx.20323.

"Chaisson insists that 'what matters ...'" E. J. Chaisson. "A singular universe of many singularities: Cultural evolution in a cosmic context," in *The Singularity Hypothesis: A Scientific and Philosophical Assessment (The Frontiers Collection)*, ed. by Eden, Soraker, Moor, and Steinhart. 2012, Springer, p. 427.

"Bejan sees the transportation networks that move people around cities and nations as trees." Adrian Bejan. *Shape and Structure, from Engineering to Nature.* 2000, Cambridge University Press.

"Perhaps Darwinian evolution is so predictable that it's boring, ..." David Toomey. *Weird Life.* 2014, Norton, p. 175: "One of the best examples is one of the earliest, in David Lindsay's 1920 novel *A Voyage to Arcturus*. The planet on which the action takes place has not only weird life, but an alternative to Darwinian natural selection, in which creatures can actually will the properties of their progeny, the result being a natural world so 'energetic and lawless' that no two creatures are alike."

"No one is a good historian ..." Victor Hugo. *Les Misérables.* 1862 (1915 ed.), T. Y. Crowell, p. 154.

"A man strikes the lyre, and says, . . ." G. K. Chesterton. *The Napoleon of Notting Hill.* 1904 (2010 ed.), Bibliolis, p. 107.

"We can read this in the genes of humans and cows." A. Curry. "The milk revolution." 2013. *Nature* 500(7460), p.c20. DOI: 10.1038/500020a.

"Paul Tough in *How Children Succeed* describes . . ." Paul Tough. *How Children Succeed: Grit, Curiosity, and the Hidden Power of Character.* 2012, Houghton Mifflin Harcourt.

"Another study found similar results for a different stress marker called CRP." W. E. Copeland et al. "Childhood bullying involvement predicts low-grade systemic inflammation into adulthood." 2014. *Proc Natl Acad Sci.* 111(21), p. 7570. DOI: 10.1073/pnas.1323641111.

"For example, how much money you make correlates . . ." J. Tyrrell et al. "Associations between socioeconomic status and environmental toxicant concentrations in adults in the USA: NHANES 2001–2010." 2013. *Environ Int.* 59, p. 328. DOI: 10.1016/j.envint.2013.06.017.

". . . even a spoonful of zinc sulfate powder." E. Ho. "Zinc deficiency, DNA damage and cancer risk." 2004. *J Nutr Biochem.* 15(10), p. 572.

"One article suggests that we supplement many common chemical deficiencies . . ." S. Loewenberg. "Easier than taking vitamins." September 5, 2012. *New York Times.* www.nytimes.com.

". . . Neanderthal children were weaned when they were a year old . . ." C. Austin et al. "Barium distributions in teeth reveal early-life dietary transitions in primates." 2013. *Nature* 498(7453), p. 216. DOI: 10.1038/nature12169.

". . . strontium levels in teeth show that Middle Eastern farmers physically moved . . ." D. Borić and T. D. Price. "Strontium isotopes document greater human mobility at the start of the Balkan Neolithic." 2013. *Proc Natl Acad Sci.* 110(9), p. 3298. DOI: 10.1073/pnas.1211474110.

"Neolithic tartar, medieval tartar, and Industrial Age tartar were compared." A. Gibbons. "How sweet it is: genes show how bacteria colonized human teeth." 2013. *Science* 339(6122), p. 896. DOI: 10.1126/science.339.6122.896.

"Man is a centaur, a tangle of flesh and mind, divine inspiration and dust." Primo Levi. *The Periodic Table.* 1984, Schocken Books, p. 10.

"Napoleon III made utensils of aluminum that were valued more than solid gold." John Emsley. *Nature's Building Blocks: An A–Z Guide to the Elements.* 2001, Oxford University Press.

". . . mercury levels in each layer fit history: . . ." O. Serrano et al. "Millennial scale impact on the marine biogeochemical cycle of mercury from early mining on the Iberian Peninsula." 2013. *Glob Biogeo Cycles.* 27, p. 1. DOI: 10.1029/2012GB004296; see also C. H. Lamborg et al. "A global ocean inventory of anthropogenic mercury based on water column measurements." 2014. *Nature* 512(7512), p. 65. DOI: 10.1038/nature13563.

". . . had up to one hundred times more lead than nearby spring waters." H. Delile et al. "Lead in ancient Rome's city waters." 2014. *Proc Natl Acad Sci.* 111(18), p. 6594. DOI: 10.1073/pnas.1400097111.

". . . it is well on the way to being metamorphosed into a new mineral." P. L. Corcoran et al. "An anthropogenic marker horizon in the future rock record." 2014. *GSA Today* 24(6), p. 4. DOI: 10.1130/GSAT-G198A.1.

"Most cities have detectable concentrations of drugs in their wastewaster . . ." C. Ort et al. "Spatial differences and temporal changes in illicit drug use in Europe quantified by wastewater analysis." 2014. *Addiction* 109(8), p. 1338. DOI: 10.1111/add.12570.

"... from being limited by nitrogen to being limited by phosphorous." I. N. Kim et al. "Increasing anthropogenic nitrogen in the North Pacific Ocean." 2014. *Science* 346(6213), p. 1102. DOI: 10.1126/science.1258396.

"So with a city, which if it is to make its mark ..." Mark Helprin. *Winter's Tale*. 1983. Houghton Mifflin Harcourt, p. 555.

"... which inadvertently shaped the evolution of cliff swallows." C. R. Brown and M. B. Brown. "Where has all the road kill gone?" 2013. *Curr Biol*. 23(6), p. R233. http://dx.doi.org/10.1016/j.cub.2013.02.023.

"One example is a kind of fox that has adapted to live in London." Alastair Bonnett. *Unruly Places: Lost Spaces, Secret Cities, and Other Inscrutable Geographies*. 2014, Houghton Mifflin Harcourt, p. 53.

"This plant can endure the harsh environment along railroad tracks." Tristan Gooley. *The Walker's Guide to Outdoor Clues and Signs*. 2014, Sceptre Press, p. 84.

"... almost all small animals died out." L. Gibson et al. "Near-complete extinction of native small mammal fauna 25 years after forest fragmentation." 2013. *Science* 341(6153), p. 1508. DOI: 10.1126/science.1240495.

"... brought with them a host of cultural and biological changes." A. Curry. "Crusader crisis: How conquest transformed northern Europe." 2012. *Science* 338(6111), p. 1144. DOI: 10.1126/science.338.6111.1144.

"... shellfish fossils from the late stone age are distinctly smaller, ..." R. G. Klein and T. E. Steele. "Archaeological shellfish size and later human evolution in Africa." 2013. *Proc Natl Acad Sci*. 110(27), p. 10910. DOI: 10.1073/pnas.1304750110.

"Also, human activity 8,000 and 5,000 years ago may have added ..." L. Mitchell et al. "Constraints on the late holocene anthropogenic contribution to the atmospheric methane budget." 2013. *Science* 342(6161), p. 964. DOI: 10.1126/science.1238920.

Chapter 12

"Play the tape a few more times, though ..." L. Van Valen. "How far does contingency rule?" 1990. *Evolut Theory* 10, p. 47.

"After three years of competition, the green lizards changed their behavior ..." Y. E. Stuart et al. "Rapid evolution of a native species following invasion by a congener." 2014. *Science* 346(6208), p. 463. DOI: 10.1126/science.1257008.

"... the genes of Californian stick bugs perching on different plants, ..." V. Soria-Carrasco et al. "Stick insect genomes reveal natural selection's role in parallel speciation." 2014. *Science* 344(6185), p. 738. DOI: 10.1126/science.1252136.

"... the genes of laboratory-grown yeast cultures that shuffle ..." M. K. Burke et al. "Standing genetic variation drives repeatable experimental evolution in outcrossing populations of *Saccharomyces cerevisiae*." 2014. *Mol Biol Evol*. 31(12), p. 3228. DOI: 10.1093/molbev/msu256.

"Giant sequoias grow on phosphorus-rich bedrock, ..." W. J. Hahm et al. "Bedrock composition regulates mountain ecosystems and landscape evolution." 2014. *Proc Natl Acad Sci*. 111(9), p. 3338. DOI: 10.1073/pnas.1315667111.

"They do so with what Brodribb et al. call 'surprising simplicity,' ..." T. J. Brodribb et al. "Conifer species adapt to low-rainfall climates by following one of two divergent pathways." 2014. *Proc Natl Acad Sci*. 111(40), p. 14489. DOI: 10.1073/pnas.1407930111.

"The fish raised on land learned to move . . ." E. M. Standen et al. "Developmental plasticity and the origin of tetrapods." 2014. *Nature* 513(7516), p. 54. DOI: 10.1038/nature13708.

"We had the sky up there, . . ." Mark Twain. *Huckleberry Finn*. 1884, Charles L. Webster, Chapter 19.

". . . Gould's quote from the first chapter, expanded: . . ." Stephen Jay Gould. *Wonderful Life: The Burgess Shale and the Nature of History*. 1990, W.W. Norton, p. 14.

". . . the explosion was *longer* in time . . ." M. S. Lee et al. "Rates of phenotypic and genomic evolution during the Cambrian explosion." 2013. *Curr Biol*. 23(19), p. 1889. DOI: 10.1016/j.cub.2013.07.055.

"More thorough comparison of Burgess Shale fossil shapes revealed more patterns, . . ." S. Conway Morris. *The Crucible of Creation: The Burgess Shale and the Rise of Animals*. 1998, Oxford University Press.

"A recent profile of Lenski claims, . . ." E. Pennisi. "The man who bottled evolution." 2013. *Science* 342(6160), p. 790. DOI: 10.1126/science.342.6160.790.

"Lenski's experiments show that evolution presses forward relentlessly, . . ." See J. E. Barrick et al. "Genome evolution and adaptation in a long-term experiment with Escherichia coli." 2009. *Nature* 461(7268), p. 1243. DOI: 10.1038/nature08480; R. J. Woods et al. "Second-order selection for evolvability in a large Escherichia coli population." 2011. *Science* 331(6023), p. 1433. DOI: 10.1126/science.1198914; a role for contingency is seen in Z. D. Blount et al. "Historical contingency and the evolution of a key innovation in an experimental population of Escherichia coli." 2008. *Proc Natl Acad Sci*. 105(23), p. 7899. DOI: 10.1073/pnas.0803151105.

". . . a study that found extreme convergence for lizards on Caribbean islands . . ." D. L. Mahler et al. "Exceptional convergence on the macroevolutionary landscape in island lizard radiations." 2013. *Science* 341(6143), p. 292. DOI: 10.1126/science.1232392.

"He has even written his own book on the Burgess Shale . . ." S. Conway Morris. *The Crucible of Creation: The Burgess Shale and the Rise of Animals*. 1998, Oxford University Press.

". . . Conway Morris has a typically rich entry . . ." S. Conway Morris. "Life: The final frontier for complexity?" Chapter 7 in *Complexity and the Arrow of Time*, ed. by Lineweaver, Davies, and Ruse. 2013, Cambridge University Press. DOI: 10.1017/CBO9781139225700.010.

"But in the very next essay, Stuart A. Kauffman . . ." S. A. Kauffman. "Evolution beyond Newton, Darwin, and entailing law: The origin of complexity in the evolving biosphere." Chapter 8 in *Complexity and the Arrow of Time*, ed. by Lineweaver, Davies, and Ruse. 2013, Cambridge University Press, pp. 162–190, quote on p. 179. DOI: 10.1017/CBO9781139225700.011.

"Meyer too focuses on the huge numbers of possible amino acids . . ." For a succinct critique of Meyer's approach, see D. Venema. "Seeking a signature." 2010. *Persp Sci Christ Faith*. 62(4), p. 276.

"John Torday has proposed another series of logically ordered events . . ." J. S. Torday. "Evolutionary biology redux." 2013. *Perspect Biol Med*. 56(4), p. 455. DOI: 10.1353/pbm.2013.0038.

". . . swim bladder genes evolved and converged four times in teleost fish, . . ." M. Berenbrink et al. "Evolution of oxygen secretion in fishes and the emergence of a complex physiological system." 2005. *Science* 307(5716), p. 1752.

"CÉLINE: Well, the past is the past . . ." R. Linklater, J. Delpy, and E. Hawke. Screenplay for *Before Sunset*. 2004, Warner Independent Pictures and Castle Rock Entertainment.

"... that genes drift about randomly inside a genome." M. Lynch and J. S. Conery. "The origins of genome complexity." 2003. *Science* 302(5649), p. 1401.

"... such as symmetry in proteins, ..." M. Lynch. "Evolutionary diversification of the multimeric states of proteins." 2013. *Proc Natl Acad Sci.* 110(30), p. E2821. DOI: 10.1073/pnas.1310980110; but for a case in which multimeric state evolved in a particular direction, see T. Perica et al. "Evolution of oligomeric state through allosteric pathways that mimic ligand binding." 2014. *Science* 346(6216), p. 1254346. DOI: 10.1126/science.1254346.

"... mammals evolve a type of wiring called enhancers ..." D. Villar et al. "Enhancer evolution across 20 mammalian species." 2015. *Cell* 160(3), p. 554. DOI: 10.1016/j.cell.2015.01.006.

"... the large-scale features of the two networks ..." F. Yue et al. "A comparative encyclopedia of DNA elements in the mouse genome." 2014. *Nature* 515(7527), p. 355. DOI: 10.1038/nature13992.

"... reconstructing the evolutionary path of hormone-binding proteins." M. J. Harms and J. W. Thornton. "Historical contingency and its biophysical basis in glucocorticoid receptor evolution." 2014. *Nature* 512(7513), p. 203. DOI: 10.1038/nature13410.

"... 81% of the time a protein will take paths in the top 30%..." J. D. Buenrostro et al. "Quantitative analysis of RNA-protein interactions on a massively parallel array reveals biophysical and evolutionary landscapes." 2014. *Nat Biotechnol.* 32(6), p. 562. DOI: 10.1038/nbt.2880.

"... a single change will open up a brand-new activity for an enzyme." M. Salmon et al. "Emergence of terpene cyclization in *Artemisia annua.*" 2015. *Nat Commun.* 6, p. 6143. DOI: 10.1038/ncomms7143.

"I am inclined to look at everything ..." Letter from Darwin to Asa Gray, 22 May 1860, DCP Letter 2814, italics original; in Francis Darwin, ed., *The Life and Letters of Charles Darwin*, Volume 2. 1888. John Murray, pp. 311–312; quoted by D. O. Lamoureux. "Darwinian theological insights: Toward an intellectually fulfilled christian theism—Part II: Evolutionary theodicy and evolutionary psychology." 2012. *Persp Sci Christ Faith.* 64(3), p. 147.

"Michael Ruse notes that in Darwin's thinking, ..." M. Ruse. "Wrestling with biological complexity: From Darwin to Dawkins." Chapter 12 in *Complexity and the Arrow of Time*, ed. by Lineweaver, Davies, and Ruse. 2013, Cambridge University Press. DOI: 10.1017/CBO9781139225700.016; p. 285.

"... an art installation for the ears, ..." Art exhibit by Janet Cardiff. "The Forty Part Motet: A Re-working of Spem in Alium Nunquam Habui 1573, by Thomas Tallis." Visited in 2008 at the Tacoma Art Museum.

"In the eighteenth century, the emphasis was on the order, as Owen Barfield writes: ..." Owen Barfield. *History in English Words* (3rd edition). 1926 (1988), Lindisfarne, pp. 178–179.

"ROTHKO: Wait. Stand closer ..." John Logan. *Red.* 2009, Oberon, Scene One.

"Then I discovered Mark Rothko through ..." Simon Schama. *The Power of Art.* 2006, Ecco.

"Darwin wrote, 'Let ...'" Charles Darwin. *Works of Charles Darwin: The Variation of Animals and Plants under Domestication in Man and Animals.* 1915, D. Appleton, p. 228.

"In *The Old Ways*, Robert Macfarlane writes: ..." Robert Macfarlane. *The Old Ways: A Journey on Foot.* Penguin, p. 192.

"Conor Cunningham writes, . . ." Conor Cunningham. *Darwin's Pious Idea: Why the Ultra-Darwinists and Creationists Both Get It Wrong.* 2010, Eerdmans, p. 148.

". . . it surprises scientists repeatedly . . ." S. Conway Morris. "Predicting what extra-terrestrials will be like: And preparing for the worst." 2011. *Philos Trans A.* 369(1936), p. 555. DOI: 10.1098/rsta.2010.0276: "the combination of primitive and advanced is deeply provoking, as is evident from the repeated employment of words such as 'puzzling' and 'surprising.'"

"The reductionist approach to evolution . . ." R. Fortey. "Shock lobsters." 1998. *London Rev Books* 20(19), p. 24.

"By contrast, I have always regarded natural history . . ." E. J. Chaisson. "Using complexity science to search for unity in the natural sciences," Chapter 4 in *Complexity and the Arrow of Time,* ed. by Lineweaver, Davies, and Ruse. 2013, Cambridge University Press. DOI: 10.1017/CBO9781139225700.006.

"Our objective was in large part . . ." Robert M. Hazen. *The Story of Earth: The First 4.5 Billion Years, from Stardust to Living Planet.* 2012, Viking, p. 201.

"Two examples that came across my browser . . ." J. Viegas. "500-million-year-old 'mistake' led to humans." July 24, 2012. *Discovery News.* news.discovery.com; and J. Frazer. "Accident of evolution allows fungi to thrive in our bodies." November 23, 2013. *Scientific American.* www.scientificamerican.com.

"Biologist James A. Shapiro gives several examples . . ." James A. Shapiro. *Evolution: A View from the 21st Century.* 2011, FT Press.

"Eminent scientists like E. O. Wilson encourage telling science as a story, . . ." E. O. Wilson. "The Power of Story." 2002. *Amer Educ.* 26(1), p. 8.

". . . narrative structures like "And But Therefore" . . ." R. Olson. "Science communication: Narratively speaking." 2013. *Science* 342, p. 1168.

"In *The Mismeasure of Man,* Gould used his measurements . . ." S. J. Gould. *The Mismeasure of Man.* 1981, Norton.

". . . Daniel Kahneman's *Thinking, Fast and Slow* . . ." D. Kahneman. *Thinking, Fast and Slow.* 2011, Farrar, Straus, and Giroux.

"In 2011 a group of anthropologists re-remeasured some of Morton's skulls . . ." J. E. Lewis et al. "The Mismeasure of Science: Stephen Jay Gould versus Samuel George Morton on skulls and bias." 2011. *PLoS Bio.* 9(6), p. e1001071. DOI: 10.1371/journal.pbio.1001071.

". . . a *Nature* editoral quickly (and inevitably) followed . . ." [No authors listed.] "Mismeasure for mismeasure." 2011. *Nature* 474(7352), p. 419. DOI: 10.1038/474419a.

"Michael Ruse in *Monad to Man,* a book about . . ." Michael Ruse. *Monad to Man: The Concept of Progress in Evolutionary Biology.* 1997, Harvard University Press.

INDEX